Climate Preservation In Urban Communities Case Studies

Climate Preservation In Urban Communities Case Studies

Edited by

Woodrow W. Clark II

ELSEVIER

Butterworth-Heinemann
An imprint of Elsevier

Butterworth-Heinemann is an imprint of Elsevier
The Boulevard, Langford Lane, Kidlington, Oxford OX5 1GB, United Kingdom
50 Hampshire Street, 5th Floor, Cambridge, MA 02139, United States

Library of Congress Cataloging-in-Publication Data
A catalog record for this book is available from the Library of Congress

British Library Cataloguing-in-Publication Data
A catalogue record for this book is available from the British Library

ISBN: 978-0-12-815920-0

For information on all Butterworth-Heinemann publications visit our website at https://www.elsevier.com/books-and-journals

Publisher: Matthew Deans
Acquisition Editor: Ken McCombs
Editorial Project Manager: Naomi Robertson
Production Project Manager: Surya Narayanan Jayachandran
Designer: Miles Hitchen

Typeset by TNQ Technologies

Contents

List of Contributors

James Barth, Professor of Economics, Auburn University, Auburn, Alabama

Samantha Bobo, Department of Planning, Rice University, Houston, Texas

Danilo Bonato, Managing Director ReMedia, Milan, Italy

Manqing Cao, University of International Relations, Beijjng, China

Elisa Castoro, Independent Writer/Photographer, Matera, Italy

Woodrow W. Clark, II, Research Professor in Economics, Pepperdine University Graziadio School of Business (PGSB), West Los Angeles, CA Campus, Los Angeles, California

Andrew DeWit, Professor, School of Economic Policy Studies, Rikkyo University, Tokyo, Japan

Athula Ekanayake, Faculty of Management, University of Peradeniya, Peradeniya, Sri Lanka

Michael Fast, Department of Business Economics, AAlborg University, AAlborg, Denmark

Rich Gibson, Sustainable Partners, Temple, Arizona

Hemantha S.B. Herath, Department of Accounting, Goodman School of Business, Brock University, St. Catharines, ON, Canada

Tejaswini C. Herath, Department of Finance, Operations and Information Systems, Goodman School of Business, Brock University, St. Catharines, ON, Canada

Naved Jafry, La Jolla, California, United States

Xing Li, Department of Business Economics, AAlborg University, AAlborg, Denmark

Lucia Elsa Maffei, Esq. Studio Legale, Matera, Italy

Melody Rong, Data Analyst, IT Health, Cardioval Company, Suzhou, Jiangsu Province, China

Garson Silvers, US Green Building Council, Beverly Hills, California, United States

Wang Weiyi, Jiaxing University, Zhejiang Province, People's Republic of China

Xi Yang, University of International Relations, Beijjng, China

Preface

This book is a result of both Elsevier Press leadership and staff with whom I have worked for over a decade now on six books, and my first book that was published by Elsevier Press as Agile Energy Systems (2004) focused on the California energy crisis to it being updated in July 2017 with a global focus. Climate Preservation (CP) is due to senior Elsevier Press (Ken McCombs and others) who supported my other books but also CP as it reflects the need for international examples from communities, cities, and nations on how to mitigate and reverse climate change, as well as enact resilience to it.

A few years ago, I took early retirement from the University of California system so that I could spend time writing and editing books on the solutions to climate change that range from public policy to technologies to economics. I have now seven books as a result of that time, from 2012 to 2017. And this year (2018), I will have three more books done and out. After CP, will be two books on Qualitative Economics (QE). It is an update on the original QE that was published a decade ago, as well as a new book about "Making Economics into a Science: Quantitative and Qualitative Economics (Q^2E), with cases in it.

All of these books are inspired by my work over 3 decades now in science and technology at Lawrence Livermore National Laboratory in the 1990s, and then public policy that I did then with the UN IPCC. From being a visiting professor at Aalborg University, Denmark, and then being asked by California Governor Gary Davis to be one of his five energy advisors at the turn of the 21st century when there was a massive energy crisis in California, had led me back into academic research and teaching.

Now I am Research Professor in Economics at Pepperdine Graziadio Business School (PGB) in Southern California, See link: https://bschool.pepperdine.edu/academics/faculty/woodrow-clark-economics-research-professor/.

Contact Dr. Clark at: Woodrow.Clark@pepperdine.edu.

Above all, I need to thank my wife, Andrea Kune-Clark, and our fifth-grade son, Paxton, for their help and support. There is more coming, so stay tuned.

Chapter 1

Introduction: Smart Green Healthy Communities Are Now the Future

Woodrow W. Clark, II [a]

Research Professor in Economics, Pepperdine University Graziadio School of Business (PGSB), West Los Angeles, CA Campus, Los Angeles, California

Chapter Outline

OVERVIEW

China has now its #13 Five-Year Plan (March 2016), which is now focusing on "the green industrial revolution" (GIR) today, not just in the future (http://www.sgcc.com.cn/ywlm/gsgk-e/gsgk-e/gsgk-e1.shtml in Mandarin and the English version from Elsevier Press http://www.amazon.com/The-Green-Industrial-).

China had already leapfrogged into the GIR (Clark, October 2014; Clark and Isherwood, 2009; Clark and Isherwood, 2007; Clark and Isherwood, 2010; Clark, 2010) along with other Asian (Japan, South Korea, and Vietnam) and

a. Dr. Woodrow W. Clark II, MA[3], PhD, is a Qualitative Economist / Managing Director, Clark Strategic Partners in Beverly Hills, California 90209: www.clarkstrategicpartners.net

Climate Preservation In Urban Communities Case Studies. https://doi.org/10.1016/B978-0-12-815920-0.00001-0

European Union (all the Nordic countries, Germany, Italy, France, Spain, and Italy) nations moving ahead. The carbon coal, oil, natural gas, and even nuclear-powered (more recently in the late 1970s) centralized energy utility industries were started by Thomas Edison in 1882, when he flipped the switch on Pearl Street substation in Manhattan. Today what began over a century ago has begun to decline.

Note that in the "smart-grid city" example below the car parked in the garage of a home is being either recharged or refueled there. Additionally, the power in the car can act as energy storage in case there is a power outage or emergency need. This kind of integrated system is beginning to occur rapidly in countries as nations and communities provide public policy along with economic incentives to create, operate, and maintain these examples of smart, green, and healthy communities.

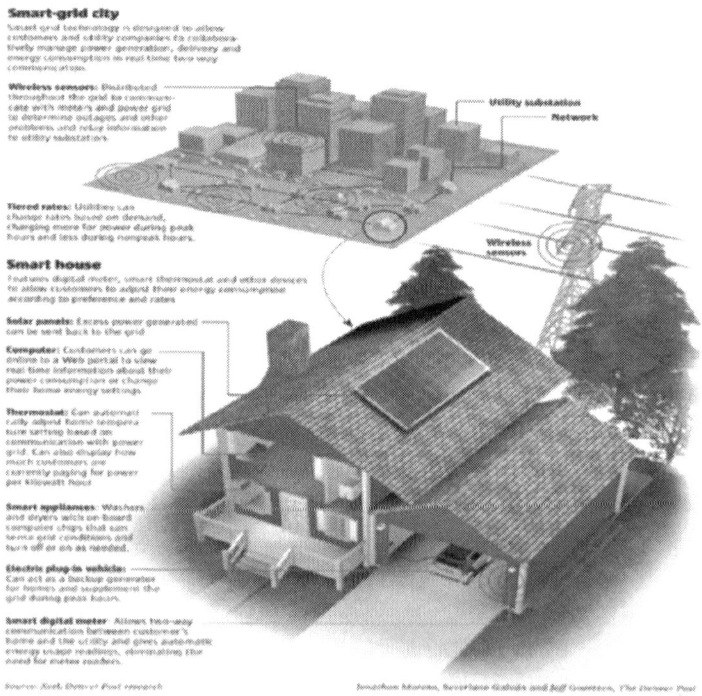

Homes need to have energy conservation and efficiency under "one roof," which is known as on-site power and part of communities needing smart green cities.

The new global energy power systems' models are "agile" (that is, "flexible") because they need to accommodate both green on-site distributed power generation and grid-connected power. Agile energy systems (AESs) was the title of my first book (Clark and Bradshaw, 2004) as it combines electricity from on-site renewable energy with electricity from traditional central grids hundreds of miles away and manages them both to meet demand and base load. Today, more and more cities (and countries) are creating AESs or "vertically" integrated energy systems. Thus Elsevier Press, the publisher of AES in 2004, asked for an update of the book, which was done in the mid-2017.

The AESs are efficient, smart, and rooted in renewable both on-site power and central grid energy power generation. Although there is usually a central grid that depends heavily on fossil fuels to generate power, agile systems allow and even encourage on-site renewable energy sources and then disperse the electricity accordingly. These distributed energy systems can be formed and operated on the local level to serve targeted communities and consumers (Clark and Bradshaw, 2004). The issue is that almost all US states allow on-site distributed power. In short, AESs are here now in the United States. And many nations have begun to create them even more rapidly than California and the United States (Fig. 1).

Regional and city-level solutions are needed to address the challenges of global warming and climate change. Rather than having centralized power plants that use fossil fuels or nuclear power to generate energy and then transmit it over power lines, local on-site generation of power from renewable sources is better for the environment, far less expensive, and much healthier

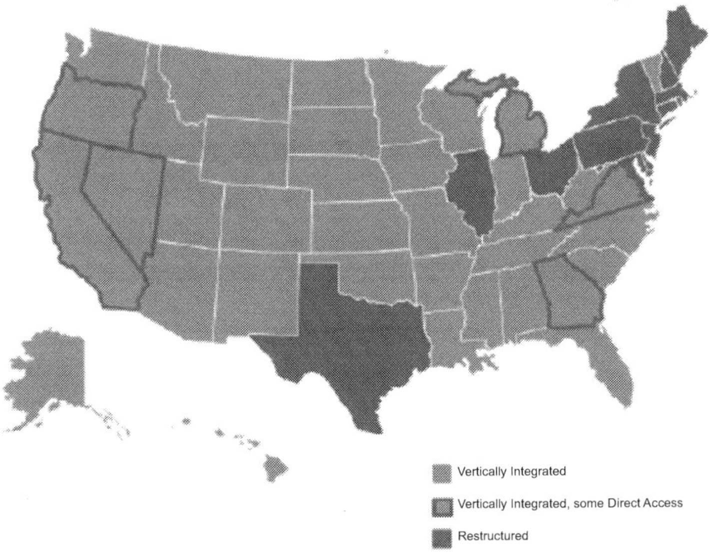

FIGURE 1 Status of electric restructuring in the United States. *Source: Advanced Energy Economy Institute "Opportunities to increase Corporate Access to Advanced Energy: A National Brief (Meister Consultants Group, August 2016).*

for the planet. All of this starts at the local, on-site level for homes, both public and private buildings with complexes. The key is technologies and systems that conserve and provide efficient power locally (http://prhlcorp.com/). As the map shows, most US states are "vertically" integrated so that on-site power is expanded (Table 1).

As this chart shows, today, the eco and green buildings of the past three decades are combined with smart technologies and systems for states and their communities, cities, and regions now combined and known as "smart green cities" (Clark and Cook, 2016). Energy needs are growing more complex as populations increase, cities expand, and power demands climb. Keeping energy use low through LED bulbs is critical for reducing energy use in buildings (www.E3EnergyExperts.com). This is needed to reduce the air and water pollution that is causing serious and costly health problems for young and old. Regulators are now implementing carbon dioxide regulations and even taxes (as was done with tobacco) to stop pollution.

Meeting the challenges of supplying energy for increasing demand, while reducing carbon emissions, calls for more complex and creative solutions based on local and national plans that have funding to implement them. The change starts with requirements to increase energy efficiency, use renewable energy generation, and create infrastructures for water, waste, and transportation that are integrated. Together these changes will help the way people live, think, and plan for all kinds of activities from using electricity to getting to work everyday. Case in point is illustrated below with the use of LED lights and how they impact energy use in buildings.

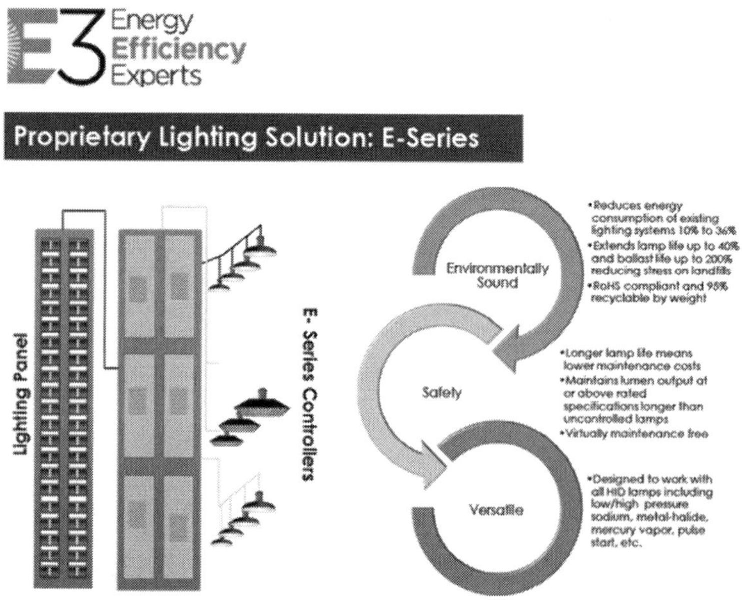

TABLE 1 States Ranked Among the Top Five States for One or More of the Identified Policies, Based on Potential to Increase Corporate Access to Advanced Energy

	Highly Ranked by Purchasing Pathway					
State	Large Offsite Purchasing	Raise DG System Cap	Enable Third Party Ownership	Enable Virtual Metering	Annual Large Corporate Consumption (GWh/year)	Corresponding Renewable Energy Capacity (MW)
Texas		X		X	106,945	40,876
California	X	X			78,504	28,909
Florida	X		X	X	49,414	19,078
Ohio				X	48,888	19,674
Indiana	X		X	X	39,876	15,842
Georgia				X	38,225	14,859
North Carolina			X		36,697	14,216
Michigan	X	X			30,608	12,317
Kentucky		X			29,845	11,830
Alabama		X	X		28,154	10,982
Minnesota	X		X		20,591	8,133

Source: Advanced Energy Economy Institute "Opportunities to increase Corporate Access to Advanced Energy: A National Brief (Meister Consultants Group, August 2016).

Energy systems are evolving and AESs are emerging. In the future, the central grid will be used for redundancy and backup purposes or act as a battery for energy storage when the sun is not shining and the wind is not blowing. These agile systems combine renewable energy, sensors, and wireless Internet connections and similar technologies that direct market mechanisms. This type of system is a new economic model that is part of making cities green and smart. In *The Next Economics* (Clark, 2012) green, smart cities and communities (Clark, 2013).

What people in all areas need to do is reduce their use of energy and have local on-site power from sources such as LEDs. For the last decade, the growth in the LED industry has been incredible. For both the consumer (lower costs for LED bulbs) and the manufactures, the costs have come down dramatically. Today, as the chart below from Goldman Sachs Global Investment shows (August 2016), LEDs lead all the other smart green industrials in term market share and growth. This does not dismiss the solar and wind industries but does make it clear that energy conservation and efficiency is critical to all areas of reducing carbon emissions and reversing climate change (Chart 1).

The value in terms of technologies with energy generation incorporated into homes and buildings along with recharging systems for electric and hydrogen fuel cars exist. As one US company, Premier Holding (PRHL), is doing this with a focus on individual homes and areas within communities (http://prhlcorp.com/). And the demand for building complexes ranging from government, office buildings, to apartment complexes is growing especially being zero emission and carbon neutral.

As a recent example, consider the new car-hailing form of transportation where the US company Uber merged with China's Didi Chuxing. As the Economist reported in August 8, 2016 (p. 49), "Uber gives app" which means in reality that the American company could not compete in China. So Uber got

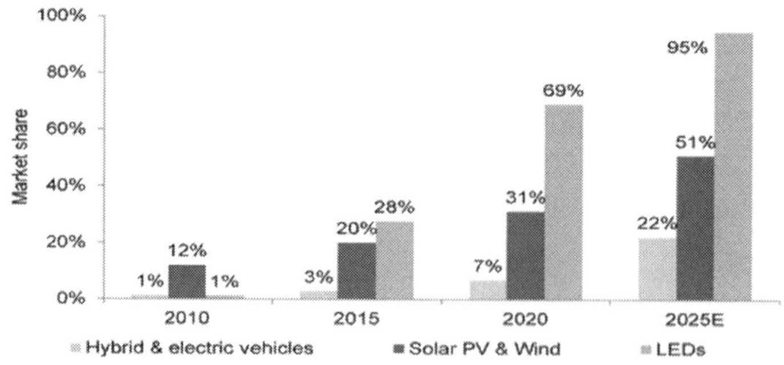

CHART 1 Compared to lighting, the transition in the capital intensive power sector is more gradual. The capacity factor–adjusted market share of solar and wind in gross capacity additions continues to grow steadily. *Source: IHS, Company data, IEA, IRENA, BP, Goldman Sachs Global Investment Research.*

17.7% ownership in Didi, with each company having a board member to the other as well as voting rights. Uber had lost $2billion over 2 years in China but will have Didi invest $1Billion into it globally (https://www.theguardian.com/technology/2016/aug/01/uber-deal-didi-chuxing-china-analysis). The *Economist* observed as other international press that this situation in China with United States could set a pattern for Google and WeChat as well as other companies (p. 50–51).

Finally, public policy (http://www.amazon.com/s/ref=nb_sb_noss?url=search-alias%3Daps&field-keywords=Agile+Energy+Systems) and economics need to be both developed, planned, and then implemented by governments (as all levels) and with the capital invested to be certain that they start and continue (http://www.amazon.com/The-Next-Economics-Environment-Climate/dp/1461449715). In the end, these economic changes come back to benefit everyone—individuals, companies, and governments.

Today, large amounts of finance and investments are going into renewable energy and smart technologies as systems that reflect the GIR.

Additionally, Goldman Sachs found recently that the LED industry has "disrupted" the Second Industrial Revolution dominated by fossil fuels for the use of energy (Chart 2).

For example, in August 2016, GE declared that it would no longer manufacturer and sell CFL bulbs. Now GE will only make, sell, and supply LED bulbs.

GLOBAL ENERGY TECHNOLOGIES TODAY

Smart green technologies range from hybrid to on-site to renewable energy and storage. The basic idea is to have renewable energy systems integrated into building as this photo above shows. Solar for energy along with energy

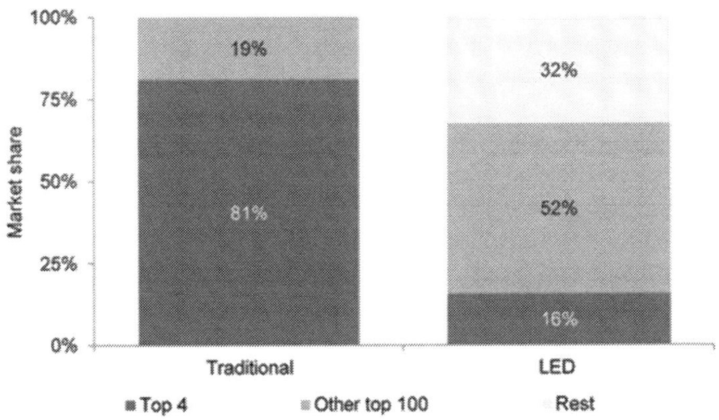

CHART 2 The shift to LEDs has already transformed the structure of the global listing industry. Vertically integrated lighting incumbents have lost market share and are now focusing on the downstream business. *Source: CSIL 2011, Goldman Sachs Global Investment Research.*

conservation from LED bulbs and efficiency through the Wi-Fi and other on-site smart systems as E3 has demonstrated in the United States through companies like PRHL (www.E3EnergyExperts.com). In this case, there is no battery or storage but a wire to a post for the power from the central grid when needed. And E3 has clearly set the pattern for the near future of the lighting industry.

As the house below shows the LED bulbs along with other smart technologies make the home a different building, and one that is free of producing carbon emissions.

Therefore, it is critical to reduce, eliminate, and replace fossil fuel use for transportation and buildings. Pressured to increase gas mileage and reduce carbon emissions, the vehicle industry is making changes. Automakers are upgrading gasoline engines and using more efficient turbochargers with computer-assisted transmissions. Ford is substituting aluminum for steel to shave 500 pounds off their F-150 trucks creating profound gas mileage savings.

SMART GREEN TECHNOLOGIES: INTEGRATED IN ENERGY INFRASTRUCTURES

Since 1949, China has had five-year plans that provide the government and population with direction from the central government about the policies and programs that need to be pursued. More recently, these five-year plans have included business development in what the Chinese refer to as social capitalism, which means that business is important, but it must consider the concerns and values of the people.

National plans need to link and integrate public infrastructure components. That way, infrastructures overlap and costs for construction, operations, and maintenance can be contained and reduced. If infrastructures can be constructed, operated, and maintained on the local level and meet regional, state, and national goals such as carbon reduction, they take on a different perspective, format, and cost structure. Economics is the key.

For the strategy, technologies are classified into seven categories: (1) innovative production process; (2) ultra-lightweight, heat-resistant material; (3) next-generation storage battery; (4) production, storage, and usage of hydrogen, etc.; (5) next-generation solar power generation; (6) next-generation geothermal power generation; and (7) immobilization/effective use of CO_2. One of the key strategies in the GIR is for new companies and older ones to merge into integrated systems such as PRHL (http://prhlcorp.com/) has done successfully in the United States.

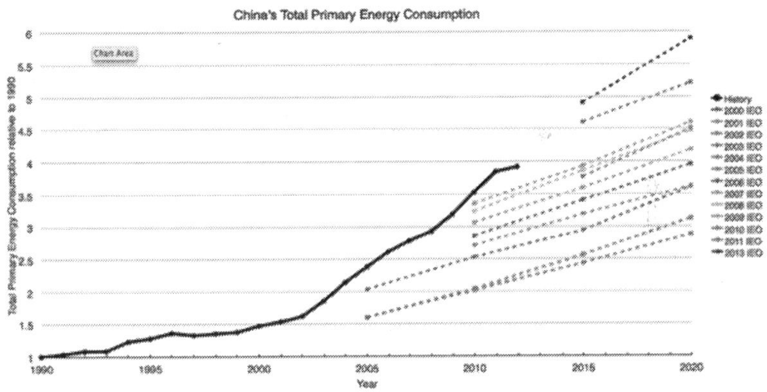

Source: U.S. Energy Information Administration.

Storage battery and hydrogen production technologies play a major role in absorbing the output fluctuation of solar/wind power. Therefore, four of the seven categories are related to renewable energy such as "quantum dot" and "perovskite," which use new structures and materials totally different from existing solar cells.

The consequences are that Big Oil is about to lose control of the auto industry as seen in the chart below. The Green Industrial Revolution (Clark and Cooke, 2014) has changed the world from getting energy from fossil fuels that pollute the environment and cause climate change, to renewable energy that come from solar, wind, water (oceans and dams) which are also smart, efficient and focused on stopping the climate from being polluted (Clark and Cooke, 2016).

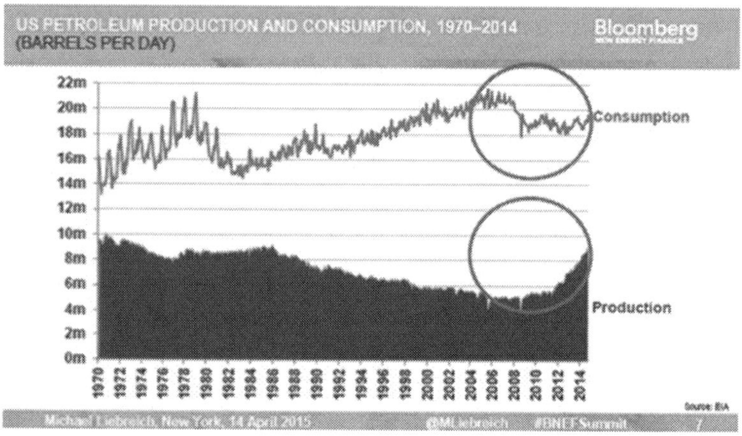

Source: EIA, 2015

As the data below show from the EU, all the extra power since the 1990s has come from wind, solar, and other renewable sources. The GIR already exists in many nations due to the increasing use of renewable energy along with storage technologies so that buildings can have reliable base load energy power.

Now the integrated energy systems from both electric and hydrogen (or other) fuel cell–powered vehicles can have storage available to homes and communities at night time. Most data track pollution and greenhouse gases (GHGs) as 40% from vehicles and about the same (40%) from buildings. The results will be reduced GHG that will also reverse and mitigate climate change.

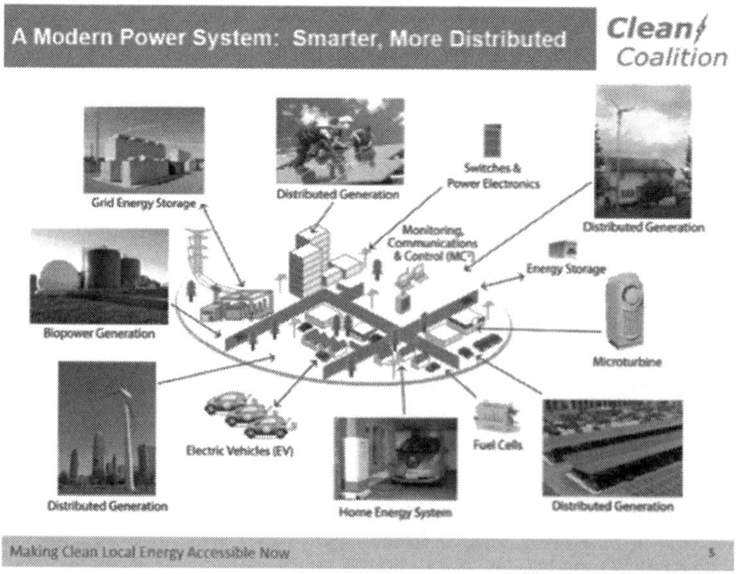

However, with the Tesla car now being all electric and some all-solar cars from Hanergy Energy Company in Beijing, China, there will be more of climate prevention push on the use of the solar energy for vehicles to be powered.

Energy efficiency can involve new appliance standards and energy codes or incentive and educational programs. Typical programs are rebates for the purchase of energy-efficient products, energy auditing services, and incentives for upgrading the operational efficiency of buildings, processing plants, and other facilities. There are also programs that encourage the purchase of new high-performance homes and others that provide design and engineering support for qualifying commercial buildings and multifamily complexes.

AGILE (FLEXIBLE) ENERGY SYSTEMS—HYBRID ENERGY TECHNOLOGIES

AESs using hybrid technologies offers substantial benefits for both energy systems to use the "green central grid power" (solar, wind farms etc) and also "on-site green power" (solar and wind on roofs etc.) which integrate the two systems in order to make them both more reliable and less costly.

The figure below shows how the central grid and on-site power should be agile in order to provide safe, green, and smart energy to entire communities. Such systems are cost-effective and reduce carbon emission and GHGs (Chart 3).

The following cases are some that show the benefits for AES hybrid technologies and their impact:

Cost Savings

- Reduced fuel costs, including storage, handling, and maintenance
- Reduced utility power consumption, especially during expensive peak hours

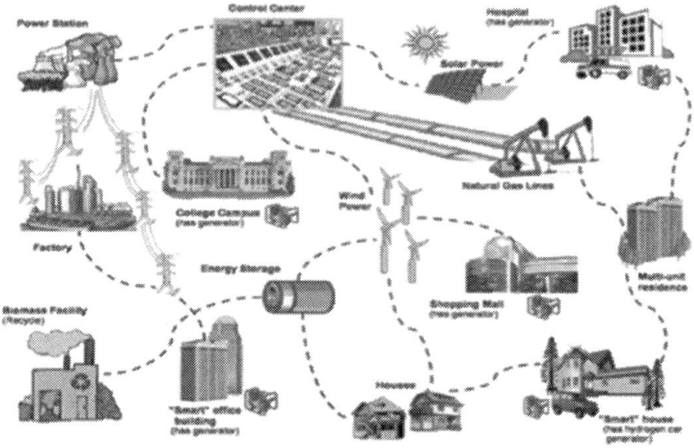

CHART 3 Agile energy systems: integrated central gird and on-site power generation with Wi-Fi and Internet (Clark and Bradshaw, 2004).

- Buy-downs, tax credits, and other incentives reduce installation cost and shorten payback period
- Reduced impact of utility rate hikes

Environmental Benefits

- Reduced GHG emissions
- Improved efficiency
- Reduced fuel consumption
- Less potential for leakage and spills

High Reliability

- Uninterrupted power supply
- Reduced risk of financial losses due to power outages
- Reduced downtime

Energy Independence

- Lower vulnerability to power outages
- Own your own power supply
- Incorporate multiple energy sources

Some examples of hybrid technologies on a regional level are as follows:

- Oak Creek Energy Systems plans to combine wind turbines with storage systems such as pumped storage or electronic storage to create a wind-driven system that can provide firm energy with some dispatchability.

- Sharp Solar Systems Division of Sharp Electronics is researching the combination of photovoltaics with electrolyzers and fuel cells and developing a control system to optimize its operation.
- SunLine Transit Agency in Thousand Palms, California, has a number of experimental fuel cell buses, and they are making hydrogen fuel for the buses in electrolyzers powered by photovoltaic arrays.

What is obvious is that the energy cost numbers begin to "add" correctly and the power supplied is cost competitive. Some of these hybrid technologies were under development in Japan with strong government support for many years (Clark and Chung, 1999). Today they show that with technologies integrated into AESs over 1–3 MW, the costs are extremely competitive.

Although chemicals and nanotechnologies are important for hydrogen vehicles and their infrastructures, there is limited information. And clearly most of it is not conclusive yet.

There are other approaches to creating hydrogen rather than through renewable electrolyzed water, sun, and wind. One is iodine sulfur reaction and hydrogen manufacturing technology through photocatalysis.

In the first Industrial Revolution, it was the steam engine and the printing press, and in the second, it was the internal combustion engine and analog communications. For the Green Industrial Revolution, renewable energy has combined with digital communications into what are known as "agile systems" (Clark and Bradshaw, 2004) whereby renewable energy, water, land, and other natural resources are integrated into one another as illustrated in Fig. 2.

Decades later, after Japan, Korea, Germany, and the Nordic countries started the GIR, the world is once again struggling with an energy crisis,

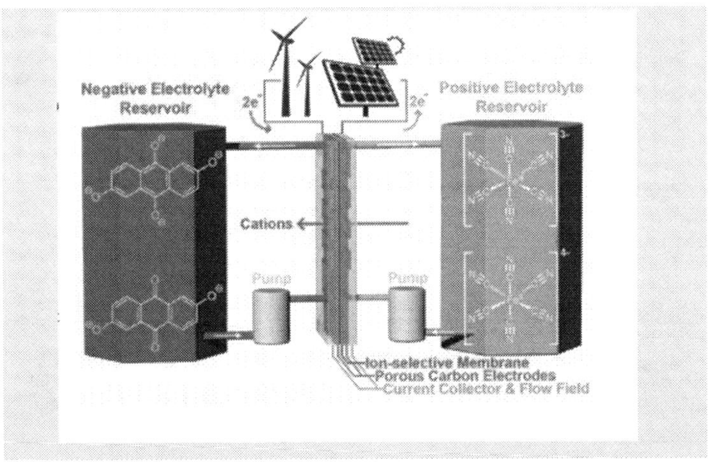

FIGURE 2 Schematic diagram of the Harvard organic flow battery.

created by a devastating earthquake and tsunami in the northeast coastal region that destroyed one of the key Fukushima nuclear power plants. From this tragedy, Japan may leap even further ahead in developing a carbonless economy as it expands renewable energy generation to compensate for the loss of nuclear power. China has watched this problem carefully and learned from it.

The key factor in the GIR is the issue of costs or "next economics" (Clark, 2012), for innovations today that can be paid for now in order to move out from the last three industrial revolutions that were primarily focused on fossil fuels for energy. As Chapter 7 will discuss in some detail, when looking at the costs for new cars that are "green" in terms of their *not* using fossil fuels such as natural gas or hybrids that combine electric with gasoline or even the sources of these fuels, the FCHV is clearly ready for the mass market, especially with the use of leases to get them on the road for everyone.

Social and environmental factors—sustainable communities, climate change mitigation, and environmental protection—are growing in importance and will soon demand far greater international cooperation and agreement. Rampant economic growth and individual accumulation of wealth is being replaced by social and environmental values that benefit the larger community. For example, the European Union is pushing for limits on the salaries of corporate executives.

Without a national policy and investment, countries and their corresponding cities cannot address basic infrastructures. Without government consensus, there can be no action, no improvement, no resources, and certainly no response to environmental degradation. Energy and infrastructure need to be integrated, and thus agile, into what many experts now call the "smart grid" (Clark and Cooke, 2014)

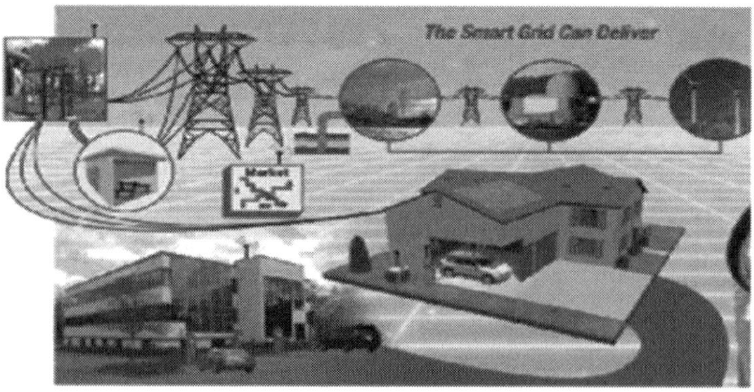

These are extraordinarily important national issues, just as important as defense or entitlement programs. To address these basic systems for the greater good, a nation needs to have plans, which are outlined and offered by the central national government.

China, not the United States, is showing real global leadership in responding to climate change. More than anything, China demonstrates how important a role the government plays in overseeing, directing, and supporting the economics of technologies and creation of employment. China's economic system is the prototype of social capitalism, which is now starting in China and is called "green development." Since the 1949 revolution, the Chinese have moved away from communism toward economic development through a series of Five-Year Plans (such as the current #13 Plan), now being referred to as guidelines.

Europeans adjusted their economies to fit the requirements of the GIR early on. Both the Scandinavians and the Germans realized that the move away from fossil fuels to renewable energy distribution would require more than neoclassical free market economics could deliver. Although the Danes and the other Scandinavians shifted national energy resources toward renewable energy power by national consensus, the Germans developed the innovative feed-in tariff process.

Energy: More and bigger problems will come with only a central electricity grid unless they are transformed into a smart, responsive, and self-healing digital network—in short, an "energy internet" as noted in the *Economist* (Spring 2004) is needed.

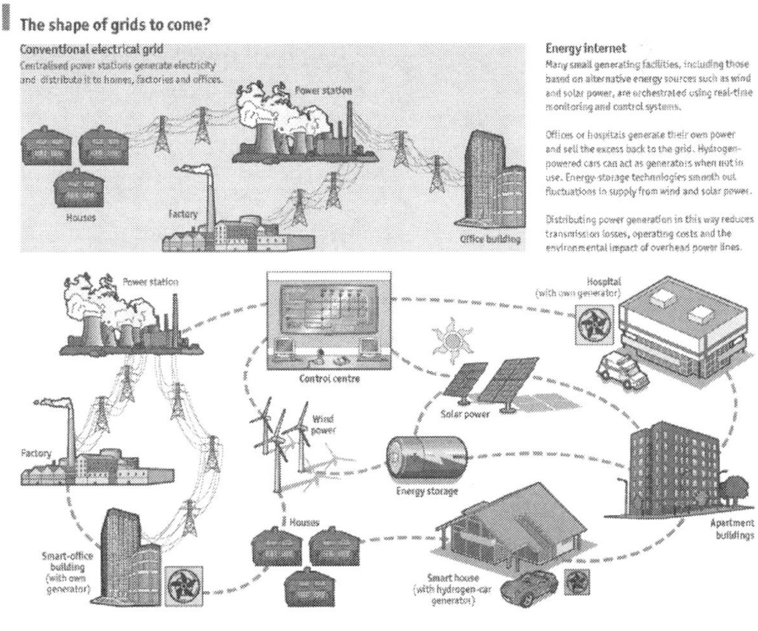

Sources: The Economist; ABB

And the agile energy grid (both central and on-site energy systems) needs to be implemented sooner than later.

The international agile paradigm is not about whether to introduce competition through markets. In reality, the emerging technologies that allow firms and communities to self-generate and self-organize energy systems guarantee a certain amount of competition. The international reality is that electric power is not evolving to a private deregulated competitive market, but rather that it is becoming a hybrid of public and private interests (again "agile" as both are needed) whereby public oversight assures that the system operates for the public good.

Indeed, this paradigm shift is to find a new nonideological approach to public sectors now in order to leverage on the local level distributed power from renewable energy sources and technologies for private company's ability to innovate and implement new ideas with the overall public interests of local citizens.

What is emerging, according to what some experts and now President XI of China labels as "green development" related energy-environment-climate change issues, are public policies and programs in collaboration with business activities (Clark and Cooke, 2014, 2015). Clearly the public has decided that it prefers environmental and health concerns over the profit-making-only interests of the market. Repeated polls of every demographic among citizens show overwhelming support of environmental protection policies and programs.

THE ECONOMICS—WHERE IS THE MONEY?

World energy consumption continues to rise as reported recently (July 2016) by the Haas Business School, University of California, Berkeley noted below.

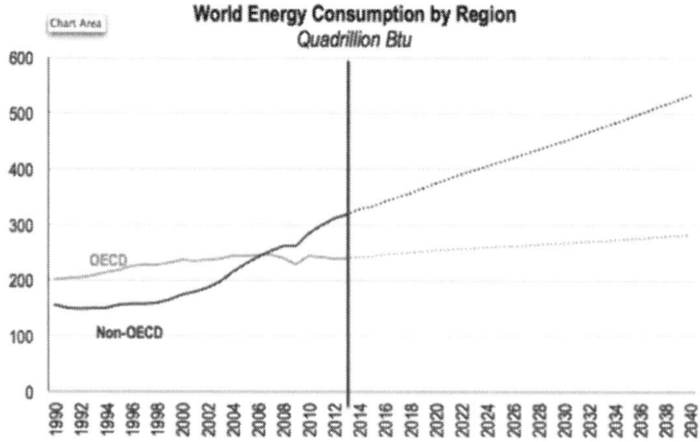

Source: U.S. Energy Information Administration.

However, there is a big difference today in the source for the energy. As noted earlier, the last century saw fossil fuels as the key source of energy for not only vehicles but also for buildings. That is changing dramatically. LED bulbs have become the "disruptive" technology along with solar, wind, and geothermal are now the key areas for energy generation around the world as solar power alone demonstrates. As the map below shows the dramatics for many reasons from the data (2016) illustrate below and noted by Mohit Anand in the Executive Summary on "Global Solar Demand Monitor" from gtm research in Q3 2016.

Source: GTM Research Global Solar Demand Monitor Q2 2016.

Then China excelled in solar demand beyond any other country. When China is even compared with other nations, the numbers become even more dramatic and compelling.

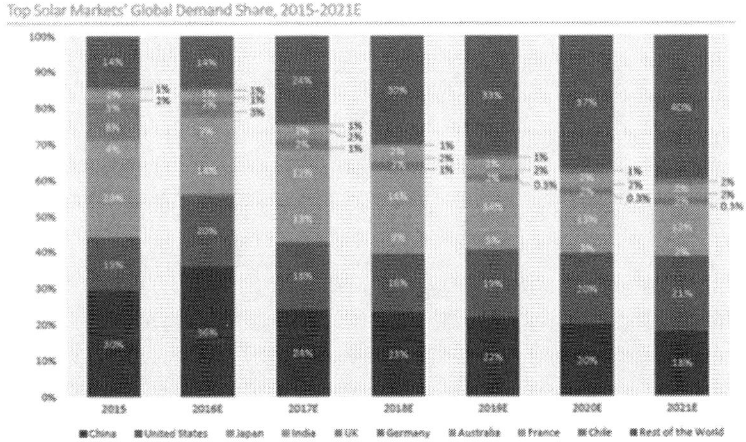

Source: GTM Research Global Solar Demand Monitor Q2 2016.

In short "Civic Capitalism" (http://www.amazon.com/The-Next-Economics-Environment-Climate/dp/1461449715) is taking place in China today as another source of solar data.

Communities need to work together and provide funding for the renewable energy technologies and their integrated systems. Some communities in the United States have local financing programs for renewable energy such as solar. Others use the "power purchase agreement," and the newest most logical funding is to include the solar and other renewable energy systems into the financial mechanism for the buildings (Clark, September 2014). New companies like PRHL are showing how this can be accomplished (http://prhlcorp.com/) and being contacted by international companies and governments especially from BRIC nations for strategies, technologies, and finance.

In the meanwhile, investment groups around the world are funding renewable energy and technology systems that are integrated as a top priority. As Wired. Com noted many USA billionaires invested in Green Tech (Nov 2015) http://www.wired.com/2015/11/zuckerberg-gates-climate-change-breakthrough-energy-coalition/?mbid=social_fb.

ECONOMICS MUST CHANGE INTO QUALITATIVE SCIENTIFIC DESCRIPTIVE ANALYSIS

Intense and constant interaction must take place between the members of the international consortium in order to commercialize the technology itself. This close collaboration supersedes industrial competitive issues and conflicts. For example, in the fuel cell case presented herein several major bus and battery or control unit manufacturers are part of the international consortium. The economic competition arises between companies in the marketing and distribution of the final zinc—air technologies on vehicles much the same way as oil, gas, and vehicles are competitive today. This is the interactionist perspective at the macroeconomic level and the subject of another book. Applying the interactionism perspective on a microeconomic level, a number of specific interactions become apparent; more details in Chapters 5 and 6.

Interactionism for Surface and Deep Representations

The first issue must clarify the surface and deep structure interactions within and between organizations forming the consortia. Linguistics is a science, as started by Noam Chomsky (1957) in the late 1950s, and has become the model for qualitative economics to be formed in order to convert economics into a science too. Using the interaction perspective outlined above, the following representation can be made of the basic organizational interactions.

The goal of the action researcher within an interactionism perspective is to establish business opportunities. Sets of representations provide guidelines for business actions as well as a road map to future interactive decisions. The representations must be stated, subject to further scientific verification, and applied in other cases. Ultimately the representation could be reduced to mathematical relational symbols. Elsewhere are several sets of rules derived from the fuel cell case (Clark, 1996).

A second interactionist result was the clarification of representations or meaning attached to words, ideas, and concepts. Consider the example of fuel cell case at the laboratory when the principal researcher and the program manager went to Europe to attend a conference on electric vehicles. They found that the definition of a "fuel cell" more accurately described the technology of zinc—air, rather as an "advanced battery." On further verification with the key staff in Australia as well as at the laboratory, the conclusion was certain.

From early December 1995, the zinc—air battery would be known as a fuel cell. Although there was scientific and technical evidence to support the name change for the zinc—air technology, the international interaction was crucial. Clearly, the technology was more of a fuel cell (that is requiring refueling rather than recharging). Furthermore, the concept of a fuel cell itself denoted high-tech advancements and therefore had more business appeal. The decision was made. A common definition was adopted and put into use worldwide. The term "fuel cell" had been agreed on at the deep structure level in order that it could be used precisely through transformation rules to the surface structure groups and organizations in everyday language usage.

A second example of creating clear and precise definitions can be seen in the creation of a new technology to parallel for refueling the fuel cells. Through numerous meetings, both within the laboratory and with the international parts, the concept of a zinc recovery unit (ZRU) with its own representations was created. Originally the principal research described the process for the zinc to be cycled through an undefined external stationary unit that pushed or spent zinc and electrolyte into an external hopper from the fuel cell. After many hours of intensive discussion among the leaders scientists (Cooper, Smith, and Tokarz particularly), two issues were clarified. The zinc was the issue and not the zinc in combination with the air. In other words, the external and separate unit needed to be concerned only with zinc processing.

The other point of clarification concerned the concept of this unit itself. Again in checking earlier documents, the process unit was called a "refueling unit" when in fact it was a "recovery unit." The difference is significant in the context of industrial ecology. Refueling implies that a raw material (zinc) has been mined and processed so that it is ready for refueling the fuel cell. In fact, the zinc is not refueled at all. The zinc is recovered because it is reformed after being used in the fuel cell. In other words, there is little loss of zinc (unlike gasoline, for example, that is burned and needs to be refueled). The ZRU, therefore, is simply an external unit that takes out the zinc from the fuel cell, reconstitutes it, and pumps it back into the fuel cell after 250—300 miles of use.

The clarification of the recovery unit (rather a refueling unit) is an enormously significant deep structure issue. Aside from the understanding and communication of the meaning into surface structure situations, the clear definition allows the scientists to actually create the ZRU. In this case, the actual ZRU had not been created and designed. Now a clear technological path could be followed. The significance of the ZRU definition will lead to patentable intellectual property among other yet unknown scientific breakthroughs.

Transformations

The relationship between the deep and surface structures occurs through transformations, which are vertical interactions between the sets of rules within each structure. In the fuel cell case, there are three "nodes" or key interactive points of contact for the rules: personal or people interactions; organizational relationships; and technology operations such as use of telephone, fax, travel, Internet, etc. Each node is noted as a "phrase marker" because it has definition and universal meaning.

As continual personal trust is being built through constant interaction within and between partnering organizations, personal nodes are triggered that move back and forth between deep to surface structures. For example, telephone conference calls (lasting hours) occur, but then a point is reached when some of the principals within the organizations need to meet face-to-face. Above, one aspect of building personal trust was described as essential for establishing networks. Such trust could only be built through constant interaction among the actors.

In the fuel cell case, personal trust was firmly established within a 2-month period between the laboratory and the power-utility company: once in Australia due to a Professional Academic Conference and once in Europe due to an International Conference. The transformational relations moved or transformed the common definition of terms and concepts to the surface structure following the set of rules outlined above. Other opportunities were deliberately created: the action researcher (Clark) as a participant observer on the laboratory team traveled to Europe on two other businesses and visited the key consortia bus manufacturer. These two meetings established an already solid interpersonal relationship through a factory tour, confidential exchange of data and information, and joint collaborative research leads to seek funding for fuel cells. Various meetings were planned in various parts of the world that would connect the key actors from each organization.

The application of economics today is in what is called "circular economics" which will be discussed in Chapter 3.

CONCLUSION

For over a century, Big Oil and then Gas with their related industries, have wrecked havoc on our environmentally fragile planet and disrupted the way humans manage their lives. Today, the loss of a major section of the West

Antarctic Ice Sheet and most of the North Pole from global warming caused by excessive carbon-generated heat appears unstoppable.

The summers are hotter and longer, and diseases like dengue fever are spreading. Asthma is afflicting children around the world in record numbers. As the ocean levels rise, coastal cities and island nations become more and more threatened with flooding and severe storm damage. It is not hard to conclude that if humans had not developed fossil fuels as a major source of energy, the planet would have industrialized slower, but been a much more sustainable environment for the future. Nor is it hard to conclude that human affairs would have turned out much differently if the oil industry had not gained so much political and economic power and come to dominate much of the world's activities.

Too bad it has taken over 100 years for it to happen when Nikola Tesla started all of the wireless world with the use of renewable energy through alternating current http://teslatech.info/ttmagazine/v4n1/seifer.htm.

Renewable energy generation is the foundation for a sustainable community and the heart of GIR. Basically, renewable energy is a source of energy that is not carbon-based and will not run out. Renewable energy sources are described as distributed energies because unlike the massive centralized fossil fuel power plants, they are spread out through many sources.

For example, solar installations can be on one rooftop or numerous as is done in Germany and Arizona and thus know as on-site power which are decentralized system that can be linked into the central power gird of a community, province, state and nation. Such distributed energy resources are found in every inch of the world—the sun, the wind, wave and tidal action, the geothermal heat under the ground, and biomass such as garbage, agricultural, and forest waste. Other renewable sources include bacteria, algae, and hydrogen when it comes from renewable electrolyzed sources.

REFERENCES

Chomsky, N., 1957. Syntactic Structures. Mouton & Co., The Hague.

Clark II, W.W., 2010. Sustainable Development Design Handbook. Elsevier.

Clark II, W.W., 2014b. The Next Economics: Global Cases in Energy. Springer Press.

Clark II, W.W., Bradshaw, T., 2004. Agile Energy Systems: Global Lessons from the California Energy Crisis. Elsevier Press and 2nd Edition as Global central and on-site distributed Energy from Elsevier in 2017.

Clark II, W.W., Cooke, G., 2014. Green Industrial Revolution. Elsevier Press.

伍德罗·克拉克,格兰特 with Anjun Jerry JIN & Ching-Fuh LIN 库克,金安君,林清富 Clark II, W.W., Cooke, G., August 2015. Green Industrial Revolution in China (Mandarin). Ashgate and China Electric Power Press.

Clark II, W.W., Cooke, G., March 2016. Smart Green Cities. Routledge Press.

Clark II, Woodrow W., Fall, 2012. The Next Economics: Global Cases in Energy, Environment, and Climate Change. Springer Press.

Clark II, Woodrow W., Isherwood, William, 2007. Special Report on Energy Infrastructures in the West: lessons learned for Inner Mongolia. PRC. Asian Development Bank.

Clark II, Woodrow W., Isherwood, William, 2009. Report on "Energy strategies for Inner Mongolia Autonomous Region". Utilities Policy Journal. https://doi.org/10.1016/j.jup.2007.07.003.

Clark II, Woodrow W., Isherwood, William, 2010. Special Issue on *Inner Mongolia, China: environmental and energy sustainable development.* and Case of "*Leapfrogging Energy Infrastructure Mistakes for Inner Mongolia*". Utilities Policies Journal (Special Issue, 2010).

Clark II, Woodrow W., July, 1996. A Technology Commercialization Model. Journal of Technology Transfer. Washington, DC.

Clark II, Woodrow W., Lead Author and Editor, 2013. The Next Economics: Global Cases in Energy, Environment, and Climate Change. Springer Press, Fall.

Clark II, Woodrow W., Lead Author/Director with RaeKwon Chung, Nov 1999. Transfer of Environmentally Sound Technologies from Developed to Developing Countries. Public Funds for technology project. Framework Convention for Climate Change, UN: Study of economics of "green" energy in six countries.

Clark II, Woodrow W., October 2014a. Global Sustainable Communities Design Handbook. Elsevier Press.

Hanergy Energy Company, Beijing China: http://www.hanergy.com/showCar/carshow.html.

FURTHER READING

Bonato, D., 2017. Circular Economics. ReMedia Company, Milan, Italy. Remedia. www.consorzioremedia.it.

Clark II, W.W., 2017b. Agile Energy Systems: Global On-Site and Distributed Energy. Elsevier Press.

Clark II, W.W., Bline, D., Demirag, I., 1998. Financial markets, corporate governance and management of research and development: reflections on US managers' perceptions. In: Demirag, I. (Ed.), Comparative Capital Systems. Pinter, London. Chapter in Book.

Clark II, W.W., Cooke, G., Fall, 2011. Global Energy Innovations. Praeger Press.

Clark II, W.W., December 2017a. Global Green Smart Sustainable Cities and Communities, second ed. Elsevier Press.

McDonough, W., 2016. Cradle to Cradle (C2C) on Circular Economics. www.i-sis.org.uk/closedLoopCircularEconomy.php.

Chapter 2

Agile Power Systems for Everyone: On-Site Distributed and Central Grid Systems

Woodrow W. Clark, II [1,a], Manqing Cao[2], Xi Yang[2]

[1]*Research Professor in Economics, Pepperdine University Graziadio School of Business (PGSB), West Los Angeles, CA Campus, Los Angeles, California;* [2]*University of International Relations, Beijjng, China*

Chapter Outline

OVERVIEW

The "Internet of Energy" (IE) is here now and real. IE is the interaction of smart distributed and/or on-site power grids together with larger historical central grid systems. These integrated power systems are known as agile energy systems (AESs), which combine both the smart and central power grids together. Not only does the information flow achieve multi-directional wireless inter-actionalism, but it also show the on-site and central energy grids value flow which achieve the interactions between people, communities, businesses and governments. The point is that these IE systems first occurred in 1893, when Nikola Tesla created and provided the night time lighting known as alternative

a. Woodrow W. Clark II, MA[3], PhD, Managing Director, Clark Strategic Partners, Southern California USA 90209. Also Manqing Cao and Xi Yang. Contact them via Dr. Clark at: www.clarkstrategicpartners.net.

Climate Preservation In Urban Communities Case Studies. https://doi.org/10.1016/B978-0-12-815920-0.00002-2

current (AC) for the Chicago World Fair—the First World Fair in history had Tesla's wireless lighting. Today communities and cities need to preserve their history and culture while also becoming smart, green, and healthy.

INTRODUCTION

People around the world daily switch on a light, lamp, and turn on their television, computers, and recharge their mobile phones and more through direct current (DC) power that Thomas Edison created and controlled at the turn of the 20th century in New Jersey and New York City and then the entire United States. However, then through the 21st century, the DC-dominated power systems provided energy flowing land-locked systems triggered by these the fossil fuel power sources for most power system. Today, into the second decade of the 21st Century, the on-site power from renewable energy sources and use of wireless systems reflect what Tesla started at the end of the 19th Century. Tesla died a poor man due to his AC energy NOT being implemented in the 20th century or even after WWII (Seifer, 2006; there are many more). Today, the AC use of wire IE is quickly becoming the AES power model around the world. The people below are the "cast of characters" back at the turn of the 20th Century whose companies then created and got control of the central grid energy systems except for Nikola Tesla as seen in the middle of the "cast of characters" (Seifer, ibid). Times have changed now from over a100 years later (Seifer, ibid)

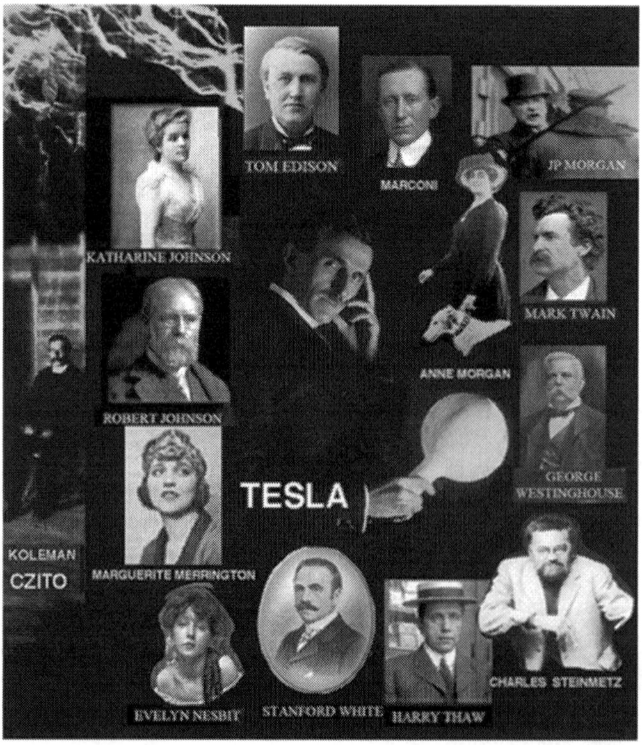

Except for Nikola Tesla seen above in the middle of pictures. Developing and poor countries are using AC from Tesla and also AES for the IE rather than the historical western DC and central power grid as the only source of energy.

Moreover, AES grid enterprises and their users are able to interact with each other. This strategy makes renewable energy power generation technologies grow and expand. Distributed generation or on-site power springs up in AES. So the network system needs to make adjustments and improvements again, heading in the direction of establishing AC wireless power that is Internet Energy (IE). Meanwhile, there are major changes in the relationship between the users and power grid. The users are not only users but also producers and coordinators of energy. Thus each joint of framework in a power grid has to have a positive effect on safety and smooth operation.

At the beginning of the 21st century, because of the old power grid and the continuous growing need of the customers, more and more contradictions came into being. Further investment in power grids and the building mode of power grid's infrastructure based on the maximum load demand were challenged. As a result, the concept of smart power grid comes into being, which logically connects the Internet technology and the physical power grid to form the modern AES edition of local distributed power grids, as seen below.

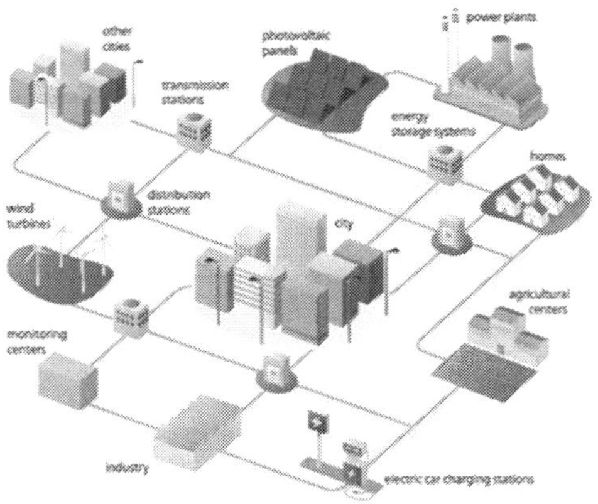

Therefore, in the Green Industrial Revolution (GIR), renewable (green and blue) energy are combined with smart digital local systems into "agile energy systems" (Clark and Bradshaw, 2004) whereby renewable energy such as solar, water, wind and other natural resources are integrated into one another as illustrated in the chart above with central energy power plants These systems are good ways to optimize the distribution of power resource and ensure the safety, reliability, and economical efficiency of electric power supply.

There are three main characters of smart power grids. First is stability. Smart power grids are still able to keep the power supply for users even when system disturbance or extreme malfunction happens. Second, is they are economical and efficient. Mainly because they increase operation and transmission efficiency to reduce cost. Third, and most importantly, they are friendly and interactive. Case in point are LED bulbs that have become known now as a "disruptive" economical technology.

The smart power grid is supposed to achieve the agile adjustment of power flowing by means of the bilateral interaction with customers. It is a good way to promote power companies and users to actively involve in the adjustment of the operation process. In other words, it meets the goals of demand side management with high efficiency and lower cost. And herein is the problem with western neoclassical economics: supply and demand have never worked in any area. The entire "balance" never existed, other than in theory from Adam Smith over a century ago (Clark and Fast, 2008; Clark and Fast 2018).

In America and Europe, the construction of smart power grids is more a focus on the community field of power distribution. The ultimate vision is to build intelligent electricity cities, more precisely, by adopting optical fiber composite cable network or power line carrier communication technology, in which the entire city is constructed. It improves the ultimate power using efficiency by systematic thinking, which includes the collection of power using information, automation of community power distribution, and the technology of intelligent housing and building complexes.

Second, these monitors, analyses, and controls of the power equipment are for the users and public electricity facilities. In China, the concept of the smart power grid is more commonly acknowledged and implemented. Because the power generation and energy demand imbalance for the geographical distribution, China's smart power grid needs more attention and change in the field of electric transmission, as well as involving the development of the small on -site and distributed power grid to ensure safety, reliability, and stability of the entire power grid for communities, cities, provinces and regions.

Around 2010, with the development of renewable energy generation technology, such as wind and solar power, communities had new demands for the construction of a power grid to meet the "friendly" need for access to renewable energy. There is one more characteristic of the new "green" power grid: using renewable energy (e.g. wind, solar, water, hydro-electric and others) that is environmental-friendly, valid, tested, honest for the distributed on-site power grid which supports the large-scale central grid access of renewable energy generation technologies, transmission and interactive systems. Therefore, the concept of energy internet locally in communities, cities, and nations is even a global energy internet strategy.

Case in point from Japan and its AES are these solar panels on top of a dam covered with green grass and more on buildings in the same area.

No central power poles or wires.

More problems come with only a central electricity grid unless they are transformed into a smart, responsive, and self-healing digital network— in short, an "energy internet" as noted above is needed which uses "green energy" for basic power sources. Therefore, the agile energy grid (both central and on-site energy systems) needs to be implemented sooner than later.

Indeed, the "green development" paradigm shift and its implementation are a new nonideological approach to public sectors in order to leverage the local on-site level distributed power from renewable energy sources and technologies for private companies (Clark, 2007; Clark and Isherwood, 2008; Clark and Isherwood, 2009). This GIR ability to innovate and implement new ideas with the overall public interests of local citizens has already started around the world.

Finally, over the last century, Big Oil, Coal, and now Natural Gas with their related industries have wrecked havoc on our environmentally fragile planet and disrupted the way humans live. Today (July 2017), the loss of a major section of the West Antarctic Ice Sheet was the size of the US State of Rhode Island. Also most of the North Pole has no ice due to excessive carbon-generated heat from global warming. Ocean water levels are rising that threaten beach communities and island nations.

The summers are hotter and longer, and diseases like dengue fever are spreading. Asthma is afflicting children around the world in record numbers. The year 2016 was the world's hottest year on record. However, the last decade has had other hot years too. As the ocean levels rise, coastal cities and island nations become more and more threatened with flooding and severe storm damage. It is not hard to conclude that if humans had not developed fossil fuels as a major source of energy, the planet would have industrialized slower, but been a much more sustainable environment for the future.

Nor is it hard to conclude that human affairs would have turned out much differently if the oil industry had not gained so much political and economic power which came to dominate much of the world's activities. It is a shame that this transversion has taken over 100 years to happen when Nicole Tesla started all of the wireless world with the use of renewable energy in the late 19th century (http://teslatech.info/ttmagazine/v4n1/seifer.htm).

Renewable energy generation is the foundation for a sustainable community and the heart of GIR. Basically, renewable energy is a source of energy that is not carbon based and will not run out. Renewable energy sources are described as distributed energies, because unlike the massive centralized fossil fuel power plants, they are spread out through many sources.

THE AGILE ENERGY SYSTEM CHANGES IN ACTION TODAY

The benefits of an agile power system are based on renewable and distributed technologies, and are discussed in details and with cases further into the book. Future energy systems must encourage the coupling of several alternative technologies into mutually beneficial hybrid power systems. Such AESs reduce reliance on long-distance transmission plans. AESs localize both the demand and supply of energy. Moreover, such systems stand in contrast to a "free market" solution which aims only to change behavior through real-time metering under the theory that if consumers know and understand their energy needs, costs, and usage, then they will cut back or alter their time of use. The facts show otherwise as noted above.

For example, consider European Union (Germany, Denmark, and Sweden in particular) and Asia with China and Japan which are referenced and seen in details later in the book. But to give an example, consider the Los Angeles, California, Community College District, also known as LACCD.

LOS ANGELES, CALIFORNIA, COMMUNITY COLLEGE DISTRICT, LACCD

Smart and sustainable communities are geopolitical areas such as cities, villages, towns, but also a small region that is primarily self-contained in terms of its purpose, technologies (ranging from renewable energy to wireless systems) with clusters of office buildings, shopping areas, living residential communities, or subsets of them like a shopping mall, walking street, or college campuses. Each of these communities are sustainable in that they can function independently with their own water, waste, transportation, and energy generation. And "agile" in that they can be connected to an energy central grid or generate their own power. Communities, furthermore, are sustainable because they strike a "harmonious" balance between economic development, job creation, and growth and their surrounding environment by protecting and preserving their local region.

These communities are smart because they use environmentally sound technology(ies) to power their own needs as well as communicate and interact within and outside their immediate regions. People use telecommunications for work, data, and information hence saving on the demands for transportation and stationary fuel supplies. In the end, smart and sustainable communities reduce the need for energy and the carbon footprint left by it. Such communities rely more on themselves than external or imported supplies and materials. College campuses are a key example and like most sustainable communities consist of three components: an overall master plan, infrastructures within the community, and finally the inner connectivity to a larger region, state, or nation.

The basic issue is the cost. How are these communities able to pay for the construction and then maintenance and updates to their facilities. LACCD has nine campuses that have from 15 to 25 buildings on each one plus athletic playing fields covering the Los Angeles County. There is also a separate main Administration building (Downtown Los Angeles in a seven-story building).

LACCD led in California and the United States for meeting both the concerns over climate change and the future needs of its students, faculty, and its boarder community by proposing to rebuild each of the nine campuses (all over 30 years old) with at least 44 new buildings and retrofitting dozens of others at LEED (Leadership, Environment, and Energy Design) building standards. LACCD is the largest community college district (2-year degree) in the world and one of the first to use LEED standards. The educational process would be enhanced, furthermore, in the design of buildings which when completed are examples of sustainable choices for "green" buildings and renewable energy power sources along with the development of an integrated curriculum focused on the balance between the environment and development.

Consistent with the nine-college campus system, and plans to build two more campuses, the LACCD Board members in 2002 furthered the academic focus for relevant and future education, curricula, research, as well as community outreach and assessment by setting a high standard for the new and retrofitted campus buildings to be "green." The Board passed a campus-wide Building Resolution in 2001 due the State of California Proposition #39 in which colleges, universities, and cities would raise bond funds from the public if over two-thirds voted for them. The voters in the LA County of Southern California approved not only the initial bond proposal of $1.245 billion to make an investment in affordable higher education but also approved a second bond issue of another $1 billion dollars in early 2003 to create a combined construction, renovation, and restoration program of $2.2 billion.

In 2005, the LACCD initiated a system-wide renewable energy central plant program and in 2007 started an "Energy Independent and Carbon Neutral" plan, which is part of the AASHE US national academic program. Another bond measure is under consideration in 2008 to further support that effort. Public policy is significant for leadership, support, and program funding. LACCD is a case study about how higher education can be a model for social and environmental change by planning and implementing programs designed to begin to solve the problems of global warming (Energy's Future: Beyond Carbon, 2006). The LACCD, with Board concurrence, identified six key policy issues:

1. Set standards for a bulk procurement program for furniture, concrete, etc;
2. Require manufacturers and factories to produce maximum recyclable content and utilize waste products in the manufacturing process;
3. Reduce long-term energy costs through an aggressive alternate and alternative energy programs;
4. Create a sustainable curriculum and the physical evidence of being in a sustainable building;
5. Work with the utility industry to foster sustainable strategies—for example, the Los Angeles Department of Water and Power (DWP) on surcharge and alternate energy support, as well as Southern California Edison on public–private partnerships;
6. Change the construction and building communities through education and contract expectations by requiring that an active green building consultant be on each project team.

The LACCD serves as a case study for how college campuses can serve as models for other such communities, towns, and cities. Most colleges and universities have the same characteristics, infrastructure needs, and environmental impacts. For example, colleges are clusters of buildings in a self-contained location. They, like other communities such as cities, shopping areas, malls, and office complexes, need power for their buildings as well as transportation, water, waste, and telecommunications infrastructures.

The major point here is the colleges did all of this without state or even local government control and oversight. In short, these local on-site distributed campuses were each able to convert them into smart, green, and healthy communities—that were and are today linked, interact, and share their experiences. Hence, many of the elements in the energy plan are now seen in the USGBC Neighborhood Program released in June 2007. One of the key differences, however, was that the LACCD emphasized also the need for curriculum and educational programs. Clearly, students needed to be trained in related and multidisciplinary fields to sustainable development including the trades such as electrical, plumbing, and software systems along with engineering and various sciences. Also staff, faculty, and community members needed to know about sustainability and its economic and business development in the broader community. The Plan includes the following:

1. Efficient renewable energy central plants
   a. Support for LEED level buildings
   b. Connecting entire campus or use of multiple central plants
   c. Financing from third parties based on campus load
   d. Campuses will be sustainable along with water, waste, transportation, and related infrastructures.

2. Performance and conservation contracts
   a. Retrofit all energy consuming elements for maximum efficiency
   b. Use conservation features in all buildings and state-of-the-art and new technologies
   c. Metering and monitoring systems
   d. Funding from private sector
   e. Aggregate through detailed technical audits
   f. Guaranteed payback within existing utility charges
   g. Funds from California Public Utility Commission
3. Energy independence and carbon neutral—renewable energy systems per campus
   a. Third party to install
   b. Solar on parking lots and roofs
   c. Buildings as integrated or add-ons
   d. DWP extension of solar/PV incentives and Edison through the California Public Utility Commission of $3.2 billion over 11 years
   e. National energy act (05) with incentives in 06 with 30% tax credit
   f. Aggregated Master Contract for buying solar and other systems
   g. Hybrid systems with storage and future technology innovation for "green" hydrogen to link with the California Hydrogen Highway (Clark, 2004)
4. Sustainable development curriculum
   a. Specific to campuses with basic courses as certificated, licenses, and degrees
   b. Career opportunity training for jobs, new companies, and advanced degrees
   c. Collaborate with unions, businesses, public, government, and nonprofit sectors
   d. Provide actual experiences on campus through building programs
   e. Sustainable development curriculum: solar, wind, geothermal, hybrids, etc., as well as new businesses, accounting, operations, and maintenance

Therefore, the results are "agile" (Clark and Bradshaw, 2004) or flexible, sustainable, and smart communities that are linked and dependent on the local and on-site infrastructures but also connected to more extensive and distant central grid ones that cover regions, cities, states, and nations (Clark and Bradshaw, 2004; Clark, 2007 and see Fig. 1 below). Thus, these agile networks have both a local and regional network that in the case of energy, for example, can be energy independent and thus more reliant on renewable energy power generation sources from local and regional resources. Hence, each of the nine LACCD campuses have developed master plans and green building objectives.

A "shared governance" program of administrators, professional educators, students, and community members that guide each campus decision-making

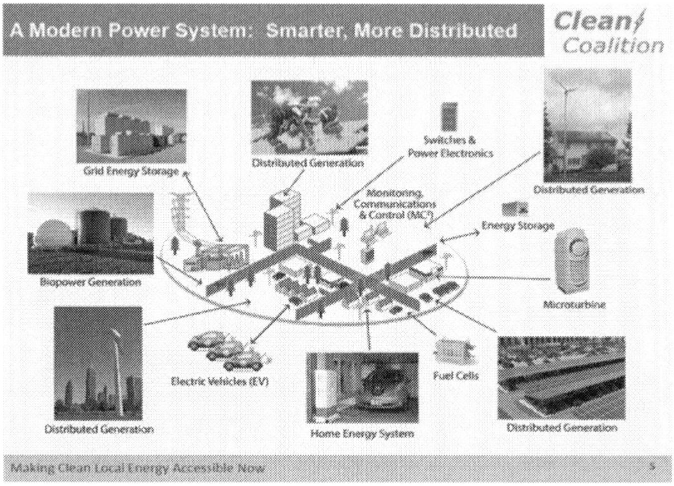

FIGURE 1 **Agile sustainable smart healthy campuses.** Below is an example of the new paradigm whereby local communities produce their own energy from renewable local sources (solar, wind, geothermal, biomass, etc.) with storage devices such as fuel cells and batteries for both stationary and mobile uses.

process. In the end, 44 buildings are in the process of being built. Contractors were hired through a rigorous competitive bid process by each campus and a construction management team was contracted to oversee the program across all campuses. In 2005, an energy team was created to concentrate on the renewable energy needs for the buildings and entire campus central plant infrastructures.

AGILE ENERGY AS HEALTHY AND SMART COLLEGE COMMUNITIES

The LACCD Board, in a bold and courageous stand to link higher education goals for research applied to the needs of society and meet the challenges of climate change. Sustainable and smart agile colleges' communities are often a threat to the established traditional approach to political governance and infrastructures operations. It represents the new paradigm (Clark and Bradshaw, 2004; Rifkin, 2004; European Union (EU) Parliament Rule, 2007). Nations, states, and cities want to control and centralize power and authority. This has been the historical pattern. Sustainable communities are a significant change in that perspective and therefore the growth of economies.

However, today with the need to meet and address the global challenge of climate change, regional and local level solutions must develop, take action, and implement politics and programs sooner than later. This issue has never been more obvious in the United States than the conflict with the decision-makers in Washington, D.C. versus states, like California. The US national government passed a "partial" and very limited national energy act (Energy Policy Act, 2005) that was primarily influenced by special interests from states advocating "biofuels" and companies that wanted to sell their products or fuels.

The US Congress tried again to pass a national energy act (2007) but was only partially successful. All this might change with a new US president and dominant political party. However, the problem with the US national government is that in the end, they might set standards (like the RPS) which are below (less than) states like California, cities, or regions like New England and New York State. Hence, any actions taken for creating sustainable and smart communities will most likely occur on local levels. By the end of 2007, over 350 US cities have now declared themselves "green." The issue will be what do they mean by green and do about it (Clark and Isherwood, 2006; Clark, 2004).

The LACCD took such action well before the state of California did. Moreover, the results are now evident in Los Angeles. Time will tell if there are results throughout the state of California and other states. Today, there is strong interest in LACCD in the European Union as it makes plans for its own "Third Industrial Revolution" (Rifkin, 2004).

The new paradigm has yet to take form in public policy in California. The college campuses that developed in the 20th century were more administrative and focused on central power plants. The development of sustainable systems, such as renewable energy sources, was slow and costly then, but is very different now in the 21st century. Yet the traditional administrative and political power plants exist today in Los Angeles, California, and throughout the United States (and globally). With no clear long-term solutions to deregulation and privatization, the result continues to be chaos, financial mismanagement, and outright fraud (Clark. & Demirag, 2002; Clark et al., 2004a) (Figs. 2 and 3).

In the end, agile sustainable college communities must develop and implement strategic plans for energy, waste, water, transportation, and telecommunications for their regions (Clark and Lund, 2007). These communities consist of college administrators, students, faculty, and staff who interact with local political leaders, business, and other groups that need to be interconnected with other college communities. Each sustainable college community must redefine, for example, the traditional central power plant and unsustainable infrastructure systems that exist today into ones that use on-site renewable energy, recycling, waste control, water, and land use as well as green building standard.

The traditional historic role (Second Industrial Revolution) of central power grids must change. Power coming from fossil fuels is bad for the

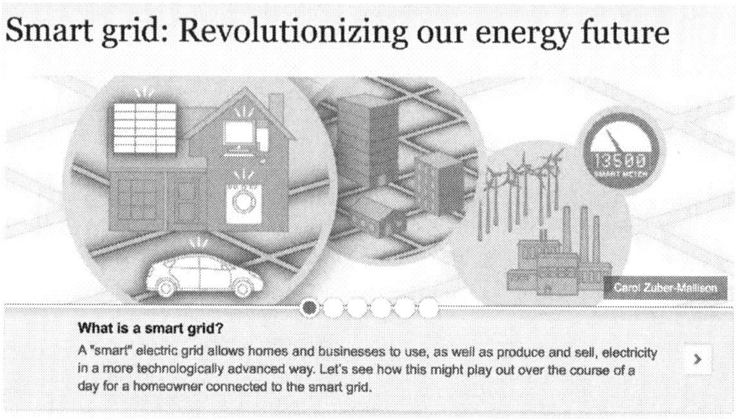

FIGURE 2 The conventional power central grid. Most of the energy generated today for all communities comes from a central plant often located miles from the end user (colleges, communities, and buildings of all kinds). Over the last 100 years, these power plants have been primarily powered by fossil fuels and nuclear power.

environment and a major cause for global warming. Downsizing, rebates, and economic incentives among other things can provide power backup and redundant power. This model is a new and different role for "public" and now "private" companies. On-site and local power is sustainable and environmentally beneficial when the energy source is from renewable energy sources combined with conservation and efficiency. Furthermore, the reduction or elimination of transportation reduces that demand for fossil fuels as well as allows and encourages people to work from home or another location where there is little or no travel. These are "smart communities."

Today, the agile sustainable and smart college communities continue to grow and are necessary for a less polluted environment and providing a "green" world for tomorrow. Many colleges in California are moving in this direction, primarily with the use of solar systems for on-site power generation. The solutions to global warming and climate change exist now and are growing on college campuses. The results will be graduates and communities who see the value (economic and environment) in making their communities sustainable with green, environmentally sound technologies for their surroundings and future generations.

FIGURE 3 Los Angeles Community College District (LACCD) Campuses. Top of the map is north and bottom is south. Location of the nine campuses over the Los Angeles County area. Two other new campuses are located near City College, one (east of it) is the former Van de Kamp Bakery Building complex and the other is the East Los Angeles (south), which is now a vacant lot.

How Do Communities and People Protect Their Historical Buildings and Communities. Case in Point Is Matera, Italy, Which Will Be the European Capital of Culture in 2019

Lucia Elsa Maffei[1], Elisa Castoro[2]

[1]*Esq. Studio Legale, Matera, Italy;* [2]*Independent Writer/Photographer, Matera, Italy*

OVERVIEW

The history of Matera is complex and fascinating: a fusion of landscapes, civilizations, and cultures. From the rock civilization over hundreds of decades to those of Byzantine and Eastern origin, to the advent of the Normans, the systematic attempt to reduce the rock city to the rules of the culture of the European city. The last eight centuries of construction and finishing of the city have tried to shape, to overcome the natural resistance of the preexisting rock habitat, determining architectures and urban arrangements of particular quality and originality, from the Romanesque to the Renaissance to the Baroque. Today, again under the banner of European urban culture, Matera faces the aspects of the challenge of renewal, sustainable recovery, and regaining of lost identity.

The city lies in a small canyon carved out by the Gravina—a rocky ravine created by a river that is now a small stream, dividing the territory into two areas, the Sassi (the ancient town) and the natural caves Park (Parco della Murgia Materana). Located in the mountains north of the Mediterranean Sea and near the Italian heel, Matera is inland on the side of a mountain.

As this photo shows, Matera was one of the first human settlements dated back to the Palaeolithic. Its architecture is magnificent and unique, consisting of the immediately visible habitations dug into the calcareous rock and the successive stratifications of houses, courtyards, ballotas, palaces, churches, and gardens, and of internal and invisible cisterns, neviere, caves, tunnels, and water control systems. The largest cistern has been found under Piazza Vittorio Veneto.

Matera preserves a large and diverse collection of buildings related to the Christian faith, including a large number of Rupestrian churches and monasteries, some of which are complex cave networks with large underground chambers. Probably the Rupestrian churches were already places of worship in the rock civilizations that preceded the Christian one.

For example, below is a picture of the Cave Church Madonna de Idris whereby the church is built both on and in the mountain site. This occurred hundreds of years ago. And the entire city has homes, offices, and businesses built into it. That is what Matera is famous for, as people do live and work in the mountain side today. Visitors from around the world come to see and experience this city and the surrounding areas as today the same historical values and their results exist in Matera. The earth is being saved, lived in, and protected for future generations.

The majestic Matera Cathedral—dedicated to Santa Maria della Bruna since 1389— the church of St John the Baptist, the Church of S. Domenico, and the Church of Santa Maria della Valle Verde on the Via Appia were built during the Middle Ages. From this moment onward a real urban nucleus takes shape, initially concentrated around the Cathedral, built in the Apulian Romanesque architectural style (the interior is on the Latin cross plan, with a nave https://en.wikipedia.org/wiki/Nave and two aisles).

The decoration is mainly from the 18th-century Baroque restoration, and a recently discovered "Last Judgement" portrait in a Byzantine-style 14th century fresco) and located at the top of the Civita hill dividing the Sassi in two: the Sasso Barisano facing east and the Sasso Caveoso facing south. Both in the Sasso Caveoso and the Sasso Barisano there are two other important churches dedicated to the Apostle Peter. San Pietro Barisano is the largest rock church of Matera.

The church of Santa Maria de Idris is located within the rocky spur of Monterrone overlooking the Sasso Caveoso, near the Church of San Pietro Caveoso and the homonymous square, offering a unique view of the city and the Gravina.

The church of Santa Lucia alle Malve—the first female monastic settlement of the Benedictine order—is located near Santa Maria de Idris in the Malve district. It is one of the most important Rupestrian churches in which there are some of the most beautiful and important wall paintings in the Matera area.

The historic center of Matera is located on a plain bordering the Sassi. As shown above, there are different levels of overlap for urban areas. Located below the central square Piazza Vittorio Veneto there are some openings that show the original level of the places, called hypogea. Dividing into continuity, the hypogea form a real submerged city connected with the Sassi. Here there are exceptional rock structures such as the large cistern called Palombaro Lungo until recently navigable; and the rock church of the Holy Spirit dating back to the 10th century AD. It is located just beyond the former convent of S. Lucia alla Fontana that divides the 18th-century church of the same name on the right and the Ferdinandea fountain on the left.

Moreover, the historic center of Matera is developed with streets connecting several public squares along what is the called "18th-century axis". In the center of Matera there are a series of important palaces and churches such as the church of St. Francis of Assisi—hypogea are also present here under the level of the homonym square. The most important noble and evocative buildings—Palazzo Gattini, Palazzo Venusio, Palazzo Malvinni Malvezzi—unraveling from Piazza del Sedile through the narrow Via Duomo come to surround the Cathedral of Matera, which has a Romanesque–Pugliese style but characterized by unique artistic elements, dating back to 1270. Here, the majestic bell tower dominates the whole amazing landscape.

Last but not least, Palazzo Lanfranchi, the National Museum of Medieval and Modern Art of Basilicata, conducts a series of exhibitions and cultural events of international importance, related to culture and history of the Matera City and regional territory

CONCLUSION: THE WORLD HAS CHANGED BY CONSERVING THE HISTORICAL PAST AND MOVING INTO THE SMART, GREEN AND HEALTHY FUTURE

AESs can (and have) created other market mechanisms at the local or regional levels, which are flexible systems rather than the traditionally central gird-oriented monopolies or new market forces, who would monopolize power generation. AESs, as they are local, allow for more public participation in setting public policy and hence provide strong public oversight.

The benefits of an AES go well beyond simply fixing the mess that deregulation has caused. AESs benefit early and quick adaptation of new technological breakthroughs. AESs favor reduced environmental destruction as they promote the health and social benefit of local communities. Finally such AESs are cost-effective solutions to a system that is now too costly on all levels and gone.

REFERENCES

California Hydrogen Highway. http://www.hydrogenhighway.ca.gov/.

Clark II, Woodrow W., 2007. The Green Paradigm Shift. Co-Generation and Distributed Generation Journal 22 (2), 6–38.

Clark, W.W., Bradshaw, T., 2004. Agile Energy Systems: Global Lessons from the California Energy Crisis. Elsevier, Oxford, UK.

Clark II, Woodrow, W, 2017. Agile Energy Systems: Global green distributed on-site and central power grid, 2nd Edition. Elsevier Press.

Clark II, W.W., December 06, 2007. Partnerships in creating agile sustainable development communities. Elsevier Press Journal of Clean Production.

Clark II, W.W., Demirag, I., Winter 2002. Enron: the failure of corporate governance. The Journal of Corporate Citizenship 8, 105–122. And Clark II, W.W., Demirag, I., 2005. US financial regulatory change: the case of the California Energy Crisis. Journal of Banking Regulation (special issue).

Clark II, W.W., Fall 2004. Hydrogen: the pathway to energy independence. Utilities Policy. Elsevier.

Clark II, W.W., Fall 2004. Distributed Generation: renewable energy in local communities. Elsevier, London, UK Energy Policy. And see Clark, W.W., Rifkin, J., Co-Chairs. Green Hydrogen Science and Technology Team composed of leading experts including Noel Laureates. www.foet.org.

Clark II, W.W., Isherwood, W., 2006. Energy Infrastructure for Inner Mongolia Autonomous Region: Five + California as a Nation-State Comparative Case Studies of Developed Countries. Asian Development Bank, Manila, PI and PRC Government, Beijing, PRC.

Clark II, W.W., Isherwood, W., Aceves, S., Berry, G., Fall 2004a. Distributed Generation: public policy in local communities. Elsevier Utilities Policy.

Clark II, W.W., Lund, H., December 06, 2007. Sustainable development in Practice. Elesvier Press Journal of Clean Production.

Clark II, Woodrow W., Michael, Fast, 2008. Qualitative Economics: toward a science of economics. Coxmoor Press.

Clark II, Woodrow W., Michael, Fast, 2018. Qualitative Economics, 2nd ed. Fall.

Clark II, Woodrow W., William, Isherwood, 2007. Special Report on Energy Infrastructures in the West: lessons learned for Inner Mongolia. PRC. Asian Development Bank.

Clark II, Woodrow W., William, Isherwood, 2009. Report on "Energy strategies for Inner Mongolia Autonomous Region". Utilities Policy Journal. https://doi.org/10.1016/j.jup.2007.07.003.

Energy Policy Act, October 2005. US Congress.

Energy's Future: Beyond Carbon, December 2006. Scientific American Special Issue.

European Union (EU) Parliament Rule (#116), May 2007. Smart and Sustainable Communities. Brussels.

LEED for Neighborhood Development: Rating System. With Congress for the New Urbanism and Natural Resource Defense Council (NRDC), June 2007. US Green Building Council, Washington, DC. http://www.usgbc.org.

Rifkin, J., 2004. The European Dream: how Europe's future is quietly eclipsing the American Dream. Tarcher/Pengin.

Seifer, Marc J., 2006. The Life and Times of Nikola Tesla: Biography of a Genius. Citadel Press, Kensington Publishing Corp, New York, NY, 1996, 1998 and 3nd Edition.

FURTHER READING

American Association of Sustainable Higher Education (AASHE), October 2006. Climate Neutral" Policy and Summit (2007), Adopted by Leadership Council. Tempe, AZ.

Bachrach, D., August 2001. Comparison of electric industry restructuring across USA. In: UC Berkeley and California Energy Commission, Governor Davis, Building for the 21st Century Infrastructure Commission, Sacramento, CA (Appendix).

Bradshaw, T.K., Winn, K., 2000. Gleaners, Do-Gooders, and Balers: options for linking sustainability, and economic development. Journal of the Community Development Society 31 (1), 112–129.

California Bio-Mass Collaborative. http://biomass.ucdavis.edu.

California Energy Commission, Future Energy Plans for California Report, Sacramento, CA. 2001. And see Clark II, W.W., January 2003. California Energy Challenge: solutions for the future. Energy Pulse: Insight, Analysis and Commentary on the Global Power Industry. http://www.energypulse.net.

California Hydrogen Business Council. http://www.CaliforniaHydrogen.org.

California Infrastructure Commission for the 21st Century, Invest in California, Sacramento, CA, 2002. http://www.ltg.ca.gov/programs/cb21/index.asp.

California Wind Consortium. http://cwec.ucdavis.edu.

Carlson, S., March 25, 2005. An engineering building as 'green' as possible. UC Santa Cruz", chronicle of higher education. Campus Architecture B11. Section B.

Chronicle of Higher Education, March 25, 2005. Campus Architecture, pp. B1–B32. Section B.

Clark, W.W., Feinberg, T., January 2003. California's Next Economy. Governor's Office of Planning and Research, Sacramento, CA, pp. 1–56. Local Government Commission web site at. http://www.lgc.org.

Clark, W., Isherwood, W., Ray Smith, J., Aceves, S., Berry, G., Fall 2004b. Distributed Generation: remote power systems with advanced storage technologies. Elsevier Press Energy Policy 32 (14), 1573–1589.

Clark II, W.W., Lund, H., December 2001. Civic markets in the California Energy crisis. UK International Journal of Global Energy Issues - Inderscience 16 (4), 328–344.

Clark, W.W., Paolucci, E., July 1997. An International Model for technology commercialization: fuel cells into vehicle process and design manufacturing. The Journal of Technology Transfer.

Cooper, J., Clark Jr., W.W., 1996. Zinc/Air Fuel Cell: a alternative to clean fuels in fleet electric vehicle applications. International Journal of Environmentally Conscious Design and Manufacturing 5 (3–4), 235–252.

Economist, May 11, 2004. Building the Energy Internet. Electronic version. www.economist.com.

Emerald Cities: the promise of green development. Special Section The American Prospect, January 2007.

European Union, Spring 2004. Boosting Innovation from Research to Market. European Union, Brussels, Belgium, pp. 65–67. And Clark II, W.W., Spring 2004. Innovation for a Sustainable Hydrogen Economy. Boosting Innovation from Research to Market, 65–67. Brussels, Belgium, European Union. www.partnersforinnovation.org. www.partnersforinnovation.org.

Executive Order S-20-04, September 18, 2006. Green Building Action Plan, Department of Finance Budget Letter 06-27 under Public Utilities Code Section 388.

Green Universities. http://www.grist.org/news/maindish/2007/08/10/colleges/.

Hebel, S., April 1, 2005. The Hard Birth of a Research University. UC Merced", Chronicle of Higher Education. Section A, pp. A1–A25.

Hvelplund, F., October 2006. Renewable energy and the need for local energy markets. Energy 31 (13), 2293–2302.

Intergovernmental Panel on Climate Change (IPCC), United Nations Environmental Programme (UNEP) and the World Meteorological Organization (WMO) established the IPCC in 1988. Four Reports (Assessments on Climate Change) were given with the last in 2007. http://www.ipcc.ch/.

Isherwood, W., Smith, J.R., Aceves, S., Berry, G., Clark, W.W., Johnson, R., Das, D., Goering, D., Seifert, R., 2000. Remote power systems with advanced storage technologies for Alaskan villages (University of Calif., Lawrence Livermore National Laboratory, UCRL-ID-129289: January 1997). Energy Policy. London: Elsevier.

Los Angeles Community College District (LACCD) www.laccd.edu.

Lovins, A., 2001. The Economic Renewal Guide: A Collaborative Process of Sustainable Development. Rocky Mountain Institute.

Lund, H., 2005. Large-scale integration of wind power into different energy systems. Energy 30 (13), 2402–2412.

Lund, H., Fall 2000. Choice Awareness: decision making at the local level. Journal of Energy Policy.

Lund, H., Munster, E., 2006a. Integrated energy systems and local energy markets. Energy Policy 34 (10), 1152–1160.

Lund, H., Munster, E., 2006b. Integrated transportation and energy sector CO_2 emission control strategies. Transport Policy 13 (5), 426–433.

Marcus, W., Hamrin, J., 2001. How we got into the California energy crisis. In: Paper given at National Renewable Laboratory, March 01. Center for Resource Solutions, San Francisco, CA.

New York, NY Brundtland Report United Nations World Commission on Environment and Development, 1987.

Porter, M.E., van der Linde, C., September–October, 1995. Green and competitive. Harvard Business Review 120–134.

Young, R., August 2007. An epiphany. web link Chronicle of Higher Education and ASSHE.

Chapter 3

Integrated Cross-Disciplinary Studies: Circular Economics With Cases in Japan

Woodrow W. Clark, II [1], Danilo Bonato[2], Andrew DeWit[a,3]

[1]*Research Professor in Economics, Pepperdine University Graziadio School of Business (PGSB), West Los Angeles, CA Campus, Los Angeles, California;* [2]*Managing Director ReMedia, Milan, Italy;* [3]*Professor, School of Economic Policy Studies, Rikkyo University, Tokyo, Japan*

Chapter Outline

OVERVIEW

"Qualitative economics" (QE) was an idea that I had with a colleague or mine (Professor Michael Fast) for a book and journal articles about 5 or 6 years ago. Then after participating in several Western Economic Association International (WEAI) conferences, I did one paper and a panel on status of economics. This was in the middle of the first decade of the 21st century. When my friend Professor Michael Intriligator (now deceased) was president of the

a. Clark was Professor of Academic Specialist, Cross-Disciplinary Scholars in Science and Technology at UCLA and Managing Director of Clark Strategic Partners. Danilo Bonato MBA and is Managing Director of ReMedia Company in Milan, Italy.

Climate Preservation In Urban Communities Case Studies. https://doi.org/10.1016/B978-0-12-815920-0.00003-4

WEAI, I talked to him about doing a special issue on *The Next Economics* (NE) which was based on QE, for one of the two journals that WEAI published each quarter. After talking with the publisher, Professor Wade Martin, we all decided on contemporary economic policy (CEP) for a special issue to be called: "Global Cases in Energy, Environment, and Climate Change: Some Challenges for the Field of Economics."

Professor Intriligator became my coeditor, and we gathered 11 papers covering this topic. In the CEP special issue, only five of the papers are to be published. Two of the papers are changed and published here in NE, along with the other six that were not accepted for the CEP. The editorial opinion about the six papers left out of the CEP, but published here, was basically that these six papers were not traditional papers in economics. Hence, the purpose of this book is to expand traditional economics by examining and providing cases of economics of this field but applied to environment, energy security, and climate change (CC) topics. The issue for everyone today is that the costs for saving the environment and solving CC are unknown and often given as an excuse to do nothing. The most common comment from the traditional neoclassical economic paradigm is that the "market" will find solutions. This book directly counters that assumption with solid data and findings in order to create economics based on science, rather than politic opinion and policies.

Without doubt, economics needs to be applied to these global societal areas. What was not done in the CEP was an analysis of the philosophical history of economics in terms of how it has impacted global environment and CC issues. The paper that addressed that issue from Professor Michael Fast and myself is revised as a chapter in this book. That chapter argues for a looking at economics in a far different manner and theoretical basis than the field has had over the last two centuries and especially the last 40–50 years. This is the next step; there needs to be a far more comprehensive examination of what, where, and why economics has come from only one western philosophical base and not others. Adam Smith was not the only economics philosopher over 200 years ago. There were others. And there were many economists who were from different cultures and wrote in different languages with positive results from their approaches to economics.

Japan and now China are cases proving that point. Later there will be more examples from both Japan and China. But consider this case from China, as it has been at least 2–3 decades behind Japan. And now it has almost leap-frogged ahead of the United States. Consider the case of the capital of China, Beijing. Most pictures of Beijing that western nations see are of a City in smog year-round. However, as Clark goes to Beijing two to three times a year now, this is what the common and current environment is there. Although this picture was taken in 2015, the atmosphere is very much like this. Clark has dozens of pictures that show the same skyline at different times of the year. He also counts how many people he sees wearing masks over their faces—the common western picture of people in Beijing. From over thousands of people in the last 4−5 years that he has counted, only 12−15 wear a mask over their face. Note also the buildings but the grass and parkland areas too.

From Dimitri de Boer (October 2015). Image by Roos mei.

As de Boer put it in the China Dialogue News back in October 2015, "Blue skies over Beijing have been a rarer sight in the past decade - but a sweeping new government document could help China to reduce pollution." That is exactly what is happening now in China.

What was the problem in China for the last 2−3 decades was that the national governments were following the western ideology of Adam Smith, and his classical economics became propagated as the truth and factual. That was wrong then and is now. Yet while Adam Smith prided himself in taking ideas and concepts from physics and mathematics, he did not use either science as the base in calculations and formulas for economics. Instead, Adam Smith used the concept of a balance between physical forces, which in the end would work in field of economics as the balance between supply and demand. However, in reality, physics is not only about balance, nor is mathematics. What the world has witnessed in the last decade is that market forces as western capitalism certainly do not demonstrate such as balance. The global economic collapse in the fall of 2008 appears to be coming back in early 2018. Science is far more than the balance of physical forces. For that matter, any science, be it chemistry, engineering, mathematics, or others, is not based on balances alone. And clearly no science has an invisible hand.

So what is science? Consider "physics (which) is often described as the fundamental science, as it seeks to understand the 'rules' or 'laws' by which the universe operates" (Perkins, 1996). What NE does is to set the stage for a far more in-depth and global investigation of economics than the CEP special issue. Consider now a summarization of the issues in each chapter of NE so that they can provide background and guidance to further study and analysis in order to create a science of economics.

THE NEXT ECONOMICS

Let's start with Rifkin, who is a well-known environmental economist, in his book, *Entropy* (1980), which discusses the extension of economics beyond the neoclassical theoretical paradigm of Adam Smith to include social issues such as the environment. Perkins, a physicist, notes that "entropy" is a scientific way to describe how "the universe is running down and getting more disordered" (Perkins, 1996, p. 3). That concept might apply to neoclassical economics, but certainly does not provide a direction in which economics must move, due to its consistent failure over the last few years to become a science.

In Rifkin's book, *The Third Industrial Revolution* (2011), he tries to make the case that economics needs to be connected to CC. And in order to do that, he argues that thermodynamics is the key, as it was the basis for making neoclassical economics into a science. Rifkin describes then how thermodynamics explains the balance between inputs and outputs as that can apply to the environment and other externalities, which make up the third industrial revolution.

While this was an attempt to explain economics as a science, and in particular to address the concerns for the environment and CC, the argument falls short. Four basic issues remain with economics, which Rifkin and other fail to address. The primary one is the acceptance of Adam Smith's theoretical basis for economics being scientific. The theories of the 17th and 18th century are dated due to the use of science in a limited manner and rooted in this historical century knowledge. Since then, science has developed and expanded with new theories and extensive research.

Second is the focus on science in a limited manner. While the traditional link between Adam Smith and Sir Isaac Newton remains the basic barrier for revolutionizing economics, it is the wrong approach rooted in the wrong assumptions; what is wrong are the particular and limited aspects of science that Smith used from Newton.

In short, science is much more than thermodynamics and the balance between forces with what economics has come to label "invisible hand." In science, there is no invisible hand. In order for something to be science (like economics should be), everything needs to be known and accounted for; repeated, and then made predictable. Economics fails to do that. As Perkins puts it, "the vast majority of these physicists had in common a few essential things: they were honest, they actually made the observations they recorded and they published the results of their discovery of the 'rules' in a form that others could duplicate and confirm. This is the basis of physics as a science." (Perkins, 1996, p. 5). That is not what economists do today or for more than half a century now. For most of them, they play the role of the invisible hand becoming visible with offering opinions, political conclusions, and even plans, which are never based on repeated analyses and predictions.

Economics as a Science

The problem with economics moves into the third issue directly. What "role model" is there for the field of economics to become a science? Economics is not a science. But there is one area of the social sciences that is scientific, particularly linguistics. Above all, economics needs to be modeled on linguistics. In *Qualitative Economics*, Clark and Fast (2008) make this point. Linguistics, particularly through Noam Chomsky and his transformational grammar work over the last 60+ years, has become the model for science in an area of research that was once considered social science. For example, see Chart 1 that shows how all languages are structured with a Surface Structure that is used in communication and also a Deep Structure that provides the meanings and definitions to words, numbers, sentences, graphics, and even more.

As above, surface structures are simply those activities and words that people hear and read from the firm. Aside from the annual report and tax statements, firms are quoted in the stock market, give briefings to analysts, and they lobby with particular perspectives to politicians. What is never known in the public are the behind-the-scenes realities of the business and its operations.

The surface structure is clearly seen in the activities of Enron, now that the behind the scenes have themselves become public since its bankruptcy in

Universal Grammar

```
T------Surface Structures-----------------------------------------------^
R      (Syntax -- the formal structure of language)                     |
A                                                                       |
N                                                                       |
S                              R                                        |
F                              U                                        |
O                            RULES                                      |
R-------Deep Structures-------------------------------------------------^
M      (Semantics -- meaning to words/sentences)                        |
A                              E                                        |
T                              S                                        |
I                                                                       |
O                                                                       |
N                                                                       |
S-----------------------------------------------------------------------v
```

CHART 1 Representation of linguistic theory.

December 2001. Below, we identify these activities as "deep structures," as they not only explain the firm's behavior, but provide predictions for its future activities. At this point, what firms want to do is satisfy their shareholders, investors, and bankers. Demirag et al. (1999) labeled this corporate behavior as "short-termism" since it was only focused on quarterly earnings statements. The motivation for corporate leadership and behavior, therefore, was the "bottom line." Bonuses and rewards are given to company executives if they can influence and enhance the bottom line each quarter.

The revelations of many corporations came about as a result of "accounting" irregularities. Nonetheless, as more and more evidence is amassed, it is clear that the corporate executives of firms either directed or agreed with the accounting schemes, since the very positive economic reports personally enhanced their own incomes and bonuses.

As also noted above, there are "deep structures" in the science of linguistics. Hence, it is important to know what is a firm, company, or business is all about. Not just their CPA and IRS accounting. How did they get there? Why? And who is involved? These are "deep structure questions." For example, most NewCo companies start with the entrepreneur. Ask most families, spouses, and colleagues of any new business venture, for example, and the answer would be that the entrepreneur is a "bit crazy." Schumpeter characterizes the entrepreneurial experience as that of a sort or "heroic figure." The "cowboy" image has now achieved the folklore status of a "cult" figure today, as with Elon Musk. However, the entrepreneurial experience is not that of only one individual.

Many entrepreneurial ventures are the results of partners, usually forming a balanced relationship. Network theory underscores that point whereby social networks either form new ventures or support them (Clark and Sorensen, 2001). Many new companies are the result of family or extended family relations. Yet within the new venture, one person tends to exhibit the basic characteristic definitions of an entrepreneur.

If following our use of transformation theory, the basic characteristic definitions of a firm can be described in a common lexicon; then a universal grammar exists for how companies actually work. With the use of theoretical considerations from linguistics, transformational rules allow explanations and predications to be made of key people as they interact with a variety of actors and in numerous situations.

The energy crisis in California has been a reaction at the surface structure level, but also offers a far more comprehensive understanding of how firms act. Through various public legal proceedings and hearings, deeper understandings of a firm's actions are now known. This is true in the context of bankruptcy, declared by Pacific Gas & Electric (PG&E) in the late spring of 2001, whereby testimony had to be given before the California Public Utility Commission. Such evidence indicated behavior by PG&E heretofore unknown.

Court testimony and legal depositions are a mechanism for getting at the deep structures of a firm. Some public interest groups have also used the American Freedom of Information Act to gather information and data often withheld by companies. This is the case now with the bankruptcy of Enron and with a number of other American companies in not only the energy sector, but also telecom companies. What will most likely occur is that a number of the corporate leaders (actors) will be found responsible for suspicious if not illegal actions taken by their companies to enhance and falsify their corporate financial performance (surface structure).

However, what does this mean for the need to research firms before tragedy (either in terms of an energy crisis or a firm bankruptcy) occurs? The issue for economics is how to get at the deep structure of a firm in order to describe, understand, explain, and perhaps predict its actions. This deep structure information about any firm is critical for any scientific consideration of economics. The aggregation or identification of principles and then rules from the actions of a firm allow economics to then be scientific.

Chomsky's books, *Reflections on Language* (Pantheon Books, 1975) and *Syntactic Structures* (The Hague: Mouton & Co., 1957), are among other books and articles that provide the details needed for documented proof on how linguistics is a science, which economics should follow. For example, Chomsky's first book in 1957 was first published in the Hague because no American or English-language publisher wanted to print a book that talked about linguistics becoming a science. Yet Chomsky's arguments and work since then turned a corner for the scientific study of languages. The point is that Adam Smith and those who interpret his work today, are still using the scientific philosophy, theory, and knowledge from hundreds of years ago, while there have been more recent breakthroughs in science within the last 50 years. Linguistics is the outstanding example.

Chomsky asks, "What is the 'science-forming capacity' that enables us to recognize certain proposed explanatory theories as intelligible and natural, while rejecting or simply not considering a vast array of others that are no less compatible with evidence?" (Chomsky, 1980, p. 250) Basically, science must describe, explain, and predict phenomena. Scientific statements are only valid if they can be replicated and proven through predictions. Modern linguistic theory led by Chomsky has been able to do just that in a nonphysical and natural science environment. The key is to extend the construction of scientific theory for languages beyond the descriptive phase and into an explanatory and predictive phase.

As described in QE (Clark and Fast, 2008) from Chomsky: by way of example, consider a typical linguistic situation. A sentence (S) contains a noun phrase (NP) followed by a verb phrase (VP) or in symbols, represented as (NP VP → S), where among the categories that figure in the categorical component are the "lexical categories," noun (N), verb (V), adjective (A), and others (Chomsky, 1980, p. 80).

Therefore, the representation of the parts of the sentence are made into symbols that allows the surface structure (spoken language) to be broken down into components. An arrow denotes the horizontal and vertical transformations (→ or $_{V}$) from the deep structure (NP → VP) into the surface structure sentence (S) and its "deep structure" or the meanings of the words, phrases, and sentences.

Today, Chomsky remains a leader in scientific thinking through linguistics as it needs to apply to societal and economic issues (Chomsky, 2012). The "environmental catastrophe. Practically every country in the world is taking at least halting steps towards trying to do something about it. The United States is also taking steps, mainly to accelerate the threat. It is the only major country that is not only not doing something constructive to protect the environment ..." (Chomsky, 2012, p. 5).

Finally, mathematics is the other scientific model for economics in both formulas and processes, including definitions of numbers, symbols, and their results. As Perkins (Perkins, 1996, p. 1) put it, "An associated feature of this simplicity or beauty is that the 'rules' can be written down very elegantly in mathematical form. This can be a problem when we first learn physics because our mathematical skills are not usually sufficiently developed."

Consider the definition of mathematics in Wikipedia:

Mathematicians seek out patterns and formulate new conjectures. Mathematicians resolve the truth or falsity of conjectures by mathematical proof. The research required to solve mathematical problems can take years or even centuries of sustained inquiry. Since the pioneering work of Giuseppe Peano (1858–1932), David Hilbert (1862–1943), and others on axiomatic systems in the late 19th century, it has become customary to view mathematical research as establishing truth by rigorous deduction from appropriately chosen axioms and definitions. When those mathematical structures are good models of real phenomena, then mathematical reasoning often provides insight or predictions about nature.

Through the use of abstraction and logical reasoning, mathematics developed from counting, calculation, measurement, and the systematic study of the shapes and motions of physical objects. Practical mathematics has been a human activity for as far back as written records exist. Rigorous arguments first appeared in Greek mathematics, most notably in Euclid's Elements. Mathematics developed at a relatively slow pace until the Renaissance, when mathematical innovations interacting with new scientific discoveries led to a rapid increase in the rate of mathematical discovery that has continued to the present day.

Part of this definition of mathematics include some discussion of the "field of mathematics" and how it became scientific. "Mathematics can, broadly speaking, be subdivided into the study of quantity, structure, space, and change (i.e., arithmetic, algebra, geometry, and analysis). In addition, to these main concerns, there are also subdivisions dedicated to exploring links from the heart of mathematics to other fields: to logic, to set theory (foundations), to the empirical mathematics of the various sciences (applied mathematics), and more recently to the rigorous study of uncertainty."

Perkins summarizes well the relationship between rules and mathematics (the core of implementing the scientific process) as applied to energy:

> *However, knowing sufficient mathematics enables us to write the rules down very simply; and very importantly, allows us to calculate many useful things. For example, knowing the rather simple mathematics of electricity allows you to calculate the rating of the fuse required in your hi-fi. An alternative is to pay an electrician $75 to do it for you! Again, knowing some of the mathematical expressions of nuclear physics permits us to calculate the electrical energy that can be generated from 1 kg of uranium. The answer is about the same as you would get from 1000 tons of coal! (Chomsky, 2012, p. 3).*

In the NE book, Clark quotes and cites each chapter to make the point about how economics can move from a field to become a science. The idea is to give the reader an overview of the main points in the chapters and the book itself. Hence, only the salient points and issues are covered that need more economic study. In particular, the goal is to stimulate further research and review of the field of economics. The time has come for just that. The purpose, in short, is to conclude what has become a pattern in economics, which leads to a global economic paradigm, if not total philosophical change in the field of economics itself. In short, NE helps make economics a science.

For example, recently an article was published (Simmons et al., 2011) on "false-positive psychology" about some research and a paper published by colleagues at the University in Netherlands that showed how research results were questionable, but, even more significantly, were used as false data for a peer-reviewed published paper. As Simmons and his colleagues put in their paper: "Our job as scientists is to discover truths about the world. We generate hypotheses, collect data, and examine whether or not the data are consistent with those hypotheses" (Simmons et al., 2011, p. 1). While this is correct, science is also about replicating the hypotheses and thus providing predictions.

Thus, when economic history examines the struggles over energy not only today in the Middle East, with the United States now engaged in the longest and costliest war in its history, but then examines US energy security in terms of other world wars in the past two centuries, the problems in terms of people, communities, and economies are staggering. There are no calculations or numbers that even come close to the results of these conflicts. What we can do, and have tried to provide some examples and cases, is ask questions and provide economic data that provides guidance and thought-provoking debate about economics itself. Now, today, economics needs to add societal issues such as health and environmental impacts into its numbers and formulas, not only locally, but internationally to understand and change the results of CC. The world is round. Hence, vast areas of land are damaged, the air is toxic, and water is polluted due to ocean and atmospheric changes from one region to another. Therefore, the numbers and calculations reported cannot even come close to being accurate. The economic costs must be comprehensive, based on hypothetical cases that are observed, examined, redefined, and tested in order to be scientific, hence accurate and predictive.

What was done in the CEP special issue is a start to make economics a science.

Economics is going through an enormous paradigm shift as it develops into a science from just being a "field of study," especially now with a focus on CC that could help to be illustrative of the entire field. In fact, applying some of the philosophical roots, analytical tools, formulas, and methodologies from the collection of these papers should provide a roadway and map into the future of economics. Certainly there will be twists and turns. Even a few dead-ends and off-ramps, but economics needs just that.

Without a doubt, economics is in a quandary while it tries to find its direction and even a new paradigm. What we have done is provide the basis for that discussion and examination into the field itself. So what is the new paradigm? Where is it going, and what will it do? We have a few initial thoughts and ideas about that and the challenges that go with it.

The *Wall Street Journal* (Bailey and Wolfram, 2012), for example, ran an article about how energy has become a key part in addressing what communities and countries can do about CC.

In the article, the concern was how renewable systems were only available to the rich due to their "hefty cost," which thus restricted and limited the "market for residential solar installations (for example) to cash-rich homeowners, restricting the potential for growth" (Bailey and Wolfram, 2012). Yet this "innovation," like many others, needs to be short term and tightly controlled.

The basic problem is that leasing contracts are stranded costs in that they must be paid back over a long period of time, with usually an additional purchase price. And the financing is higher due to third parties providing the financing at a price above the actual purchase costs to the person leasing the energy system. What is more useful to do is to think and consider other finance models: for example, the use of mortgages for including energy (renewable power, storage, refueling, gird models, etc.), just as water, waste, electric along with air conditioning and heating are done today within the costs for most buildings. Such financial innovations are now being included in current laws being considered and passed in California. Now the costs for renewable energy are another asset valued in a building that are bought and sold with the building.

One of the significant factors in looking at CC is the continuing role of the United Nations and all nations to find resolutions. The fact that yearly meetings are being held to find answers to CC had not resulted in much. Part of the reason concerns the nations involved. Another key part is the politicians and to whom they report when they are back home. But in the end, it is economics. For example, the UN IPCC (Intergovernmental Panel on Climate Change) has been debating if CC is a result of humankind or the natural evolution of our planet. In 2007, the UN IPCC stated clearly the problem was humankind. However, in that debate was the underlying conflict and theme rooted in economics: can humankind adapt (that is, live with) or mitigate (that is, stop) CC? In short, what are the basic economics and costs associated with one or the other of these areas?

Now let's move into the core issues surrounding CC: economics and the people who control or oversee groups, companies, and countries who benefit either through adaptation or mitigation or both. That core basis for debate within economics leads the chapters. Consider, however, a very important philosophical perspective about economics that sets the forum for a dramatic shift in economics and could lead to the field of economics becoming a science.

According to Adam Smith, the founder and most well-known creator of modern western economic theory that is often noted as being foundation of "classical economics," the use of taxes for both controlling consumption and considered financial resources can be: "Sugar, rum and tobacco are commodities which are nowhere necessaries of life, [but] which are … objects of almost universal consumption and which are therefore extremely proper subjects of taxation." (Toedtman, 2012). The application of this basic neoclassical principle that is never mentioned in the modern economic theory over the last 4 decades is applicable to economics of CC.

The chapters in this book discuss the "basic problems in the field of economics" as it impacts externalities such as CC, environment, and national security. In that regard, one of the fundamental problems rests with how to analyze economics and the environment. Dole points out several economic approaches that are meant to solve environment problems. He concludes after reviewing cap and trade (CAT) economics that has become popular with environmental policy makers, numerous NGOs, and businesses. Part of the issues that President Obama faced with his attempt to create a national energy program was with "cap and trade" economics. Dole notes in "Market Solutions for Climate Change" that command and control (CAC) works in many ways. CAC is fairly successful in Nordic countries and China where government oversight and even control of economics is accepted and successful.

However, as Dole points out, CC is questionable in the United States, primarily for political reasons. After reviewing market solutions to CC and especially the current focus in California on CAT, Dole argues that the only economic way to mitigate CC is through a carbon tax. Adam Smith would agree with that, but today's political interruption of his economic paradigm disagrees with that. Hence, in the classical economic paradigm, laws need to be created and managed that make polluters pay for their impact on the environment through pollution and emissions. The issue of economics becoming a science is not solved, but the classical economic paradigm of Adam Smith would agree with that policy.

Li and Clark look into what these new approaches to economics might actually do on the microlevel, but also how to characterize the new emerging economic paradigm through national planning such as "energy economics in China." Both authors have worked in China and find that their way to characterize all these concerns over the environment and economics should be seen and described as "social capitalism." Below are some social ideas (CC and the environment) and economic applications for financing renewable energy systems in order to make them affordable. The key issue is how renewable energy, one of the more significant, concrete ways to solve and slow CC, can be economically viable for people in their homes, work, and public service roles. In short, to mitigate CC, affordable technologies and systems must be available now and not just in 10–20 years.

The second section of articles focuses on the "next economics," starting with an overview of what this means for moving economics into a science through a different historical and philosophical perspective that applies to the environment, energy, and CC. The other chapters in this special issue discuss issues and ideas in this area, but there are missing pieces. The critical one is the health costs to people, with CC surrounding everyone's daily life.

THE GREEN INDUSTRIAL REVOLUTION: AS GLOBAL GREEN DEVELOPMENT

Then, as Clark presents in his NE book chapter, there is a green industrial revolution. Here, the difference in the new economic paradigm is that more than economic numbers must be placed in the equitation. One of the key concepts concerns communities being sustainable. More will be said about that below. But at this point, the key issue is that communities of every kind are configured through infrastructures. Today, most economic analyses are focused on one area or another, like energy, transportation, water, etc. Few economic analyses study the overall integration of the entire community infrastructure systems. Even more significant are the needs of areas not usually considered in economics like the environment, CC, and health issues.

Awareness of these "externalities" to economics has become prominent with the reports from the UN IPCC and debates among nations over the Kyoto Accords. What has become obvious in all of these discussions about the environment is that the western nations are dependent on neoclassical economic theories. However, these economic theories fail to be able to take into account the perspectives of all western nations and certainly leave significant economic questions in others.

Fast and Clark next present a new philosophical approach to economics altogether. In his book with Clark, QE, in 2008, they both provided a challenge to economics from the basic ground level: philosophical roots to western economics. In that book and this paper, the fact is that there are other western philosophical theories, some of which are close to Asian theories and philosophers, that challenge the roots of neoclassical economics. However, moving into the modern 21st century, the legacies of the 20th century economic theories are questionable. The subtitle of QE, originally drafted and written in the late 1990s and revised in the mid-2000s, is "toward a science of economics," exactly the same concern of *The Economist* special issue with the Bible melting on the cover about "modern economic theory" (July 16, 2009) failing by it documenting significant concerns about western economic theory in mid-July 2009 after the global economic collapse 9 months earlier in October 2008.

If indeed, economics is not a science and perhaps an art or social study, then how can it become a science? As the QE points out and documents at length, science is a not a matter of numbers and formulas. Instead, science of any kind concerns creating and testing hypotheses, one after another. The use of numbers and statistics is not as significant as the mathematics and philosophical basis on which the hypotheses and data is centered. Hence, the science of economics does not exist.

Also if that is the case, how can economics become a science? This is the challenge. What QE did was look at various sciences and social "sciences" to which Fast cites in his article as there being only one: linguistics. The creation and growth of modern linguistics can be directly attributed to Noam Chomsky at Massachusetts Institute of Technology. His work for almost half a century has made linguistics a science—one in which the field of economics needs to research and learn from. Fast presents some of those structural forms of linguistics that make it a science. QE presents some case studies with these applications from linguistics to make it a science of economics. See Clark and Lund (2006) for looking at economics in the context of sustainable global communities, and then see Clark (2007) as to how new economics focused on the environment and energy applies to emerging Asia nations (Clark, 2007) and China in particular (Clark and Isherwood, 2007, 2009, 2010).

Each of these chapters and many more now as well as those in the three sections of the book cover important economic topics in CC. What they all conclude is that economics cannot be narrow and hence limited. Science is certainly not that way. The focus on sustainable communities (Clark, 2009, 2010) provides a basis for cases and baseline data from which further hypotheses can be made, data gathered, and concrete scientific economic data derived. The result is that sustainable communities are one basis for the economic understanding of CC. If humankind is to resolve CC and go beyond adaptation to it, then there must be hard scientific economic ways to mitigate and stop CC.

Critical to understanding energy are its economics. As Matteson takes the topic on a regional level, using California as a case, he then moves into the national needs for the United States to produce a national energy plan. His data provides a basis for prediction. And the results are concerning, as they provide comparative data on the impacts of energy production to the environment. Professionally, Matteson spent most of his career managing the energy system for the University of California, Berkeley, and then the entire University of California system with its 10 campuses. He was able to see what the energy demands were for the campuses.

Then California Governor Brown went to the meeting in Beijing, China, in June 2017 before the United Nations met for the G20 Summit in Germany. Clark was interviewed live on June 11 from Los Angeles, California, for an Energy Conference in Beijing before the UN Paris Accord meetings hosted by Germany in July 2017. Clark foresaw the United States pulling out of the UN G20 so that the "nation-state" of California (Clark, 2017c,d) would stay in the UN G20, which is now G19. Over 170 other states, provinces, and cities around the world agreed with California, with some west and east coast American states joining California and the UN G19.

What are particularly concerning are those energy needs and the sources for the energy as the shift from coal plants and power transmitted from long distances in other states. Since California has no coal resources and limited natural gas, the predominant energy resources for the current state's economic growth are renewable energy power from solar, wind, hydro, run of river and ocean. However, there are significant barriers to the consumption of renewable energy sources due to the control of the power being transmitted primarily through central energy producing systems controlled by a few large power utilities.

While Matteson had been an advocate of energy deregulation power generation, he saw what happened to California in the early part of the 21st century. The solution for the future, from his data, was not just centralized power generation from a small group of utilities, either as government or private businesses, but on-site power for buildings, residents, complexes like shopping malls, offices, and academic institutions. The near and long-term future needed to plan for this shift from central power plants to local power generation. This was the same conclusion in 2004 and then the update in 2017 of Agile Energy Systems (Elsevier Press) about the need to have a combination of both current central power plants and on-site power systems.

Grose then outlines and discusses in detail the chapter on "achieving economic gains," the "core green economy" needs to be adaptive within the rest of the economy. The chapter reviews conventional economic issues (15 in all) as segments to the "core green economy" ranging from efficiency, transportation, and infrastructure to water, waste, and agriculture to advance materials and buildings, but not including renewable energy generation. The analogy to the IT industry for the green economy is similar to Rifkin's recent book (2011). And some of the ideas, including the need for public policy, specifically with the case of California, are useful.

But the key issue in her chapter on the link is also to achieving these economic gains with "the setting of environmental goals." The case of California in its energy crisis at the turn of the millennium is important. Providing public policies on conservation and efficiency along with setting a renewable energy portfolio standard and clean car act, among others, are important. However, public policy(s) needs more, as the last decade has demonstrated: financial and monetary support. In short, the policies need money in order to implement them. These do not have to be public finances and/or incentives but some actions that correct the problem, as California had, in the energy sector, with the public policy for "deregulation," which caused a major economic crisis in California that continues today and now is in the United States and other nations (Clark, 2017a,b,c,d).

The issue of adaptation is the key. Does it do enough? When Rifkin uses "entropy" in that way to relate economics from science to the environment and CC, it too does not go far enough. While the United Nations debates, even today, what to do about CC, the basic problem is between those nations who want to adapt to it against others who argue that CC must stop or be mitigated and reversed. The latter view of global climate and environmental perspectives sets the stage of what must be done, since vast regions of the world are continuing to experience severe atmospheric and ocean changes.

Adapting to these changes needs to be done for survival, but mitigating them is the only real solution. This conflict in philosophy is the same as the consideration of economics today in areas that are new to the field. However, if the more comprehensive and global perspective is taken in economics, then economics becomes a science. It must be scientific to get results, replicate them, and construct rules that are measurable and evaluated again and again and again. This science of economics brings into it other critical areas and the actual solutions to CC. That is, what can people, governments, and businesses do about it. There is no "invisible hand" that applies to science or the real world.

As Clark and Li provide in their chapter on "social capitalism" and this book, there is a need to think of basic economics in a new and different way. "Social capitalism" is a big part of that way of thinking, and how to make economics scientific. In other words, making money is fine within economics, but economics still needs to connect to society and social issues like the environment. Interestingly enough, *The Economist* appears to be interested in these more historical models of economics as well. In a special issue on state capitalism (Economist, 2012), *The Economist* talked about how emerging nations were merging government and market economics in order to construct and build their economies.

While *The Economist* did not agree with that economic model, although it had been done historically in all western nations, the conclusion was that it worked in the nations today known as BRIC (Brazil, Russia, India, and China). In fact, *The Economist* added Chile, Argentina, and some Eastern European nations as well. Each of these nations are rapidly becoming powerful economic powers that challenge the past 100 years of both EU and North American hemisphere domination.

China has become the most significant and outstanding case in point, as Clark and Li discuss, but also as Clark and Cooke (2011) provide a global perspective on energy innovations. The Chinese 5-year plans are examples of what they do as a nation implementing social capitalism. In short, over 1.3 billion people have a national and then local plan to use a roadmap. In modern, 21st-century China, those 5-year plans are indispensable for businesses and international policy and corporate leaders.

So what are the economics that need to be examined?

THE ECONOMIC PARADIGM CHANGE: STOP AND REVERSE CLIMATE CHANGE

Many communities, cities, and other organizations such as academic institutions and private sector business recognize the need for policies that direct their facilities and infrastructures to be "green" based upon some criteria, such as the US Green Building Council (USGBC) certification for achieving LEED (Leadership in Energy, Environment, and Design) from basic criteria to higher standards. Individual buildings are to be "net-zero" carbon emissions. Many organizations are seeking to make their entire facilities "energy independent and carbon neutral." Only since June 2007, the USGBC has created "community" or LEED neighborhood standards. This set of criteria reflects the broader concerns for clusters of buildings with designs that are integrated with basic infrastructure needs.

Developing dense, compact, walkable communities that enable a range of transportation choices leads to reduced energy consumption. "Communities" thus has a broad definition because they can range from college campuses to cities, towns, and villages that are self-sustaining and provide for multiple uses ranging from housing, education, family events, and religion to business complexes, shopping streets, malls, and recreational activities. Thirdly, a sustainable smart community is a vibrant, "experiential" applied model that should catalyze and stimulate entrepreneurial activities, education, and creative learning, along with research, commercialization, and new businesses.

Communities can be sustainable in that new economic and social programs can be created or recreated. However, the difference today is the need for sustainability to include renewable energy resources such as wind, solar, biomass, ocean, geothermal, and "run of the river" (not large hydroelectric dams or nuclear energy plants) energy sources that do not impact the environment negatively, but instead are green and clean. Therein lies a "paradigm change" to all agile energy systems, which combines local renewable power and fuel resources with grid-connected ones.

Because of global concerns, many nations and now regions, states, communities, and cities have developed their own policies to increase renewable energy power generation as part of the solution to respond to the threat of CC. Since the primary infrastructure sectors that impact global warming are energy and transportation, they must be examined in order to find ways to reverse the warming of the Earth. One key element in achieving such goals is to consider how renewable energy can impact and change the transportation sector to be more environmentally sound and sustainable. Several different technologies have been put forward, but in practice, no single technology can solve the problem on its own. Many different contributions have to be combined and leveraged to coordinate with parallel activities in the energy sector. The Pew Charitable Trusts reported that since 2005 (by 2009), there was a 230% increase in clean energy investments (Hershauer et al., 2010, p. 4).

On the local or regional level, sustainable and smart communities must have three components: Primarily, there is the need for a master strategic plan for infrastructure that includes energy, transportation, water, waste, and telecommunications, along with the traditional dimensions of research, curricula, outreach, and assessments. Secondly, there is the array of issues pertaining to the design, architecture, and siting of buildings and overall facility master planning, which must be addressed from the perspective of "green," energy-efficient orientation and be designed for multiple use by the academic and local community.

The California energy crisis was the tip of a much larger problem in California, as well as in the United States and other industrialized nations. While private companies took over much of the state's energy-generation capacity, related and similar issues confront California and other nation-states such as infrastructures for water, waste, and transportation that are separate but interconnected sectors (Clark and Bradshaw, 2004). These sectors remain in a crisis mode but are ignored and unattended, since there have been few visible crises in them. However, the impact of hurricanes and storms as a result of CC on communities in the southern part of the United States and globally has begun to make the public more aware and ready to take constructive action, starting with the need to conserve and use energy efficiently.

Even though California is one of the states that uses the least amount of electricity per capita, the impact of its dominant use of fossil fuels (about 58% from coal and natural gas) negatively impacts the environment and pollutes the atmosphere. It is far broader than energy deregulation. It is attitude, lifestyle, ethics, policy, psychology, etc. The simple policy of deregulation is but the microtip of an incredibly complex series of issues about global warming, waste, and misuse of natural resources, etc.

A part of the solution that came from the California Energy Crisis (2000–2003) is the creation of such as what I have called "agile energy systems" in which communities have clusters of buildings, colleges, local governments, residential divisions, shopping malls, and office buildings that have their own "on-site power generation systems" (Clark and Bradshaw, 2004). The book Agile Energy Systems is about how energy must be generated and distributed to both central grid (usually a public or government managed company) that gets energy from coal, natural gas, hydro-electric dams dams and even nuclear plants. Today now there are solar "farms" and wind farms that generate large amounts of energy but that need to be sent over wires (and now wireless) to the consumer. However there needs to be on-site power (one building, shopping mall and family home etc.) called "distributed" energy that is usually solar or other renewable energy source. So agile (which is science for "flexible") energy systems are both. But the agile system approach, which has become reality throughout California today, is to have a combination of local on-site energy generation (e.g., solar systems, combined heat and power, use of biomass and other renewable sources for energy) along with the central grid power generation.

For some clusters of buildings, like colleges, the use of solar for power is significant during the day, but on nonsunny days and at nighttime, the central grid becomes the power source for the community. A key component to buildings today is their design (such as LEED standards) so that they are environmentally sound. The design and construction of buildings and clusters must be addressed as the shift from a centralized to a decentralized or a combination of centralized and decentralized energy production.

The place to start is with small, relatively self-contained communities or villages within larger cities and regions. The issue is to get communities off their dependency on central grid–connected energy, since most of these power-generation sources come from fossil fuels like coal, natural gas, and nuclear power. Local on-site power can be more efficiently used and based on the region's renewable energy resources such as wind, solar, and biomass, among others. This model is now being realized in Denmark where many communities are generating power with wind and biomass combined to provide base load. Denmark has a goal of 50% renewable energy generation (primarily from on-site and local resources) by 2020. The country is well on its way to meeting and perhaps exceeding that national goal (Clark and Lund, 2006).

For example, the EU has committed EUR 2 billion with a matching EUR 3 billion over the next 5 years. The "challenge" is to have the public and private sectors become partners. These "civic markets" must collaborate in developing sustainable and smart communities. Sustainable communities include the facilities, land, and the infrastructure sectors that intersect, such as energy, water, IT, waste, and environment, and they must have public sector involvement and oversight to set goals, policies, and programs and provide finance.

The basic issue always is money and finance: economics.

Creating sustainable communities of any kind means finding the funds to pay for the technologies, systems, and operations. These areas are skillsets that must be included in the development of any sustainable community (Clark, 2010). While articles in the special issue address various levels and approaches to the economics of sustainable technologies, they fall into some methods set often by industry to pay from products like renewable energy such as solar panels. Solar panels are a good example.

Two roads have been set to improve energy efficiency since with the residential sector at 36% electricity consumption in the United States. One is that the government has created demand-side management to increase adoption of (1) efficient appliances (e.g., Energy Star program), and (2) building practices (e.g., LEED programs). In 2005, the US Congress placed an emphasis on more efficient lighting, which leads to the use of lights such as LED (light-emitting diode).

The second road is to reduce load on the grid, such as through better monitoring and feedback from meters.

A smart grid (Jerry, 2010) monitors and controls energy use, which reduces and conserves energy use, thus reducing the load demands. In 2007, the US Congress passed and established the official American national policy, the Smart Grid. California, 3 years later (2010), enacted a law that ordered the state's investor-owned utilities to develop smart grid plans by the summer of 2011. Now these plans are being implemented.

The new economic paradigm follows the critical strategy for both roads to the meet reduction and efficient use of energy with its impact personal behavior. The old paradigm did not connect individual behaviors with energy use. Today, with smart grids and their impact on personal behavior well documented, there is a need to connect the energy savings from meters and other energy-efficient resources with utilities and policy makers who "will need better information so that these savings" will have more beneficial financial return for the overall economics of central grid energy demand, renewable energy, and infrastructures, including those same demands for on-site and building power use.

There are currently many pressing climate, environmental, and economic issues facing Europe and the rest of the world now that will impact our children and grandchildren (http://www.amazon.com/The-Green-Industrial-Revolution-Engineering/dp/0128023147/ref=sr_1_3?ie=UTF8&qid=14). That is why the European Commission launched a new, comprehensive research and innovation program in 2013 named Horizon 2020 that will build a more solutions-oriented approach to CC. As a matter of fact, it is of particular importance to not only develop a program that is convincing from a research perspective, but also one that is relevant in terms of the goals of the European Union, including prosperity, quality of life, sustainability, growth, and jobs. These are the main reasons why another version of this blogpost has already been published in Italy. Now it is time to bring this topic and ideas to the United States and other English-speaking nations and then also to other nations around the world. Stay tuned in the European Union, western hemisphere, Asia, and Africa, as well as island nations, as we will be blogging there too with this article soon.

Circular Economics

One of the key objectives of Horizon 2020 is to position Europe as a frontrunner in the development of a "circular" and "green" economy that focuses on sustainability. At this point in history, the European Union needs to enhance its international competitiveness through resource productivity and improving its position for low-impact, resource-efficient technology, products, and services in order to achieve its sustainable development goals. This is a dramatic change from the conventional neoclassical economics of Adam Smith (Clark and Fast., 2008) and needs focus on more than the "market."

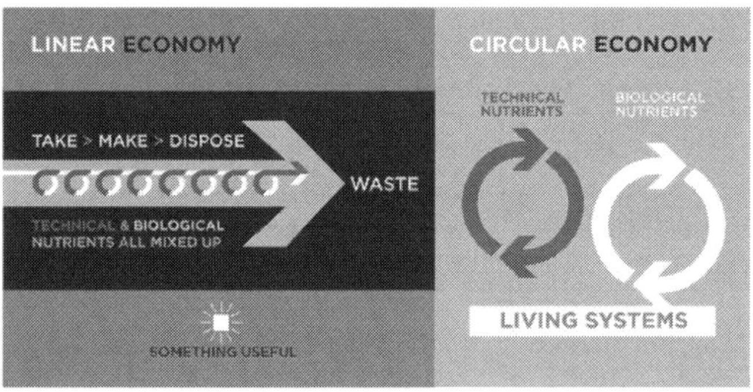

Neoclassical conventional economics versus circular green economics (http://ec.europa.eu/growth/industry/sustainability/ecodesign/index_en.htm).

The European Union (and the world) is facing a series of systemic crises ranging from economic, environmental, natural resources, social, etc., to population growth, conflicts, and wars. Eco-innovative solutions are required to tackle almost all of these systemic challenges in order to transform the economies and to make the way of life of European citizens more sustainable. The *Green Industrial Revolution* was published in 2015 http://www.amazon.

com/The-Green-Industrial-Revolution-Engineering/dp/0128023147/ref=sr_1_3?ie=UTF8&qid=14.

This dramatic green revolution has started in the European Union and is now in China and other areas of Asia: http://www.apo-tokyo.org/publications/apo_news/september-october-2016/.

A sound, systemic economic model for eco-innovation within a broader perspective is needed that considers all innovative solutions, which are sustainable from a triple economic, social, and environmental point of view or paradigm change called the NE www.springerpress.com/Clark. The circular economy (www.ellenmacarthurfoundation.org/.../euro) is the best approach to solve environmental problems. CE makes the most sense and is now a policy for the entire EU due to Belgium's presidency of the council of the European Union, Flemish Environment Minister Joke Schauvliege who in 2014 became president of the EU's Environment Council. Since Schauvliege was appointed collaborations between EU nations have grown and spread around the world. See: http://www.eutrio.be/pressrelease/schauvliege-calls-sustainable-materials-management.

Circular economics was created by an American design architect for the Ellen MacArthur Foundation (2017). Case in point from EU Commission Staff Working: "Report Critical Raw Materials (CRMs) and Circular Economy" January 16, 2018.

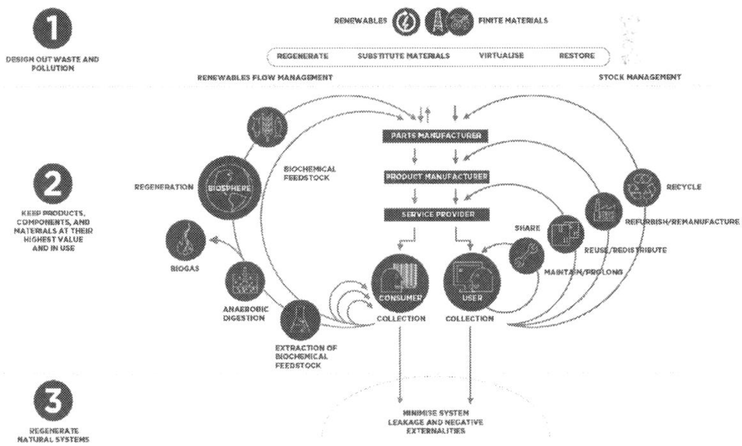

One of the key priorities for Europe is in the circular economy context. For example, Remedia (www.consorzioremedia.it) is related to the sustainable supply of nonenergy and nonagricultural raw materials. See below Remedia model throughout the European Union.

Started in 2005, Remedia has a long-term vision to tap the full potential of secondary raw materials and to boost the innovation capacity of the EU secondary raw materials sector. This dynamics of the circular economy then turn into a strong, sustainable pillar of the EU circular and green economy. Industries (and governments) must be active in addressing societal environmental challenges, which increase benefits for every person and community.

This can only be done by gaining relevant knowledge about secondary raw materials in Europe. It also requires that stakeholders, including relevant authorities, raw materials and downstream industries, research communities, and society work toward the same objectives for the reuse and recycling of all aspects of products, manufacturing, shipping, and sales to the customer. Thus people will know what they are buying and be more concerned with the use and therefore the reuse of their products purchased.

There is a crucial need to improve awareness on critical raw materials and to monitor the dynamic development of the global situation, as too many companies still ignore the fact that their products are endangered by the lack of monitored and controlled European resources.

Circular Economy: Work in Progress

What exactly can the public and leadership expect? The experts who are working on the new package are well aware that, at present, about 90% of the European industry turnover (sadly one of the best performances in the world) is based on linear economic models of procurement, production, consumption, and disposal of matter. This neoclassical economic model is now defunct and proven to be only an ideological model that does not work in today's European Union, let alone around the world for other nations (www.springerpress.com/Clark-Fast).

That is one reason why Clark and Fast have been asked to update their book on QE in 2008 and now an update to a second edition of QE in the fall of 2018, 1 decade later from Springer Press. The "invisible hand" never existed in the past and needs to be monitored for the future in a scientific manner. Governments have always been involved in nations and their development on all levels. "Civic or social capitalism" is needed now more today than ever before (http://www.amazon.com/The-Next-Economics-Environment-Climate/dp/1461449715).

Furthermore, it is recognized, at least in principle, that the only alternative to ensure a viable safe future for new generations of European citizens (and refugees) is that of a rapid transition to an industrial system that is based on the circular economy paradigm providing economic, family, and cultural support for everyone. Businesses and governments need to rethink production cycles so as to eliminate the concept of waste through optimized models for the reuse of products, disassembly, and recycling of goods.

The circular patterns to which the industry should converge will have to specialize and focus on environmentally positive products, goods, and services in order to address the different issues related to fast-moving consumer goods from those of durable goods (http://www.earthaccounting.com). Hence, the European Commission Critical Raw Materials Matrix in its 2017 report on Clinical Data Management System (CDMs) (January 2018) uses the quantitative methodology (with now qualitative methods) that applies two criteria: the economic importance and the supply risk of the selected raw materials.

- Economic importance: this analysis is achieved by assessing the proportion of each material associated with industrial megasectors at an EU level. These proportions are then combined with the megasectors' gross value added to the European Union's GDP. This total is then scaled according to the total EU GDP to define an overall economic importance for a material.
- Supply risk: in order to measure the supply risk of raw materials, the World Governance Indicator was used. This indicator takes a variety of influences into account such as voice and accountability, political stability and absence of violence, government effectiveness, regulatory quality, rule of law, and control of corruption.

Critical minerals and metals are essential to low-carbon and environmental technologies, such as solar photovoltaic, wind, lighting, and electric vehicle industries. Disruptions of the supply chain can be dramatic in times of crisis from the environmental, social, and economic point of views. This is the reason why there is a need for a global picture in terms of raw materials beyond conventional neoclassical supply and demand into collaboration with countries outside Europe. Today, products, especially technological and social media ones, are global from their creation to distribution and sales. The use of Netflix, YouTube, Facebook, and other forms of communication need to be part of the circular economics that reuses and recycles the end products into new ones. The entire process needs to be enhanced.

Success will be seen and thus be able to achieve a number of improvements in access and availability to raw materials, sustainable management of resources within a greener and more circular economy, and world-leading innovations both in technology and business practices to meet the growing demand and secure the supply. This effort should also result in changes in consumer behavior that are more consistent with the sustainable use of our raw materials. Thus, circular economics works with industry on environmental strategies such as these below from Remedia:

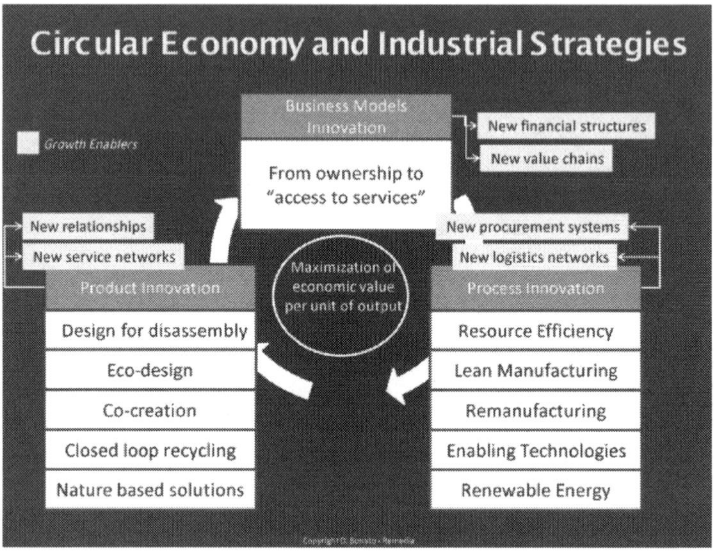

Copyright © D. Bonato, 2015.

The practical applications are for economics to correct CC problems; Hoexter discusses the Feed-in-Tariff (FiT). This approach to financing renewable energy, especially large systems, has worked well in Germany. Now, Spain, Japan, Canada, and other nations have created FiT plans as well. The United States is behind. Some cities in the United States have them, and California adopted a mild form of the FiT in the fall of 2009. But in general, the United States is far behind in part because such programs have been the most successful for large arrays of solar-concentrated systems and wind farms. These large systems transmit power and hence threaten the power generation from energy and utility companies. Smaller economic programs and hence renewable energy systems for homes, buildings, and communities such as colleges, resorts, shopping areas, and retirement complexes are needed.

The NE book also covered some new economic studies in the European Union. Italy became a good place to report on wind energy in particular, as there are many wind systems being installed today with more advanced and detailed information. N. Cardinale, Rospi, Cotrufo, and T. Cardinale present a study of wind energy in Matera City on its Matera's Murgia Park in southern Italy between Basilicata and Puglia in terms of "economic-environmental performance in the Mediterranean area." The authors review the history of wind turbines, which date back to sailing in 2500 BC, and windmills first built during the 7th century in Persia that came to Europe 500 years later. However, it was not until windmills moved from the Mediterranean to the northern European regions that the technology expanded and grew.

Wind turbines have now spread around the world, and Denmark has become the recognized leader in the technology, industry, and impact. Lund and Ostergaard provide an extremely good case of a town in Denmark (Frederikshavn) using 100% renewable energy by 2015. With only a few years to go, they are over halfway there. Wind is one of the core renewable technologies accounting for over two-thirds of the power generated. As N. Cardinale et al. put it, "The wind energy represents, among renewable sources, the one with the highest potential of use, as it is an absolutely free resource exploitable by using a simple turbine without the high cost of installation."

In Italy, there was a 40% increase in wind power from 2008 to 2009, generating over 4898 mW of power. Southern Italy represented 88% of the national installed capacity. In the study of four wind turbines for the park in Matera, the authors found that in a 20-year period of time for the life cycle of the wind turbines is best suited for "micro wind turbines," since the wind speeds are not as strong as those need for standard large wind turbines. The micro–wind turbines with horizontal axis and a 6-kW power have a cost that is "close to thermal power plants and nuclear third-generation power plants with the (same) environmental benefits." Such results for regions and areas that need power systems that can use renewables like wind (and solar) are good for the environment and far less costly than other power systems such as nuclear power plants.

The case of what the government system can do to lead and implement energy conservation and reduction is important. In California, for example, the Caltrans (California State Transportation Authority) has the responsibility for buildings and roadways. What Prey discusses is the research in LED lights since 1991 and then the extensive use and results of LED lights for streetlights and bridges over a decade ago (1998) that set the specs for the LED manufacturing industry (national product specifications) and the standard for conservation and efficiency throughout California and today, the entire United States.

THE GLOBAL PERSPECTIVE AND INTERACTIONS

In the early part of the 21st century, "the energy/carbon footprint for the statewide owned/maintained traffic signal upgrade exceeded 93% reduction from the incandescent baseline." For Caltrans alone, that meant a 13-MW grid load was eliminated very day. And the return on investment (ROI) was within 3 years. This led to California, then other states, and finally the US Congress in 2005 with its National Energy Act, banning incandescent traffic lamps throughout the United States. By 2011–2012, the California state "moved the development and discussion phase to that of a deployment phase." The NE book provided many chapters from a global perspective on what to do.

In this same regard, Prey argues that the states, like California, which is now a "nation-state" (Clark, 2017b), can often take the initiative in terms of innovations and program implementation for other environmentally sounds and economical technologies such as HVAC (air conditioning) systems from high electricity—based to thermal-based absorption ones that use solar thermal pretemp to reduce the need for carbon-based fuels. The same can be done, argues Prey, with hydrogen for transportation by using ammonia compounds to transport hydrogen for use in energy centers and vehicles. In short, the numbers work for a positive (short-term) ROI and for reducing carbon and other emissions.

Nuckols emphasizes the need for coalitions both within the United States and internationally, as the need to address environmental, land, and economic issues has expanded greatly. He notes, for example, the Apollo Alliance founded in 2003, which has evolved into the blue—green partnership, which bring together traditional labor unions and environment groups. From that, 3 years later, a Blue-Green Alliance emerged, with blue-collar labor unions and the green environmentalists actively working together. By 2011, both groups merged to "continue to partner in battling for jobs in the green economy."

Resource efficiency, conservation, and new technologies are not enough. They all must be connected, which is the basis of circular economics in practice.

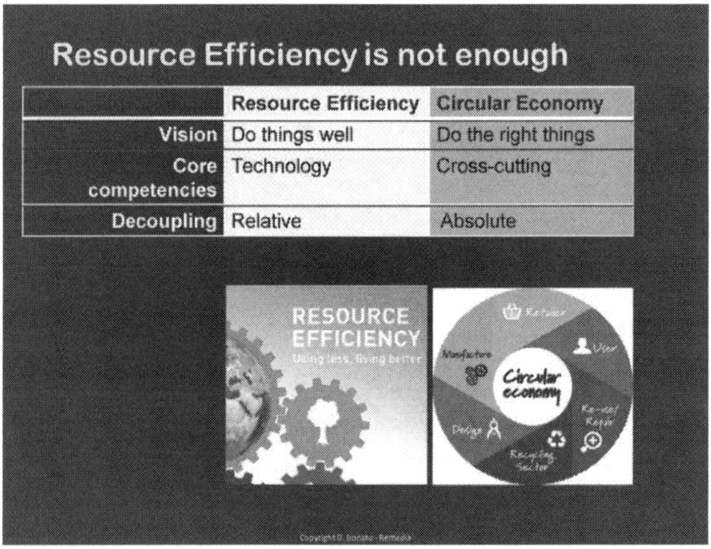

Copyright © D. Bonato, 2015

One area of focus for these and related alliances is agriculture. The concern today is the economics of agriculture due to increased competition. Yet the concern for the environment and pollution has become more dominant and costly. While some of the changes are needed in particular for large agribusinesses, it seems that the environmental concerns and even costs are borne by the smaller ones through alliances in order "to generate cooperative dialogue that leads to a reduction in the exposure of toxic elements, both in global communities and particular worksites."

The international social movements have also been playing a significant role, since the concern over ethical issues and relationships are predominant. "These activists feel strongly that a healthy community and workplace sets the stage for resilient job creation and sustained economic growth." Schumacher, as Nuckols points out, is "divergent" and only solved by "higher forces of wisdom, love, compassion, understanding, and empathy." In short, a higher level of concern must be made into action. Science, as Nuckols concludes, in our world today cannot be significant on its own, because "science on its own can give no reasons for sustaining humankind."

Nijaki talks about a "green economy as sustainable," but that goes beyond traditional concepts of development and growth. In short, she argues that "aggregate economic growth alone (measured as GDP) in terms of measures of productivity, skills, and wealth may be an oversimplification of true economic development." Basically, she defines growth as both a distribution issue of the benefits and costs of growth and also the currently more extensive definition to increase the quality of life. This chapter covers a more extensive and broader definition of economics that helps to frame "green" growth.

Furthermore, Nijaki reviews conventional and standard economic theory, especially as it applies to development and growth. Some economists even took the classical economic theory a bit further to argue the need for technological innovation and change to stimulate and push development and growth, but expanded the concept into new products, human capital, and production methods. In the last 4 decades, the emphasis that placed demand-side economics at the center of all growth was replaced by Milton Friedman and his supply-side economics, based in neoclassical economics, which still holds control of economic teaching, research, publication, and even public policy. However, Thomas Friedman had pushed the three phases of globalization, which morphed from being a world that was flat in the early part of the 21st century to a partial round world again due to CCs in the atmosphere, oceans, and even land masses.

CONCLUSION: WHAT IS UP AHEAD IS MOVING NOW

Today (2017), there are global moves into a "green development" future now. Not in another decade or two. The clock is ticking.

Thus, "economic growth may sometimes be in conflict with measures that protect equity considerations and quality of life goods." But in the end, the "green economy aims to widen the view of economic growth or progress through an integration of environmental considerations in the development process." It reframes growth as "green growth," and thus limits development by taking into account quality of life considerations that are hinged on environmental quality today and into the future. In this way, the metrics for evaluating development choices and their successes is changed to one that seeks to reference the long-run environmental effects of economic action and inaction.

The key is how "green economy" is defined, measured, and scientific. Companies today are seeking to be labeled "green" while they and consumers are looking for green products and services. Naijaki provides a framework for green growth as part of NE but covering four areas in some detail: (1) diversity in sectors, (2) practitioners versus producers, (3) regulation centric, and (4) small and startups. In the end, not all growth and jobs will be green, which she attributes to local and regional government policy makers.

Government involvement is only part of the need in economics for it to be scientific and concerned with societal issues. And in fact, government is really a major part of the solution. The supply-side economics argues for "market forces" and related neoclassical economic theories with an invisible hand called government. This approach to economics is not scientific since there factors that create hypotheses, observations, data, and repeated experiments in order to set rules for future measurement, analyses, and evaluation. Government needs to be present for the objective oversight of the economics of science much like a physicist would be in and out of the laboratory.

In the conclusion of this book, Clark leaves it to every reader to decide, interact, and work with others in their community as well as gather ideas from around the world. The point is to know that the readers have a good sense of where economics is going now and into the future. Enjoy. This is just the beginning of NE. There is a lot to do.

REFERENCES

Bailey, E.M., Wolfram, C., June 18, 2012. A whole different kind of innovation. The Journal Report: Innovations in Energy, Wall Street Journal. www.wsj.com.

Chomsky, N., 1957. Syntactic Structures. Mouton & Co., The Hague.

Chomsky, N., 1975. Reflections on Language. Pantheon Books, NY.

Chomsky, N., 1980. Rules and Representations. Columbia Un. Press, NY.

Chomsky, N., May 8, 2012. Plutonomy and the Precariat: On the History of the US Economy in Decline, pp. 1–5. TomDispatch.com.

Clark II, W.W., 2007a. Eco-Efficient Energy Infrastructure Initiative Paradigm. UNESCAP, Economic Social Council, Asia, Bangkok, Thailand.

Clark II, W.W., December 06, 2007b. Partnerships in creating agile sustainable development communities. Elsevier Press Journal of Cleaner Production 15, 294–302.

Clark II, W.W., 2009. Sustainable Communities. Springer Press.

Clark II, W.W., June 2010. Sustainable Communities Design Handbook. Elsevier Press.

Clark II, W.W., November 8, 2017a. One of the Keynote Speaker on "The Global Green Economy", for National Council of Green Energy. Ministry of Environment and Ministry of Economic Development, Rimini Fiera, Italy.

Clark II, W.W., August 2017b. China and the nation-state of California. Beverly Hills Courier 11.

Clark II, W.W., June 6, 2017c. Interviewed Live from Los Angels to Beijing on "Clean Energy" during Under2 Conf in Beijing. On CGTN Media. www.cgtn.com. https://www.youtube.com/watch?v=HhYZWBBkZKw&index=3&list=PLt-M8o1W_GdSNhViAVk4Fq8IItW8PHBVf.

Clark II, W.W., 2017d. Agile Energy Systems: Global On-Site Distributed and Central Grid Systems. Elsevier Press.

Clark II, W.W., Danal Bonato on Circular Economics in HuffPost in USA (English) and EU (in Italian) http://www.huffingtonpost.com/woodrow-clark/

Clark II, W.W., Bradshaw, T., 2004. Agile Energy Systems: Global Lessons from the California Energy Crisis. Elsevier Press, London, UK.

Clark II, W.W., Cooke, G., November 2011. Global Energy Innovation. Preager Press.

Clark II, W.W., Fast, M., 2008. Qualitative Economics: Toward a Science of Economics. Coxmoor Press.

Clark II, W.W., Isherwood, W., 2007. Energy Infrastructure for Inner Mongolia Autonomous Region: Five Nation Comparative Case Studies. Asian Development Bank, Manila, PI and PRC National Government, Beijing, PRC.

Clark II, W.W., Isherwood, W., 2009. Report on energy strategies for inner Mongolia autonomous region. Utilities Policy. https://doi.org/10.1016/j.jup.2007.07.003. Woodrow W. Clark II, and William Isherwood, (Authors and Co-Editors). Special Issue of Utility Policy Journal: China: environmental and energy sustainable development. Winter 2010.

Clark II, W.W., Lead Co-Author and Co-Editor W. Isherwood, Utilities Policy Journal. Special Issue on *China: environmental and energy sustainable development*. Winter 2010.

Clark II, W.W., Lund, H., September 2006. Special Issue on "Sustainable Development: the economics of energy and environmental production". Journal of Cleaner Production.

Clark II, W.W., Olav Jull, Sorensen, Dec 2001. Entrepreneurship: theoretical considerations, International Journal of Technology Management. Inderscience, London, UK.

Demirag, Istemi, Iqbal Khadaroo, Clark II, W.W., 2009. The institutionalization of public-private partnerships in the UK and the nation-state of California. International Journal of Public Policy 4 (Nos. 3/4), 190–213.

Economist, July 16, 2009. Modern Economic Theory. Special Issue.

Economist, January 12, 2012. State Capitalism. Special Issue.

Hershauer, James C., George Basile and Scott G. McNall, 2010. The Business of Sustainability: Trends, Policies, Practices, The Pew Charitable Trusts, Praeger Press.

Jerry, J., June 2010. Transformational relationship of renewable energies and the smart grid. In: Clark II, W.W. (Ed.), Sustainable Communities Design Handbook. Elsevier Press, pp. 217–232.

Perkins, L.J., 1996. Senior Physist, "What Is Physics and Why Is it a 'Science'?" Lecture at University of California Physics Seminar. University of California, Berkeley, pp. 1–3.

Rifkin, J., 1980. Entrophy. Palgrave Macmillan, New York, NY. Sekisui House, 2016. CSV Strategies, 2016: https://www.sekisuihouse.co.jp/english/sr/datail/__icsFiles/afieldfile/2016/07/29/all.pdf.

Simmons, J.P., Nelson, L.D., Simonsohn, U., May 23, 2011. False-positive psychology: undisclosed flexibility in data collection and analysis allows presenting anything as significant. Association of Psychological Sciences 1–8. Sage Publications, Accepted.

Toedtman, J. (Ed.), May 30, 2012. Tax Attack: Is it Time for a Sugar Tax?. AARP Bulletin (website reference).

FURTHER READING

Clark II, W.W., 2003a. Science parks (1): the theory. International Journal of Technology Transfer and Commercialization 2 (2), 179–206. Inderscience, London, UK.

Clark II, W.W., 2003b. Science parks (2): the practice. International Journal of Technology Transfer and Commercialization 2 (2), 179–206. Inderscience, London, UK.

Clark II, W.W., 2008. The green hydrogen paradigm shift: energy generation for stations to vehicles. Utility Policy Journal. Elsevier Press.

Clark II, W.W., Cooke, G., 2016. The Green Industrial Revolution. Elsevier Press, 2014 and shorter earlier version in Mandarin.

Clark II, W.W., Li, X., 2003. Social Capitalism: transfer of technology for developing nations. International Journal of Technology Transfer. Inderscience, London, UK.

Lund, H., Ostergaard, P.A., 2009. Sustainable towns: the case of Frederikshavn – 100% renewable energy. In: Clark II, W.W. (Ed.), Sustainable Communities. Springer Press, pp. 155–168.

Nularis. Patent Pending. Light Emitting Diode (LED). Clark II., W.W., Brown, W., Fram J., February 12, 2012. US Patent Office, Washington, DC.

Rifkin, J., 2006. The European Dream. Palgrave Macmillan, New York, NY.

Rifkin, J., 2011. The Third Industrial Revolution: How Lateral Power is Transforming Energy, the Economy and the World. Palgrave Macmillan, New York, NY.

Wikipedia, July 2012. Mathematics. en.wikipedia.org/wiki/Mathematics.

Xing, L., Clark, W.W., 2009. Crises, opportunities and alternatives globalization and the next economy: a theoretical and critical review. Chapter #4. In: Xing, L., Winther, G. (Eds.), Globalization and Transnational Capitalism. Aalborg University Press, Denmark.

Chapter 4

Infrastructures for Green, Smart and Healthy Communities

Xing Li[a,1], Woodrow W. Clark, II[b,2]
[1]*Department of Business Economics, AAlborg University, AAlborg, Denmark;* [2]*Research Professor in Economics, Pepperdine University Graziadio School of Business (PGSB), West Los Angeles, CA Campus, Los Angeles, California*

Chapter Outline

OVERVIEW

Now China's leadership under President XI is struggling to look for new sources to sustain its desire for economic growth. One of the most important shifts in China's "green development" orientation is to lead China into a

a. Xing Li, Ph.D., is a Professor at AAlborg University, AAlborg, Denmark.
b. Woodrow W. Clark II, Ph.D., is a Qualitative Economist.

Climate Preservation In Urban Communities Case Studies. https://doi.org/10.1016/B978-0-12-815920-0.00004-6

transition from export-oriented growth to a new model based on consumption and outward investment. So, at the heart of the "One Belt, One Road" (hereafter OBOR) strategy, which covers about 65% of the world's population and one-third of the world's GDP, lies the idea to create viable economic belts: (1) a land belt that includes neighboring countries surrounding China, especially those countries on the original "Silk Road"[1] through Central Asia, West Asia, the Middle East, and Europe and (2) a maritime belt that links China's port facilities with the African coast, pushing up through the Suez Canal into the Mediterranean. The OBOR seems to revive the ancient Silk Routes with a 21st-century twist.

Being an ambitious project, the OBOR initiative aims to link interregional trade across some 60 Asian and European countries along a new Silk Road. It is China's most important and grand strategic initiative and a crucial component of President Xi Jinping's foreign policy. The OBOR attempts to provide a new prism to view the world economy and the global balance of power, not a conventional prism to see the world through the Atlantic with the United States on one side and Europe on the other. Global attention on the OBOR strategy has been linked with China's proactive engagement in financial "minilateralism" in recent years as an attempt to create global alternative financial institutions, such as the BRICS Bank,[2] the AIIB,[3] and the New Silk Road Fund.[4]

1. The conception of the "Silk Road" refers to the ancient Silk Road, a historical network of trade routes started during the Han Dynasty (206 BC–AD 220) between Europe, China, Africa, and many other countries on the Afro-Eurasian landmass. It had both land and sea routes where trade between China and regions and countries along the routes was vibrant with goods and cultural exchanges.
2. China is the largest stakeholder of the BRICS Bank, contributing 41% of the total 100 billion US dollars capital base, at the sixth BRICS summit in Fortaleza, Brazil. According to the official published *Fortaleza Declaration*, "the BRICS Development Bank is designed to mobilize resources for infrastructure and sustainable development projects in BRICS and other emerging and developing economies".
3. It refers to the "Asian Infrastructure Investment Bank" (AIIB), an initiative led by Chinese government in 2015. It is an international financial institution with a pivotal aim to support and finance infrastructural projects in the Asia–Pacific region.
4. During the APEC Summit in Beijing in November 2014, China's President Xi Jinping announced that China would contribute 40 billion US dollars to set up a Silk Road Fund at a dialogue meeting on strengthening connectivity and improving cooperation in the country's neighborhood. According to Xi's explanation, "the new Silk Road Fund will be used to provide investment and financing support to carry out infrastructure, resources, industrial cooperation, financial cooperation, and other projects related to connectivity for countries along the 'One Belt and One Road' initiative".

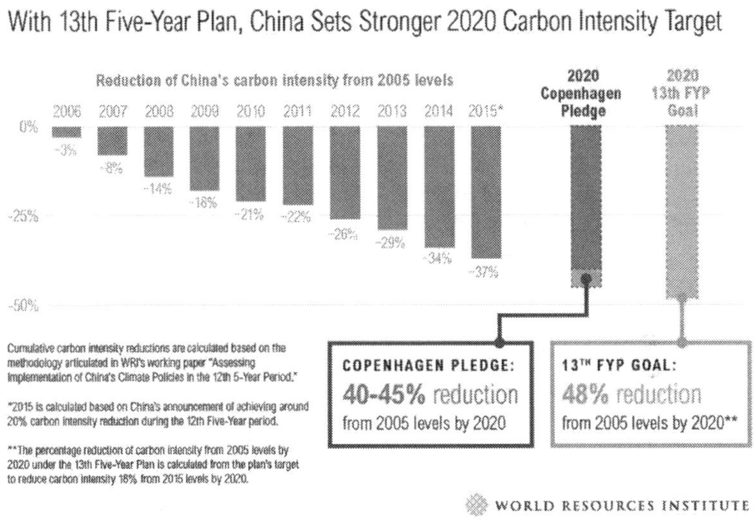

Furthermore, the concepts that President XI spoke about at the UN G20 Conference in China on September are "green development" and bringing back the "Silk Road." Both policies are critical to China and other nations.

One interesting new "green development" under Chinese President XI is in the southern part of China, city of Shenzhen, which mirrors the US Silicon Valley in Northern California near San Francisco. This area and region in China is known as the "Calichina" where Chinese companies do not compete with the United States but rather "share" companies and economic eco-friendly strategies. As noted by a professor in Australia recently, "If Chinese millennials are anything like American ones, they are likely to value access to the App Store more than ambitions of geopolitical domination. And if that is the case, the world has little to fear from a China-centered world order, now or in the future" (Babones, 2018).

In the global scheme of things, the renewable energy train has already left the station and not even the United States will be able to derail it. In 2015, renewable energy set new growth records both for investment and new capacity. Investments reached nearly $286 billion, more than six times more than in 2004. In addition, for the first time, renewable energy accounted for more than half of all new power generation capacity added in the world, according to the Global Trends in Renewable Energy Investment (2016).

The first practical solar cell was invented in the United States. Until the 1980s, almost all solar cells were built in the United States. President Ronald Reagan slashed federal funding for solar energy by about two-thirds, rolling back the progress made under the Carter administration.

"It is the Reagan Administration's contention that much of the Federal work in solar energy supports near-term development and commercialization and, hence, can be reduced or eliminated," according to a study by the Congressional Research Service. "While the Administration contends that it is, the Solar Energy Industries Association predicted that the budget [cuts] proposed for 1982 would collapse the industry."

Reagan's budget cuts did not kill the solar industry so much, as he sent it overseas. Case in point is Hanergy Company in Beijing that created and produced thin solar films such as that on its corporate buildings. The six-story buildings of the Hanergy company are all solar covered and have over 43% of its energy coming from solar cells. See the Hanergy Research Center Building below:

In the 2013 book *Let It Shine: The 6000-Year Story of Solar Energy*, John Perlin explains how Herwig's prediction has come to pass. "Starting in 1994, the Japanese government (as Chapter 9 points out) offered a generous subsidy to encourage individuals to place solar cells on their rooftops, resulting in the placement of photovoltaic systems with a capacity of 3 kW on 539 rooftops.

By the time the program ended in 2004 almost 400,000 rooftops had photovoltaics. The initiative helped Japan become, by 2002, the world leader in the installation of grid-connected systems and the manufacture of solar cells. Rooftop capacity after the first year of the program totaled 1.6 MW. Eleven years later, more than a gigawatt (a billion watts) had been installed."

Germany has gone even farther than Japan, installing more than 30,000 MW worth of solar photovoltaics. To put this in perspective, about 4000 MW worth of solar PV has been installed in the United States.

Aside from Reagan taking off the solar roof on the White House that President Carter put on, was not the only casualty of President Reagan's energy policies. Reagan's energy agenda also proved to be ruinous for America's long-term competitiveness in advanced battery technologies from solar to others including wind, water, and geothermal.

"If Reagan had continued the programs the Jimmy Carter administration started, we'd be a lot further ahead," Robert Huggins, a professor of materials science at Stanford University, told Seth Fletcher in Bottled Lightning. "But that didn't happen."

Instead, US companies ceded the advanced battery industry to Japanese competitors such as Sony and Panasonic. As Elton Cairns, a leading expert on advanced battery technology, told Fletcher:

> *Throughout the eighties, the Japanese really kept at the R&D. It took a lot of systematic work to develop a commercial lithium-ion cell: testing carbons for the negative electrode, working on procedures for making the positive electrode, finding out what would be the best electrolyte. But they did it rather quietly. Most people not inside those efforts didn't fully realize what was taking place in terms of moving toward commercialization.*

Ultimately, American battery companies such as Everyday and Duracell have been "kind of got left in the dust." It remains to be seen whether Trump will slash federal funding budgets for energy technology R&D. He has plenty to slash.

CONCEPTUALIZING THE INTERNAL–EXTERNAL NEXUS OF THE OBOR STRATEGY

The post-Mao development strategy adopted during early periods of the economic reform in the late 1970s was heuristically underlined by Deng Xiaoping's maxim, "watch cautiously, hold our ground, meet the danger calmly, hide our capacity, while also get something accomplished" (冷静观察、稳住阵脚、沉着应付、韬光养晦、有所作为). Deng's words implied an implicit strategic thinking, namely to concentrate on China's internal development while waiting for the right time when the country is ready to assert itself in the global sphere and make an accomplishment. The flag slogan

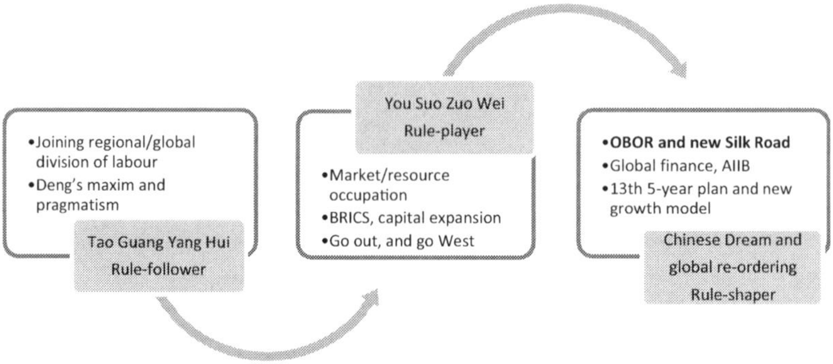

FIGURE 1 The internal–external nexus in China's development strategy.

under which Beijing projected development objective during the past decades was "peaceful ascent" (和平崛起)[5] or "peaceful development" (和平发展).

Fig. 1 shows the internal–external nexus of China's economic development during the different periods in which China transformed itself from being a passive rule-follower (recipient of global industrial outsourcing) to being an indispensable contributor to global economic growth, and finally to becoming a proactive rule-shaper in global governance. Internally, the OBOR initiative is a further extension of China's "go out" and "go West" policy strategy, aiming to focus more on the quality and equality of economic growth rather than growth "at any cost." Externally, the OBOR initiative, apart from economic consideration, can be seen as Beijing's foreign policy tool to build its leadership and to provide an internal–external channel for Chinese industrial and capital outward expansion by leveraging China's comparative advantage in infrastructure development, financial power, and manufacturing capacity.

FIRST PERIOD: JOINING REGIONAL AND GLOBAL DIVISION OF LABOR

The first phase of the "Tao Guang Yang Hui" strategy in the early period of China's economic reform marked Beijing's willingness to accept the regional division of labor and to be part of the regional flying geese economic order

5. The concept of "peaceful ascent" was formulated by Zheng Bijian in a series of speeches and articles from 2002 onwards during which he functioned as a spokesman for Chinese government viewpoints. The notion aimed to counterweight two worldwide perceptions on China—"the threat of China" and "the collapse of China." However, the word "ascent" (崛起) can also imply "challenge" in a revisionist manner, causing the concern led by the United States–led world order. Therefore, the term "peaceful development" is more often applied today to replace "peaceful ascent."

based on dynamic comparative advantage. The concept of the East Asian "flying geese" pattern of regional economic integration was coined by Japanese economist, Kanane Akamatsu in the mid-1930s, who developed the theory of a multitier hierarchical "flying geese" model in which industrialization could be promoted and spread from developed countries to the less developed countries as costs in those economies rose. The hypothesis of the flying geese pattern suggests that a group of nations in this region are flying together in layers with Japan at the front layer. The layers signify the different stages of economic development of various countries. The flying geese model implies a life cycle of industries by which older technology and knowhow are passed down the chain of developing economies. It also entails a product cycle theory of trade and investment.

Since its economic reform in the early 1980s, China's production and labor relations within region was also part of the flying geese model. However, what is unique about China is the nation's vast size, and its internal unbalanced development level between regions indicate that the Chinese economy is flying in different layers simultaneously and has multiple chain production and labor relations with many countries (Li, 2016). This situation was obviously an advantage because it enabled China to cooperate with other countries in almost all industrial sectors. On the one hand, China was capable of cooperating with Japan and the East Asian Next Industrial Economics (NIEs) in developing high-tech industries while it could inherit labor-intensive industries from them. On the other hand, it could also export capital, intermediate products, and some of its labor-intensive light industries to less developed countries in the region. Based on these closely linked economic relations, Chinese economy was increasingly integrated with the regional economy (Fig. 2).

The transition from the traditional vertically integrated utility to the emerging "agile power system" is a struggle between two different perspectives on how to increase the public good—on the one hand "privatization" and the other hand "civic markets." Deregulation was proposed under the premise

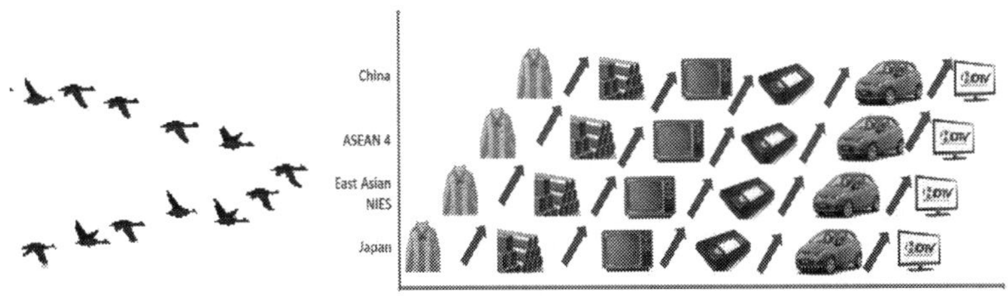

FIGURE 2 The flying geese model of East Asian division of labor.

of privatization, which is the neoclassical economist view that competitive markets reduce prices and allow maximum choice, while providing optimum conditions for technological innovation. The belief has been a cornerstone of Prime Minister Margaret Thatcher's public sector reforms during the 1980s as well as the philosophy behind relaxing regulations in many other industries from banking to airlines. In all these cases, proponents not only look at the declining prices and innovation but also have to acknowledge gaps in service and occasional scandals such as the collapse of the American Savings and Loan institutions. The failure of the California deregulation can be added to the list of failures.

The philosophical premise that the public good rests at the core and is the theoretical basis for any analysis of market options is key in any discussion of the energy sector. The public good is a concept linked to community-wide, shared benefits by people that transcends special individual or group interests, and hence benefits everyone collectively. Put in another way, public good is a shared benefit that in aggregate exceeds those that might otherwise be retained as exploitive benefits for only a few individuals.

At the turn of the 21st century in California, these same environmental problems in China and other nations impacted the public good in another way: health which is a very big externality cost around the world as this map shows (2010) but worse today (2018):

Sources of Air Pollution-Related Deaths in 2010

A growing amount of literature in "environmental justice" documents the health costs for communities surrounding power plants. San Francisco has been active in meeting the concern through its Bay View Hunters' Point community action group which seeks to close two local power plants and replace them with renewable energy generation. The concern then for public good translates into high medical costs for society. Community-based action groups (Citizens for Better Communities, among others) have provided technological expertise and contact networks to provide solutions to local community clean energy needs.

Some of these energy-environmental problems can be addressed by regulatory prohibitions that reduce or eliminate some of the offending behaviors. However, the record both of utility regulations and environmental laws shows that much progress can be made by requiring that technologies be developed and installed that reduce pollution, even if the market would not voluntarily select them.

In addition, a range of environmental problems need to be addressed by experimental technologies, and considerable investment in alternatives that may cost more at the early stages, until the economies of scale make them more price competitive. The requirement for including advanced technologies, that are not immediately competitive in the energy sector mix until they reach enough volume that they become cost-effective, is one of the key objectives of progressive regulatory agendas. Moreover, California has pursued this strategy during the energy crisis through several civic market partnerships, collaboratives, agency and commission programs as discussed later in Chapter 7 on advanced technologies and economics.

Other technological issues also benefit the public good in ways that need public encouragement and regulation to assure that the technological advances are provided. The three most important of these are renewable energy generation (grid, distributed generation and on-site), peaking power, and conservation. The public sector through government programs such as regulation, standards, and codes as well as public procurement policies can assure that these benefits are met.

California has demonstrated this approach since 2001 numerous times. For example, to conserve energy, the California Consumer Agency (CCA) launched a public awareness campaign, called "Flex Your Power" that has proven to be so successful that the public and private sectors reduced energy consumption by an average of 12% over the critical summer months of 2001.

The same agency has spearheaded a "Sustainable Public Building" program whereby state, and now local, government buildings are installing solar PV systems, net metering, and conservation devices. Procurement policies aimed at purchasing and installing such systems have had a direct impact at encouraging the marketplace. All these programs have helped drive down the consumer price as the market becomes cost competitive.

The equity issue also remains one for nonmarket regulation. The problem in power is that it needs to be available to a range of people who cannot afford it but who need it for their safety. Similarly, some groups of people have very strong needs but the cost of service is greater than they can afford at least initially. For example, rural populations were last to get power because of the high cost of providing lines and distribution networks for power over much longer distances on average than in urban areas, and rural populations tended to be poor. This led to public rural electrification programs in most countries. Both of these issues contribute to public well-being, safety, and security.

SECOND PERIOD: THE EMERGENCE OF CHINA-CENTERED REGIONAL ECONOMIC ORDER

Since the 21st century the region witnessed a gradual shift away from the vertical "flying geese" model to a new horizontal modeling of regional economic integration. The rise of China as a new regional economic hegemon shows a unique and complex relationship with the flying geese dynamics. On the one hand, it can be argued that China's domestic development in the past 3 decades also resembled certain flying geese characteristics manifested by its coastal and inland economic interactions, while its economic success can be seen as a result of its integration in the intraregional flying geese dynamics. On the other hand, China's vast size and its unbalanced development levels in different internal regions with their respective levels of industrial stages and comparative advantages enabled China to cooperate with the rest of the countries of the region and even the whole world—core, semiperiphery, and periphery countries—both vertically and horizontally.

Today there is a consensus in East Asia that China's economic power, especially its growing domestic market, has been a major engine for growth, boosting regional trade growth and economic integration. China is now virtually the number one trading partner for most East and Southeast countries, and only in 2013 did the United States take over China's position as Japan's largest trading partner (due to the China–Japan crisis over the territorial disputes). This situation is not only an East Asian phenomenon but also a global one: in 2013, China overtook the United States as the world's largest goods trader. As noted in an RT.com article "As recently as 2006, America was the larger trading partner for 127 countries, versus just 70 for China. By 2011, the situation had changed drastically: 124 countries for China, 76 for the US" (quoted by *Forbes*, October 19, 2014). The same situation has been taking place in East Asia (see Fig. 3).

Today, virtually every East Asian country counts China as its largest trading partner and its largest market (Fig. 3, left side). There is no doubt today that China is increasingly looked upon as the locomotive of the region's economic growth and development. If the analytical perspectives are extended to depict the regional "economic order," then recent studies indicate that the

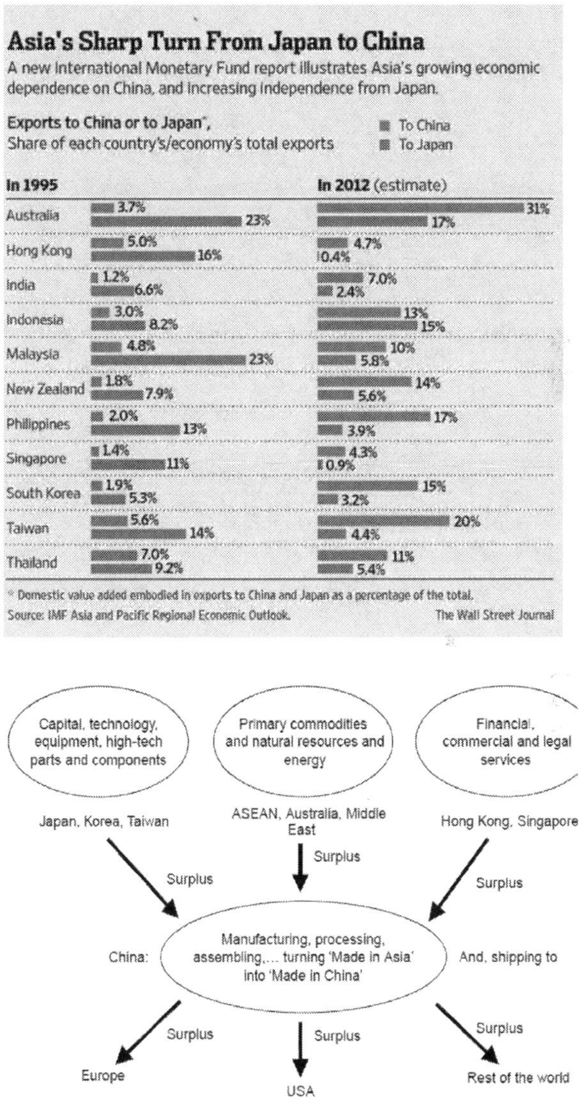

FIGURE 3 The return of China's historical position as the region's gravity. *Left side, The Wall Street Journal, May 13, 2014.*

Chinese economy is playing a pivotal role not only as an engine of growth for East and Southeast Asia but also as a hub for regional economic integration (Fig. 3, right side):

> *China has in fact become an important regional 'integrator' through its many global and regional production networks. China's exports (over 50% being processing trade) embody raw materials, parts and components, technology and equipment, and financial and economic services from different Asian economies, converting 'Made-in-Asia' into 'Made-in-China' products for the world market.*

THE THIRD PERIOD (OR "WAVE"): THE OBOR INITIATIVE AND CHINA'S INTERNAL ECONOMIC RESTRUCTURING

To explore the domestic factors behind the OBOR strategy, we can find the internal imperativeness for the Chinese economy to look for a new growth model fueled by consumption and services instead of fixed asset investment. China's high economic growth during the previous government was based on a massive state-led fiscal stimulus projects that was able to shield China from the destructive effects of the global financial crisis. But, such an investment-driven growth model also intensified economic imbalance and wasteful spending. In addition, since 2006 many of China's production sectors have been facing the problem of overcapacity, and the OBOR initiative is to transfer overproduction and overcapacity to new markets in the neighboring countries.

What China needs to do and is doing now (see Appendix on China) is move into the "third wave" which realizes that the most attractive firms are those that create permanent jobs, contribute to community well-being, and stimulate the conditions for new growth, and these firms typically are interested in other factors than simply lower costs—highly trained and skilled workers and superior quality of life for executives and workers as a primary example. Thus if the first wave was about attracting footloose firms, and the second was about reducing costs for all business, the third wave is about creating a favorable context for sectors of the economy that can grow most rapidly in a local community, creating business environments where innovation and success flourish.

The tools of third wave economic development differ dramatically from the tools that are used in the first and second waves. The processes are cumulative however. All economic developers will mobilize incentives for firms looking to come to their community, and they will try to offer second-wave community improvements to businesses when they can. What third-wave economic developers add is that they see networks of firms, not just individual firms. They seek to find the intersections between firms and to build regional collaboration that will lead to clusters that are more competitive than if the firms operated alone.

Whereas first-wave developers viewed themselves in competition with other communities, the third-wave developer looks for potential cooperative arrangements where all firms win. Leadership and the ability to broker deals where firms cooperate are the hallmarks of the third wave; conflict resolution is often necessary to form the trust that allows cooperation. Interestingly, third-wave development strategies are more complex and skilled than the efforts associated with the earlier strategies, but they are generally inexpensive. Building partnerships is complex, but the purpose is to create value in partnerships rather than to spend a lot of money.

For communities looking at the changing energy situation and trying to figure out the economic development implications for themselves, this constitutes a new opportunity. New energy investment brings value and jobs to the local economy, although most communities have not wanted large centralized power plants because of their negative environmental impacts. Moreover, large power plants employ relatively few people to build and operate, with fuels coming from outside the local region. Renewable and distributed plants on the other hand are sometimes more labor intensive and do not send lots of local money outside the local economy. Today, communities are in a competition to become the next "silicon valley" of wind, fuel cells, or the hydrogen economy.

The new energy system is good for local economies because of jobs and cash flow. The strategy by which local communities benefit requires intervention, however, from an economic development perspective to broker the relationships and build the partnerships that allow new agile power systems to be established in a community and to contribute jobs and wealth. Economic development and sustainable energy development based on renewable sources and conservation are not in opposition but in fact are symbiotic. The case study of a San Francisco neighborhood that discovered this helps to prove the point.

Green development is being led by the Asian economies of the Republic of Korea (ROK), Japan, Singapore, and PR China as well as some European economies such as the Nordic countries and Germany (see Chapters 3 and 6). The United States is lagging behind as a nation, although some of its states such as California, Oregon, and New York and the New England area are moving ahead. Asia takes an early lead.

The oil embargoes of the 1970s pushed Japan and the ROK toward economic and energy policies that eventually led to their development of the GIR. By the 1980s, many Asian and European countries had realized that their futures were not rooted in the same carbon-intensive economies that had built the Western economies of the Second Industrial Revolution, as they now became dependent on oil imports and started environmentally damaging programs to obtain oil and natural gas. These Asian and European nations wanted more energy security and developed national policies and programs to reduce their growing dependency on foreign fuels.

China started the same thing, in association with Singapore, at the turn of the 21st century by creating the Tianjin Eco-city. All buildings in the Tianjin

Eco-city comply with the Green Building Evaluation Standards, a unique benchmark that Singaporean and Chinese expert teams developed. The smart green city has a sound water management system that allows residents to drink directly from their taps, while sewage water is treated to provide a supplementary supply. As seen here, there are more and more cities that are "smart, green, and healthy" which are both affordable and good for the environment.

Clean, renewable energy sources such as solar water heaters and geothermal heating systems are used in the city to supplement traditional en-

Model city in Asia (2016).

ergy supplies. Also, a collective system of waste management and recycling is integrated with waste disposal and incineration processes to regenerate energy as well as minimize the strain on landfills. Thailand, despite being toward the other end of the economic spectrum in Asia, is also taking significant concrete steps toward a smart green city. Its capital, Bangkok, has a population of over 8 million people. Threatened by an increase in extreme weather and heat waves due to climate change, the city has introduced a number of ambitious environmental, local public health infrastructure, and health programs.

Europe's energy strides in Europe, while Denmark and other Nordic countries shifted national resources toward renewable energy power by national consensus, and Germany developed the innovative feed-in-tariff (FiT) financial process that started in the early 1990s under the national government that recombined East and West Germany into one nation.

Germany's FiT was part of the 2000 Energy Renewable Sources Act, formally called the Act of Granting Priority to Renewable Energy Sources. This remarkable policy was designed to encourage the adoption of renewable energy sources and to help accelerate the move toward "grid parity," making the price of renewable energy the same as that of the existing power from the grid. Under FiT, those generating eligible renewable energy, either homeowners or businesses, are paid a premium price for the renewable electricity that they produce. Different tariff rates were set for different renewable energy technologies, based on the development costs for each resource. By creating variable cost-based pricing, Germany was able to encourage the use of new energy technologies, such as wind power, biomass, hydropower, geothermal power, and solar photovoltaic, as well as to support the development of new technologies. Funding some of the green market for economists and others argued that there was no need for regulation; the government would act as "the invisible hand." Nothing could be further from reality. Without national and local policies in place, countries cannot address the issues, and there can be no action, no improvement, no resources, and certainly no response to environmental degradation.

There is not, and never was, such a thing as the government being an "invisible hand" in capitalism. Governments have always been involved and need to be even more involved for a better future. The key for local, state, provincial, and national governments is to have each of the major infrastructure components, i.e., energy, water, waste, telecommunications, and transportation, linked and integrated.

That way, the overlapping costs for construction, operations, and maintenance can be contained and reduced. If these basic infrastructure components can be constructed, operated, and maintained on the local level and meet regional, state, and national goals for carbon reduction, they take on a different perspective, format, and cost structure. Plans need to have financing, and vice versa. For example, the most significant result of the German FiT was that it stabilized the renewable energy companies and reduced the financial risk of energy investment. By guaranteeing investors compensatory payments down to the last pfennig per kWh, the FiT program created a secure climate for investment. The program covered up to 20 years per plant, with the exception of hydroelectricity installations, which required longer amortization periods. The law also offered means for altering the compensation rates for future installations, if necessary.

So the green development is not all about costing money. The GIR is about climate change mitigation, renewable energy, smart grids, and environmental sensitivity. But achieving the benefits of green development like a wave of new smart green technologies, business enterprises, and green jobs, will require substantial public and private financing. The green development economy will be needed to accelerate the necessary changes and stop climate change.

PARTNERSHIPS TO CREATE CIVIC MARKETS IN "GREEN DEVELOPMENT"

The problem, however, is that the private sector has taken over these "mechanisms" and will try to control and manipulate it as is its typical approach to the market. A number of private "exchanges" have been created and more are being formed. The government must oversee and regulate these exchanges, especially as they begin to conduct business. However, an even more compelling and significant infrastructure development is on the near-term horizon in infrastructure development: hydrogen and all solar powered vehicles.

Surprisingly, the problems confronting all new massive innovations (like the Internet) are not insurmountable. Hydrogen and fuel cells are well into the marketplace. Government is leading the charge. In California by late 2002, three major efforts were underway: one with the South Coast Air Quality District in the Los Angeles' region; another inland in the Palm Springs area with Sunline Transit; and the last in Northern California with Alameda-Contra Costa Transit District (east of San Francisco).

The transit systems have led the building of a hydrogen infrastructure as the government can exercise both planning and procurement (noted below). Civic markets are just that role whereby the government and industry collaborate together for mutually common and supportive ends. See various plans by these public organizations as well as private sector companies supplying hydrogen systems and major automobile companies providing vehicles and stationary fuel cells. The future is now.

In terms of how agile energy systems work and are organized, for example, the plans of regional transit systems are one of the driving forces. The CFCP is an excellent example of how agile systems work in an area related to energy: transportation and air pollution. The State has a long history of air quality problems, and there is no need to review them here. But it is important to note how public–private partnerships have helped to resolve the problems. The background of air quality problems in California rests in the regions but became institutionalized into the California Air Resources Board (CARB) in the early 1990s. The CARB set up zero-emission vehicle regulations (ZEV)—hence requiring automobile manufacturers to "sell" ZEV cars into California under an annual formula.

The ZEV regulations basically saw electric vehicles as the future form of transportation for the State. The key to the success of the program was to have the public sector oversee the regulations for vehicles sold in California by working with the carmakers to produce such vehicles. By the late 1990s, it was clear that electric battery vehicles were not the "only" solution to the problem. One clean alternative would be "fuel cell" vehicles (Cooper and Clark, 1996; Clark et al., 2002) and especially those that ran on hydrogen (Rifkin, 2003). Resistance from the carmakers, especially the American companies, was overwhelming as they attempted to stop the California deadlines.

However, the carmakers had made great technological advances and with the collaborative efforts of the CFCP were able to not only find compromised technological solutions but also regulatory avenues to get the ZEV into California. While the time lines have been pushed back slightly, the public—private effort has worked successfully. The same is now true for the State Fuel Cell Collaborative (SFCC) where companies and public sector leadership appears to be moving clean fuel cells into the market place. Much of the progress in both mobile and stationary fuel cells is on the local and regional level.

Government and the private sector can collaborate rather than be adversaries. This civic market approach has been very significant with the fuel cell partnership and other collaborative efforts in the State to derive practical approaches to implementing public policy: in this case with ZEVs and in other cases with energy and environmental policies. Porter (1995) not normally recognized for the positive role that government can play in stimulating new technologies, let alone good public policy, makes the point that the Dutch government passed some rigid standards for tulips in their country that not only protected their domestic market but stimulated international market growth.

Now today we have the Tesla car from Elon Musk who got a significant amount of his money to build the car from the US Department of Energy, and at the end of 2017, his company was over \$670 million in debt. Yet he launched his XSpace Ship in January 2018.

Falcon Heavy blasts off Launch Pad 39A at the Kennedy Space Center on Tuesday February 2, 2018. *Joe Raedle/Getty Images.*

Consider China Today Moving Ahead Into a Smart, Green, and Healthy Nation

Melody Rong
Data Analyst, IT Health, Cardioval Company, Suzhou, Jiangsu Province, China

OVERVIEW: VIRTUAL POWER PLANT

Transferring from a central grid to on-site power, virtual power plants are literally distributed energy sources. They are designed to cut down cumulative power demand during peak hours, supply renewable energy, and prevent intervention on the grid. Due to the problems and concerns with nuclear power, China is working with the United Kingdom on what to do to make the ones that they have safe and secure. However, the nation is moving away from nuclear power and into renewable green energy systems for both on-site and central grid power. The graph below shows how these systems need to and will be interacting.

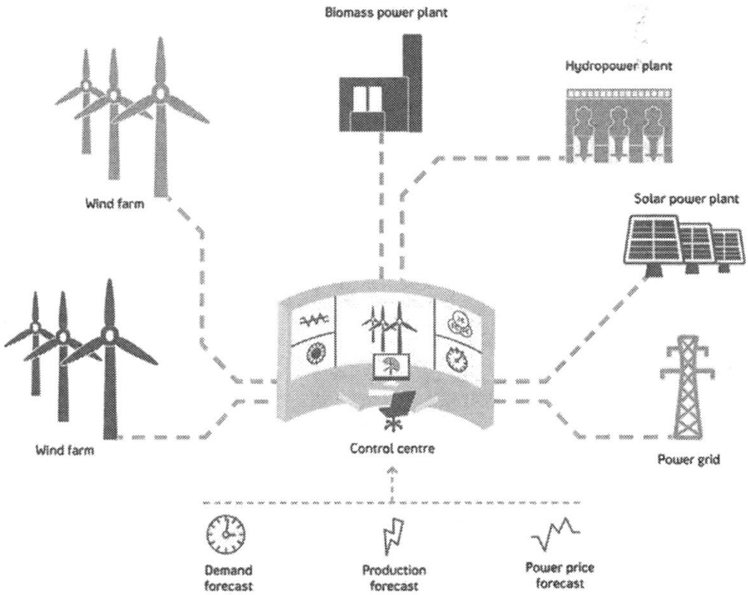

A graph of the virtual power plant. Graphics: Statkraft.

China has been advancing in distributed power and will increase renewable energy usage. The 12th Five-Year Plan in 2012 planned to produce distributed power from solar energy with a capacity of 10 GW.

China's largest energy utility, the State Grid Corporation has encouraged households and companies to utilize distributed power by providing credit for extra power they deliver to the grid since March 2013. The grid is connected to on-site power at less or equal to 10 kV, with capacity of each on-site power being less than 6 MW.

Moreover, distributed power technologies solve issues from natural gas flaring and can stimulate transformation from oil and gas industry.

China has been leading in investment on renewable energies with $89.5 billion in 2014. 19.81 FW and 10.61 GW have been installed in wind and solar energy, respectively, from both central plant and distributed power. China is aimed at installing 100 and 200 GW for solar and wind, respectively, by 2020.

SOLAR ENERGY

China is expected to develop 85.00 GW solar energy by 2035. The estimated investment on solar power is $107 billion from 2011 to 2035.

Incentive plans of renewable are applied. For solar energy, investment has been corporate in constructing solar photovoltaic (PV). Distributed solar energy produces power with a value of $0.25, while the state grid generates electricity with a value of $0.12.

China has also developed new policy to decontrol the distributed solar power market and centralize payments from on-site power.

Distributed PV projects in 2014 include Coca Cola with 10 MW PV, Haier with 20 MW PV, Galanz with 50 MW PV, and Midea with 35 MW PV.

Below is a chart illustrating connected solar energy in China from 2005 to 2020.

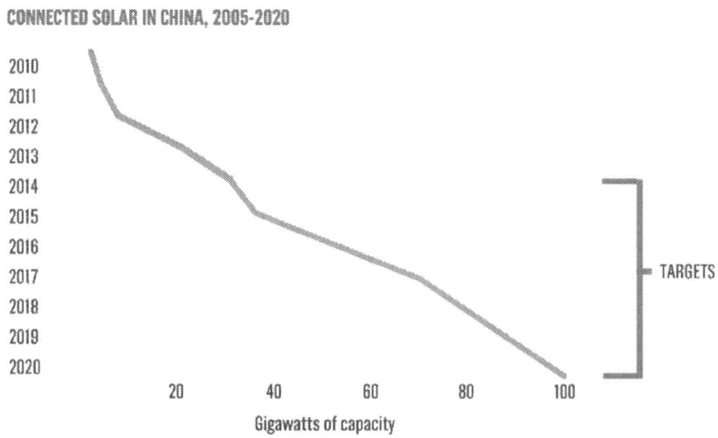

Source: NEA, China State Grid, Azure International (Note: 2015 goal of 17.8 GW in total (centralized + distributed). By October 2013 cumulative centralized capacity exceeded 15 GW.).

Case in point is Zhejiang Province, Yiwu Grid-Connected Photovoltaic System was the largest system in 2008 and had 1295 kW.

WIND ENERGY

In 2011, China has had 65 GW wind capacity, of which 47 GW is electricity—grid connected. State Grid of China is aimed at installing grid-connected wind energy capacity with 150 GW by 2020. China is believed to have 800—1000 GW wind capacity by 2050. The estimated investment on wind energy is $808 billion for 2011 to 2050.

Below is a chart illustrating connected wind energy in China from 2005 to 2020.

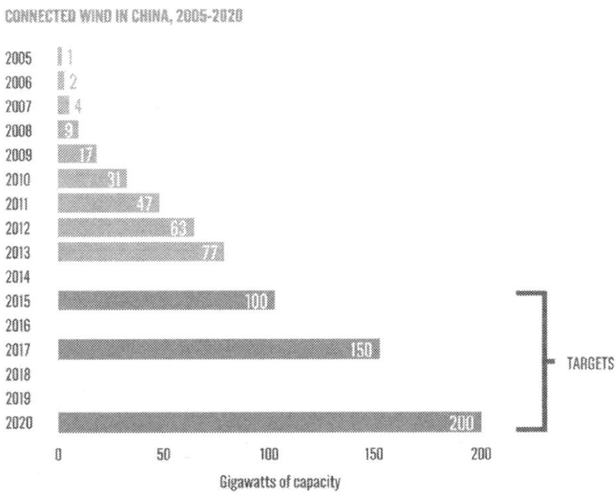

Source: National Energy Administration (NEA), China Wind Energy Association (CWEA), Azure International.

DISTRIBUTED WIND POWER HAS BEEN DEVELOPING AND BENEFITS FROM GRID INTERCONNECTION

A modular heliostat matrix concentrating thermal system was constructed in Shanghai. The heat conservation ratio has been up to 70%−80% with an operating temperature of 350−400°C. Power supply is stable and not intermitted due to its heat storage technology, which has a working coefficient analogous to coal.

Biomass

Biomass is composed from living organisms and photosynthesis. All organic wastes are able to produce liquid, solid, and gas energies, and get rid of organic pollution.

China has 720 million tons of biomass sources so far and plans to have 30 GW capacity for biomass energy by 2020.

In Henan Province, Zhoukou has built a biomass gas project based on a hybrid anaerobic digestion. Raw materials were livestock and urban catering wastes. There were 10 major fermentation tanks. Each tank had a volume of 2500 cubic meters and generated a total of 35 million cubic meters biomass fuel gas in a year. This project has reduced 420,000 tons of annual greenhouse gas emissions. Below is a picture of Zhoukou Anaerobic Biomass Fuel Gas Project.

Geothermal Energy

Geothermal energy produces electricity by tapping hot steam and then bringing it to the ground surface. China has much potential in renewable energies. Geothermal power reservation in China is approximately 135 billion tons of standard coal, which has around 3.16 billion tons reserve. Geothermal reduced greenhouse gas emissions by 19.87 million tons in 2008 and contributed to RMB 7.092 billion.

The BeiyuanJiayuan No. 6 Residential Quarters Geothermal Heating Project in Beijing is an example. It is the largest geothermal project in China. It uses 406,000 square meters floor space and has three geothermal wells. The system supplies heating for winter and cooling for summer in all targeted areas. This project is able to replace more than 8100 tons of fuel coal per year.

Below is a picture of BeiyuanJiayuan Residential Quarters.

Another project is practiced in Tianjin—the Historical Cultural Street Geothermal Utilization Project. It applies temperature differences and heat pumping to implement a cycled model of recharging in summer for heating in winter, and recharging in winter for cooling in summer. The recharging rate is 100% due to the stability of water level and quality of wells. This project conserves energy and benefits the environment. Below is a graph of Geothermal Utilization Project at Historic Cultural Street in Tianjin.

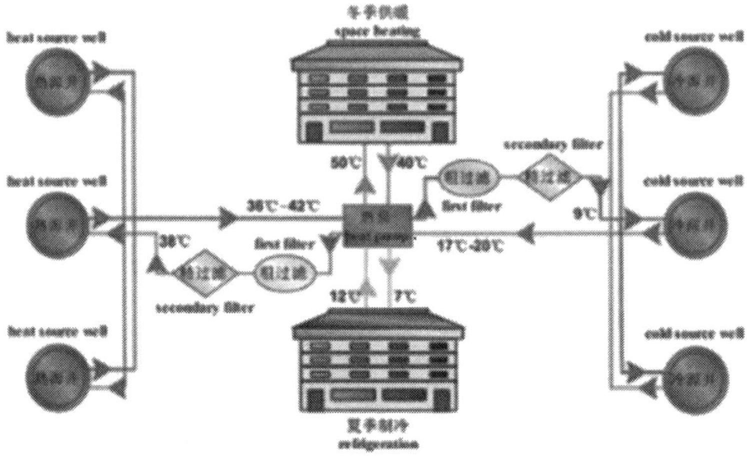

Water Dams—Large and Small

China has also employed hydroelectric power through Three Gorges Dam Project in the Yangtze River which is in the Xilingxia Gorge in Hubei Province. The Project finished establishment in 2008. The dam is the largest hydroelectric power plant worldwide with 185 m height and 2309 m width. The plant gives approximately 85 TWh energy output each year, which is equivalent to one-tenth of the demand in China. There are 32 major power generators from the dam and produced 18,300 MW in total during the year of 2008. The estimated cost of the project was $29 billion. Technology transfer was applied from foreign groups.

Construction of the dam is controversial as it can have negative effect on the reservoir, break the balance of upstream, take away nutrients from fish downstream, and damage the ecosystem. On the other hand, other countries have carefully reviewed the project. Japanese Ministry of International Trade and Industry argued that the project would help manage flooding issues and reduce emissions from coal.

The energy has been distributed to two cities through three 500 kV DC transmission lines. Directions include the East China Grid with 7200 MW capacity, the South China Grid with 3000 MW capacity, and the Central China Grid with 12,000 MW capacity.

Wave Power

Wave power is another renewable energy that has been widely used for water pumps and desalination plants. It is generated when electricity generators are operating on the ocean surface. Multiple factors all influence wave energy output, such as wavelength, wave speed, wave height, and water density. Below is a graph presenting the wave power principle.

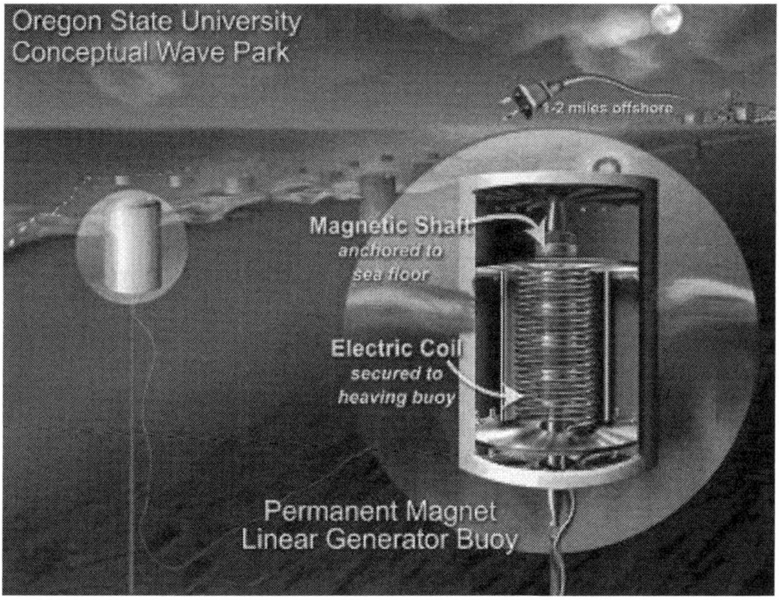

Source: Nicolle Rager Fuller, NSF.

While practical wave power stations are limited worldwide, China planned to launch its first generator plants in 2015 with Suzhou Eco Wave Technology, a subsidiary of Tel Aviv–based Eco Wave Power as shown in the photo below.

$800,000 was invested on the planning by the company and Chinese government. The power plant is connected through grids and has 100-kW capacity.

Moreover, Eco Wave implements its "Power Wing" technology to monitor water level change and wave motion to overcome severe conditions such as storms.

Run-of-River

Run-of-river is one form of hydropower. It is produced from elevation differences of river by introducing part of a river with into a pipe, and then spinning a turbine to power electricity with the water force. Enough rain and snowpack are desired to generate run-of-river power. Facilities manage the water fluctuation and supply electricity continuously.

China has leapfrogged hydropower generation. It has installed a total of 319,370 MW hydropower until 2015 and 1,126,000 GWh power was produced last year. One example is the Yugur Run-of-River Hydro Project.

The project has a capacity 13 MW with two turbines of 6.5 MW. It applies the Heihe River located in Gansu Province to provide power for the Northwest China Grid with up to 50,000 MWh in 1 year.

In addition, Hanergy, a Chinese solar energy company also proposed to implement dam power from the Amur River in August 2016. The Amur River is also known as Heilongjiang River in Chinese. It is 2800 km long and has a catchment area of 2.1 million square kilometers.

CONCLUSIONS

Therefore, the ultimate goal of the grand OBOR strategy aims to redirect the country's domestic overcapacity and capital for regional infrastructure development to continue to keep Chinese industry and production robust, to keep low unemployment rate through retaining a decent GDP growth rate, and to resolve China's internal overinvestment problem by exporting some productions, such as steel, cement, and aluminum, which used to be the driving sectors in China's own industrial and infrastructural development, to the construction of overseas infrastructural projects, such as railways, sea ports, highways, airports, etc. Externally, it aims to pave the transmission belt for Chinese goods and services to enter new markets and improve trade and relations with Southeast Asia, Central Asia, and European countries. It is an effective way to boost China's regional influence.

The OBOR initiative will provide more opportunities for the development of China's less developed border regions. China also intends to explore new investment options that preserve and increase the value of the capital accumulated in the last few decades. OBOR has the potential to grow into a model for an alternative rule maker of international politics and could serve as a vehicle for creating a new global economic and political order.

The OBOR initiative is also included and is inextricably linked with China's new 13th Five-Year Plan (2016–20) for the first time. The Five-Year Plan functions as a national guide for countrywide development and investment strategy throughout the period. It marks a dramatic shift of China's growth model from capital accumulation–led growth to innovation-led growth together with integrated urban–rural development and green development.

The Plan echoes the OBOR initiative by urging metal, steel, and heavy machinery companies with stunning overcapacity (particularly those with declining domestic demand) to find overseas markets. The regional perspectives of the Five-Year Plan, seen from the OBOR perspectives, show a long-term economic strategy to boost economic integration across the belt-road countries and regions.

NOTES

1. According to the Copenhagen School, securitization is a process whereby a securitizing agent tries to establish, socially, the existence of a *threat* to the survival of a unit. When a subject is securitized, it comes out of the scope of *normal policy* and moves into the scope of *emergency policy*, which usually legitimates the use of force (Buzan et al., 1998).
2. This is a Chinese idiomatic expression which means "to hide brightness, and to nourish obscurity." Its underlined notion reflects an implicit strategic choice, namely to wait for a time when China is ready to assert itself in the global sphere and is ready to make a challenge.
3. This is also a Chinese idiomatic expression which is semantically connected with the "Tao Guang Yang Hui" aphorism (see note 2). It stands for "striving for achievement" and "making a difference" after a period of accumulation of physical and psychological strength.

REFERENCES

Babones, S., January 25, 2018. America Has Little to Fear from a China-Centered World: Washington Post, World Post. https://www.washingtonpost.com/news/theworldpost/wp/2018/01/25/china-us-relations/?utm_term=.5f2c4812396b.

Buzan, B., Wæver, O., de Wilde, J., 1998. Security – A New Framework for Analysis. Lynne Rinner Publishers, Inc., Colorado.

Clark, Woodrow, W., Emilio II, Paulocci, John, Cooper, 2002. "Commercial Development of energy – environmentally sound technologies for the auto-industry: the case of fuel cells". Journal of Cleaner Production, Special Issue.

Cooper, John F., Woodrow W., Clark, 1996. "Zinc/Air Fuel Cell: an alternative to clean fuels in fleet electric vehicle applications". International Journal of Environmentally Conscious Design and Manufacturing 5 (3-4), 49–54.

Li, X., 2016. Understanding China's economic success: "embeddedness" with Chinese characteristics. Asian Culture and History 8 (2), 18–31.

Rifkin, Jeremy, 2003. The Hydrogen Economy. Penquin, New York, NY.

FURTHER READING

BBC Mundo, 2014. Los Impresionantes números del Gran Canal de Nicaragua. In: 22 diciembre. Available at: http://www.bbc.com/mundo/noticias/2014/07/140708_economia_canal_nicaragua_cifras_nc.

Breslin, S., 2010. China engages Asia: the soft notion of China's 'soft power'. ETHOS 8, 5−11. Available at: https://www2.warwick.ac.uk/fac/soc/pais/people/breslin/research/sbreslin-china_engages_asia-ethos_issue_8.pdf (February 13, 2017).

Broomfield, E.V., 2003. Perceptions of danger: the China threat theory. Journal of Contemporary China 22 (35), 265−284.

Chan, H., April 12, 2016. Will the Beijing consensus follow the Washington consensus to Oblivion? IPP Review. Available at: http://ippreview.com/index.php/Blog/single/id/97.html.

China's Clean Energy in Progress. Ministry of Science and Technology of P.R. China. Retreived from: http://www.china-un.org/eng/chinaandun/economicdevelopment/kj/P020100618014743196837.pdf.

China's Fast Track to a Renewable Future. RE100 China Analysis, April 2015.

China International Hydropower Association, May 2016 (Retrieved from hydropower.org).

Duarte, P., 2017. Pax Sinica. Chiado Editora, Lisbon.

Ekman, A., 2016. China's Multilateralism: Higher Ambitions. European Union Institute for Security Studies (EUISS). Available at: www.google.dk/url?sa=t&rct=j&q=&esrc=s&source=web&cd=1&ved=0ahUKEwiHmtGR4KvWAhWBfxoKHWkZCIgQFggqMAA&url=https%3A%2F%2Fwww.iss.europa.eu%2Fsites%2Fdefault%2Ffiles%2FEUISSFiles%2FAlert_2_China.pdf&usg=AFQjCNHUxRv0jmJGkDD42fN99k_JVyw9fA https://www.google.dk/url?sa=t&rct=j&q=&esrc=s&source=web&cd=1&ved=0ahUKEwiHmtGR4KvWAhWBfxoKHWkZCIgQFggqMAA&url=https%3A%2F%2Fwww.iss.europa.eu%2Fsites%2Fdefault%2Ffiles%2FEUISSFiles%2FAlert_2_China.pdf&usg=AFQjCNHUxRv0jmJGkDD42fN99k_JVyw9fA.

Financial Times, 2017. China Encircles the World with One Belt, One Road Strategy. Available at: https://www.ft.com/content/0714074a-0334-11e7-aa5b-6bb07f5c8e12.

Hanergy Solar Giant Wants to Dam the Largest Free Flowing River of Asia?, August 20, 2016. River without Boundaries.

Hubbard, P., 2008. Chinese concessional loan. In: Rotberg, R. (Ed.), China into Africa: Trade, Aid, and Influence. Brookings Institution Press, Washington.

Ikenberry, J.G., 2008. The rise of China and the future of the west. Foreign Affairs 87 (1), 23−37.

Ikenberry, J.G., 2011. The future of the liberal world order: internationalism after America. Foreign Affairs 90 (3), 56−68.

Kautsky, K., 1914. Ultra-imperialism. Available at: https://www.marxists.org/archive/kautsky/1914/09/ultra-imp.htm.

Kynge, J., July 29, 2015. China renminbi goal needs open markets. The Financial Times. Available at: https://www.ft.com/content/ee74deea-15d1-11e5-be54-00144feabdc0.

Lenin, V., 1948/1917. Imperialism, the Highest Stage of Capitalism. Lawrence and Wishart, London.

Mearsheimer, J., 2006. China's unpeaceful rise. Current History 105 (690), 160−162.

Mearsheimer, J., 2010. The gathering storm: China's challenge to US power in Asia. The Chinese Journal of International Politics 3 (4), 381−396.

Nuclear Power in China, September 20, 2016. World Nuclear Association.

Ordoñez, S., February 25, 2015. Who's behind the 'Nicaragua grand canal' - and why? CNBC. Available at: http://www.cnbc.com/2015/02/25/nicaragua-grand-canal-chinese-built-waterway-generates-lots-of-questions.html.

Owens, B., 2014. Chapter VI. Five Countries That Illustrate the Potential of Distributed Power. The Rise of Distributed Power. Ecomagination. General Electric Company.

Pan, C., 2012. Knowledge, Desire and Power in Global Politics: Western Representations of China's Rise. Edward Elgar, Cheltenham, UK.

Prospects for Distributed Combined Cooling, Healing, and Power Systems in China. L. Feng, Y. Wang. Falcon Group Inc. Institute of Engineering Thermophysics Chinese Academy of Sciences.

Pu, X., 2012. Socialisation as a two-way process: emerging powers and the diffusion of international norms. The Chinese Journal of International Politics 5, 341–367.

Ramo, J.C., 2004. The Beijing Consensus. The Foreign Policy Centre, London. Available at: http://www.google.dk/url?sa=t&rct=j&q=&esrc=s&source=web&cd=1&cad=rja&uact=8&ved=0ahUKEwi4nNOO36vWAhWSLFAKHXKDBIYQFggqMAA&url=http%3A%2F%2Ffpc.org.uk%2Ffsblob%2F244.pdf&usg=AFQjCNE7hBtQCbvssHinJhCqBRz-xpr82A.

Riley, C., Mullen, J., September 15, 2016. China will Own Part of New U.K. Nuclear Power Plant. CNN Money.

Ritzer, G., 1996. Sociological Theory, fourth ed. McGraw-Hill, New York.

Romana, H., 2010. Personal Interview by Paulo Duarte. Lisbon.

Scaling Wind and Solar Power in China: Building the Grid to Meet Targets. DB Climate Change Advisors. Deutsche Bank Group.

Shambaugh, D., March 6, 2015. The coming Chinese crackup. The Wall Street Journal. Available at: https://www.wsj.com/articles/the-coming-chinese-crack-up-1425659198.

Shen, S., February 6, 2016. How China's 'belt and road' compares to the Marshall plan. The Diplomat. Available at: http://thediplomat.com/2016/02/how-chinas-belt-and-road-compares-to-the-marshall-plan/.

The New Green Grid: Utilities Deploy 'Virtual Power Plants', August 1, 2016 The New Green Grid: Utilities Deploy 'Virtual Power Plants'. Admin. August 1, 2016. Green Energy Solutions.Greenenergyinvesting.net.

Three Gorges Dam Hydroelectric Power Plant, China. Projects. Retrieved from: power-technology.com.

Tiezzi, S., April 10, 2015. How China seeks to shape its neighborhood. The Diplomat. Available at: http://thediplomat.com/2015/04/how-china-seeks-to-shape-its-neighborhood/.

Udasin, S., March 26, 2015. Eco Wave Power to Establish its First Chinese Power Station (New-Tech. Israel News. Jerusalem Post).

Various News Releases, 2016, Azure International.

Wang, Y., 2016. Offensive for defensive: the belt and road initiative and China's new grand strategy. The Pacific Review 29 (3), 455–463.

Wave Power, November 2, 2016. AENews (Retrieved from alternative-energy-news.info).

Yugur Run-of-River Hydro Project. The Carbon Neutral company.

Chapter 5

Policies, Partnerships and Plans: Case of China

Xing Li[a,1], Woodrow W. Clark, II[b,2]

[1]*Department of Business Economics, AAlborg University, AAlborg, Denmark;* [2]*Research Professor in Economics, Pepperdine University Graziadio School of Business (PGBS), West Los Angeles, CA Campus, Los Angeles, California*

Chapter Outline

---

a. Xing Li, Ph.D. Professor at AAlborg University, Denmark.

b. Woodrow W. Clark II, Ph.D. Managing Director, Clark Strategic Partners, Los Angeles, CA, USA.

Climate Preservation In Urban Communities Case Studies. https://doi.org/10.1016/B978-0-12-815920-0.00005-8

OVERVIEW OF HISTORICAL BACKGROUND

The historical transformation from Maoist social-communist China to Dengist capitalist China since the end of the 1970s represents a sharp contrast as to China's national policy objectives, political agenda, economic development, and more importantly, development strategy. The Dengist China took the capitalist mode of economic development based on privatization of ownership and the means of production and distribution to the marketization and allocation of resources, including the total acceptance of economic inequities and political privileges.

The basis of this form of capitalism was initially to place emphasis on Western market–oriented economics and advanced technologies as the essential productive forces, along with the promotion of the interests of the privileged, professional, and entrepreneur classes, to also include the commercialization of welfare and social security (Chan, 2009). The 1970s was the period in which Western capitalism, promoted by the United Kingdom and the United States, became the norm for over 3 decades (Economist, 2009).

More significantly what China has witnessed in the past 3 decades is an economic growth path based on increasing demand for energy consumption based on an annual gross domestic product (GDP) growth of 10%. By the end of the first decade of the 21st century, China had outpaced the United States in clean technology investments (Scientific American, 2011). According to M. L. Chan, China at the end of 2010, ranked number No. 1 in clean technology investments, whereas the United States fell to No. 3 (Chan, 2010). However, China also become the No. 1 nation in carbon and air pollution surpassing the United States, even though the per capital air pollution in China is less than that of the United States.

By the end of 2008, China ranked fourth in the world in wind turbine manufacturing and installation with 12 GW of power (Zhen et al., 2009). Today Chinese companies control over half of the annual US $65 billion wind turbine market. Could China become the No. 1 economy in the world, surpassing the United States (Time, 2011)? With its GDP at $5.88 trillion in 2010, China surpassed the GDP of $5.47 trillion of the former No. 1 economy Japan. The Chinese GDP per capita of $7518 is much lower than $33,828 of Japan, but some economists predict that by 2030, China will have a GDP of $73.5 trillion compared with $38.2 trillion of the United States and Japan will be in the third place with $8.4 trillion. The world economics has changed as John Ross, an economist, noted "This has produced in China the greatest economic achievement in human history." Yet there is a human and an environmental economic cost, which are known as eternalities. Also, some Americans are skeptical and fearful of the Chinese economic challenge. A recent series of meetings in the US Congress have focused on "China's energy and climate initiatives" with a subtitle of "successes, challenges, and implications for US policies" (US Congress, 2011). The presentations were varied and presented no conclusions. Today, the challenge has affected China, and its current government is dedicated to do many things about it.

During the socialist period (after World War [WW] II), due to historical reasons, China took a development course with a commitment and goal by emphasizing human capacity, economic equality, and balance, thereby mobilizing social and economic resources in pursuit of a self-reliance development strategy. No matter how the socialist strategy is interpreted and assessed from today's perspective, it was historically the only possible option if China wanted to sustain its economic development, national security, and independence. The strategy of self-reliance emphasized the primacy of internally generated *independent* development not only at the national level but also at the provincial, regional, and local levels. The national Five-Year Plans started under Mao continue today. These internally generated *independent* development plans were reflected in the institutional structure of each unit,[1] where the multifunctions of party leadership, production planning, medical service, and resource demand and supplier controls were closely integrated. Self-reliance under such a socioeconomic structure in which politics, economics, service, and supply existed together at the unit level, created a sustainable society based on its own resources and needs and under the overall guidance of the state's Five-Year Plans.

CHINA'S NATIONAL FIVE-YEAR PLANS, HISTORICAL EVOLUTION AND TRANSFORMATION

The Five-Year Plans or *The Five-Year Plan for National Economic and Social Development* are national goal-setting economic and policy position papers derived over a year or two from high-level and local committees. At the national level, the Five-Year Plans outline national key construction projects and infrastructure plans and administer the distribution of productive, manufacturing, and business sources; growth of academic and educational institutions; along with individual sector contributions to the national economy. The plans also provide large sums of national funding for implementation. Aside from giving the nation, business, government, and foreign interest a "road map" about Chinese policies, the plans map the general direction of future development including specific measureable policies and targets. The last Five-Year Plan was from 2006 to 2010 and officially called the *11th Five-Year Development Guidelines*.

Now the new current 13th Five-Year Plan from 2016 to 2020 is being enacted by China's President XI. More details will be given in Chapter 5.

In the context of the 12th Plan, energy, particularly for homes and buildings, self-reliance also meant that the focus for renewable energy was primarily technologies that could help in the heating and cooling of buildings. Solar thermal systems were developed by state-operated companies in the 1990s and then spun off into private-public-owned firms. Sundra is a case in

1. A unit, in Chinese "Dan Wei," refers to any functional organization, for example, a ministry, a university, a company, a factory, etc. A "Dan Wei" was self-managed "miniwelfare state" combining supply, demand, and welfare.

point with solar thermal systems that appear all over China on homes (Kwan, 2009). Now their market has grown worldwide as examples from colleges in the United States illustrate (Eisenberg, 2009).

The Chinese choice of a self-reliance and self-sufficiency development path was projected as a potential "ideological threat" by the capitalist Western nations, because the central goal of the socialist politics with specific plans from the central government were seen as an attempt to challenge the capitalist ideology of competition that would reduce costs but unfortunately, also lead to an inequitable hierarchy in the world order (Downs, 2000, 2004, and 2006; Jiahua et al., 2006; Kaplinsky, 2006; Liu, 2006; Li et al., 2010; and today, US Congress, 2011 among others).

Seen from the interpretation of world system theory, the socialist self-reliance and self-efficiency policy aimed at transforming the basic logic of capitalism into "social capitalism" (Clark and Li, 2003). However, in reality it was actually designed toward a nation-wide mobilization for industrialization for the purpose of catching up with the core advanced capitalism that rewards people with money, no matter from where or by what means the funds were acquired.

This is because "socialist" (not "communist") nations like the Nordic nations, Germany, Iceland, and others were still operated within, but redefined for the capitalist world economy, and the dynamics of capitalism was capable of distorting and limiting national economic planning, leading to the constraints of their policy options (Chase-Dunn, 1982, 1999). The creation now "qualitative economics" at the turn of the 21st century has provided a new economic paradigm as chapters in this book document. Nevertheless, such a socialist project based on self-reliance that existed outside the United States has led the capitalist world economic system. In other words, it was merely an ideological challenge without being able to construct a sustainable alternative model or paradigm based on recent historical evidence, to replace the Western capitalist system.

In addition, the Chinese self-reliance policy was also a response to both the internal constraints of socioeconomic backwardness and the external pressure by the US-led economic blockage and trade embargo after WWII and the Korean War as well as from the failure of China's dependent relations with the Soviet Union since the Republic was founded in 1949. What should be pointed out here is that China's emphasis on self-reliance and independence in the Chinese energy industry was primarily derived from the lessons learned due to the Sino-Soviet split in which China was deprived not only of the Soviet technicians and specialists who were helping China develop its industries but also of around 50% of its oil supplies that were imported from the Soviet Union (Downs, 2000, p. 11–12). The discovery of the Dajing[2] oil field at the end of the 1950s was declared to mark the end of

2. Daqing was the largest oil field in China, and it is located in Heilongjiang province. It was discovered in 1959 and the production started in 1960.

China's external oil dependence both for defense and civilian applications. Such was the case in the modern market-oriented economy after Mao in the 1980s.

The post-Mao transformation of policy orientation from socialism to market capitalism with its objectives has its roots in the change of the regime's perceptions of the external environment of international political economy since the 1980s and especially due to the end of the Cold War in the early 1990s. The perceptions were generated from the following conceptualization of international relations. (1) The superpowers, including their respective alliances, were exhausted in their endless competition, leading to a situation where no major serious conflict was likely in the future even with the almost decade-long US military involvement in the Middle East and the 2011 military warplane presence in Northern Africa.

On the other hand, there would be the emergence of nonconventional security challenges whereby the Chinese government—controlled oil and gas companies bought international oil and gas producing and transport companies. (2) Economic development became the key objective for all nations, and economic power thus emerged to become more important and relevant than traditional military strength. Thus, soon, China had to face economic interdependence by increasing its global economic presence fuel supplies. (3) The post—Cold War US-based and US-controlled world order, which is an American-centric new world order, would likely remain for an unknown period of time. Therefore, China should, in the words of Deng, "observe calmly, secure our position, cope with affairs calmly, hide our capacities and bide our time." (4) There would be growing global competition for natural resources and hence for energy security. It was this last area that began after the turn of the 21st century, when China's unknown global challenges were still being defined and hence significant tasks needed to accomplished moving from totally government-controlled industries to ones that were collaboration or joint ventures with foreign companies and often owned by a majority of Chinese workers (Clark and Isherwood, 2010).

Consider the case of BRICS (Brazil, Russia, India, China and South Africa) as Professor Xing Li published in the fall of 2017:

The BRICS phenomenon in which the rise of emerging powers is coupled with a global financial crisis has caught global attention. Research and academic communities are being written widely that the world order is experiencing as a rebalancing of power and reordering. After almost a decade since the birth of the "BRICS" acronym (Brazil, Russia, India, China and South Africa), what is the situation today? The situation in Brazil is particularly discouraging. The Brazilian economy is heavily reliant on commodity exports (oil and gas especially as other countries like Venezuela), and hence went into went into recession. South Africa is now in an historical recession but not due to the fossil fuel industry.

The Russian economy has for many years pulled back and suffered from collapsing oil, and now even more due to natural gas prices. Furthermore, the Ukraine crisis turned Russia–West relations back into the Cold War era. Due to a devaluation in currency and Western sanctions restricting its trade, Russia is experiencing its worst recession in 6 years. India's economic growth is far from robust, as the country has a relatively high level of unemployment, along with its "caste system" still in place so that vast numbers of poor people are suffering from comprehensive poverty and backward infrastructures.

Despite the fact that in 2017 China also experienced its slowest pace of economic growth (6%–7%) in recent years, the country is the world's second largest economy and is becoming the biggest global stabilizer in the BRICS nations. China is dominating all the new institutions in which the BRICS interests are claimed to be presented. However, the Chinese state-led and investment-driven growth model also intensified economic imbalance and wasteful spending as well as overbuilding and the demand for energy primarily from fossil fuels. In addition, since 2006 many of China's production sectors have been facing the problem of overcapacity. Over the last century, China has changed dramatically, especially in the last 4 decades. Now the new Chinese central government in Beijing under President Xi has created a new "One Belt One Road Initiative" that aims precisely to transfer overproduction and overcapacity to new markets in neighboring countries and regions.

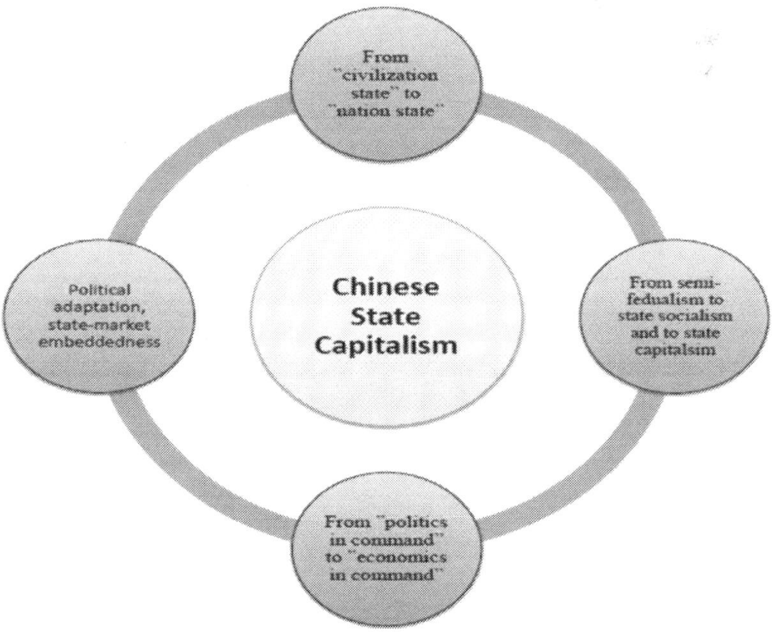

Historical evolution and factors of Chinese state capitalism (Li and Shaw, 2018).

The "Silk Road" model from centuries ago has now in the 13th Five-Year Plan turned to air and water to move goods and services from and to China. This global policy matches the program from the central government for "green development," which will reduce the air, water, and carbon emissions that have now made China the No. 1 nation in the world, after the United States. Even though there is less pollution per person than in the United States, the impact on people, their lives, and their health is costly in many ways for China. Most emerging powers (BRICS and others) in general and China in particular have followed a very opportunist foreign policy to find a balance between defending their national interest and resisting the "hegemons" of the existing order. Shown below is the Silk Road previously (in red) and what it will be again (blue for water as well as air) today with China's leadership:

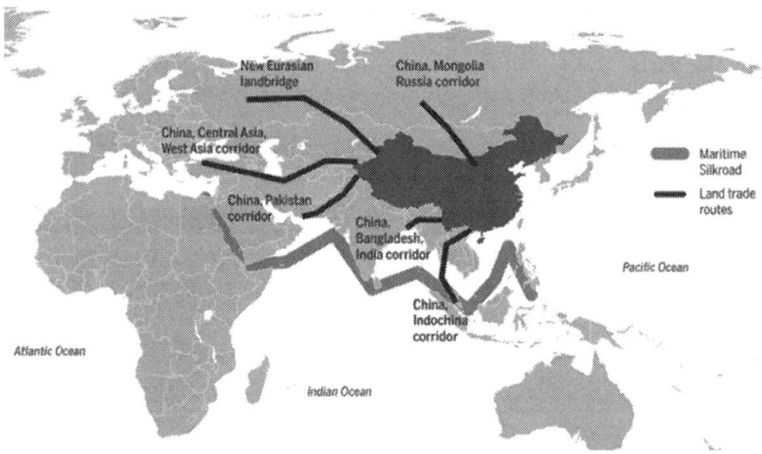

The world is witnessing neither the repeat of history manifested by the world wars (apart from some signs of great power rivalry) between the rising and the existing powers nor the harmonious integration and accommodation of the latecomers into the established system. So, a different but interesting question is: what is the nature of the relationship between the BRICS/emerging powers and the existing world order? To rephrase it:

Is the nexus between the US-led/Western-dominated existing world order and the emerging world reorder brought about by the rise of the BRICS/emerging powers more intertwined than conflictual?

The transition from "hegemony" to "interdependent hegemony" occurred due to the global financial crisis in 2008 with the decline of the hegemonic dominance of the US-led world order. That and the rise of the emerging powers has successfully penetrated into some power areas in terms of economic competition, capital accumulation, political and economic influence, as well as technical and material capacities.

China in particular is performing outstandingly in terms of its global share of high-tech manufacturing commodities, financial competitiveness, as well as international aid and overseas investment. It is argued that the world order is entering into a new and dynamic era of interdependent hegemony. This dynamic global change is now showing that the sources to feed and maintain the areas of structural power and monopoly are no longer dominated exclusively by the United States/West, as they are largely dependent on the inputs from emerging powers.

Notwithstanding that the concept of "hegemony" is an important tool for understanding and analyzing politics and international relations, "interdependent hegemony" is a better concept in describing, understanding, and analyzing the world order in transformation. The concept of "interdependent hegemony" implies a dialectic process of mutual challenge, mutual constraint, mutual need, and mutual accommodation. It symbolizes a dynamic situation in which the existing system's defenders and the new emerging powers are intertwined in a constant interactive process of shaping and reshaping the world order, whereby nation states, global governance, transnational actors, civil societies, and interest groups are incorporated into the dominant project of transnational capitalism.

Driven by the above-mentioned new strategic understanding, China has been pursuing a global foreign economic policy that was directed at creating a stable and peaceful environment for its economic growth through active engagement with the West and with the surrounding Asian nations (Clark and Cooke, 2011). This strategy has become China's globalization focus for a new or next economy (Li and Clark, 2009). China grasped opportunities for increasing international trade and foreign direct investment, and, more importantly, for securing access to natural resources and energy supplies through its own international trade and investment in the resource-rich regions such as Africa, Latin America, and Southeast Asia, and in recent years, Central Asia. China's global policy strategies under an active role of the state have been seen as effectively making it one of "globalization's great winners" (Thøgersen and Østergaard, 2010).

In short, China is no longer an "emerging nation" but was a founding member of BRIC (Brazil, Russia, India, and China) but has leapfrogged technologically, economically, and politically (Clark, 2010b; Clark and Isherwood, 2010) into The Green Industrial Revolution (GIR), which Clark and Cooke had published in Mandarin (2014) and then more comprehensively in English (2015), which moves an economy off the dependence on fossil fuels to renewable energy, smart green grids, and advanced storage technologies, leading to sustainable communities that reverse carbon emissions and reduce pollution, which started in Asia and the European Union over 2 decades ago (Clark and Cooke, 2011).

China, by the end of the first decade of the 21st century, moved actively into the GIR. The Pew Charitable Trusts (SJBJ, 2011, p. 8) quoted in Apr. 2011 with global data. For example, China reported $54.4 billion in clean tech investments for 2010, which was an increase of 39% from the year before and thus led all nations. Germany was second with $41.2 billion, with the United States third at $34 billion or 51% over 2009. Italy was fourth with $13.98 billion or up 124% from 2009, whereas the rest of the European Union ranked fifth with $13,48 billion, which was off 1% from 2009.

China's remarkable achievement in economic growth was made possible by its growing involvement in the capitalist world system. Steven Chan verified this fact as he told the story of SunTech becoming the world's No. 1 solar manufacturing company in 2010 (Chan, 2011). However, China remained in charge with caution and intense controls from the central government. It did not, for example, experience the deep 2008 global economic recession. In other words, China's economic growth is inseparable from its increasing dependence on global markets, with some estimates suggesting that more than 40% of its gross national product is derived from international trade.

In other words, China's rapid economic growth has been driven by exports with the assistance of foreign investments and joint ventures that have dominated the most dynamic sectors of the economy. Its market-driven growth encourages more concessions to induce capital flows and growth in unlimited possibilities of expansion and more structural changes to meet the demand of the overwhelming pursuit of external markets and resources (Lo, 2011). In addition, its integration with the world market is followed by overdependence of the productive forces on the fluctuations of the world market. The most affected area is China's energy demand and supply. Chart 1 illustrates the problem in China today with most of its energy demand coming from the supplies of coal, natural gas, and nuclear power. To summarize part of a recent series of presentations to the US Congress (April 2011):

> *China is now the global manufacturing leader of most renewable energy technologies, and the largest user of clean energy. In 2010 alone, clean energy investment in China topped $154 billion, while approximately 77 gigawatts of old fossil fuel power was shut down.*

THE DEBATE OVER ENERGY- AND CONSUMPTION-BASED ECONOMIC GROWTH

During the past 3 decades, China's GDP has enjoyed an average growth rate of 9%−10%. Since 2002, China's energy consumption has been growing at a faster rate than its GDP growth. As the following chart shows, China has had

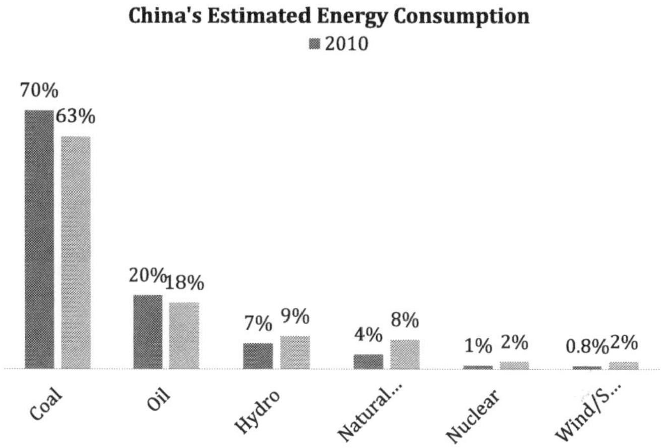

CHART 1 Future projection of China's energy consumption. *APCO Worldwide (2010, p. 7).*

the fastest economic growth than any nation in the world over the last 40 years (World Bank World Development Indicators, China National indicators of Statistics, December 2017).

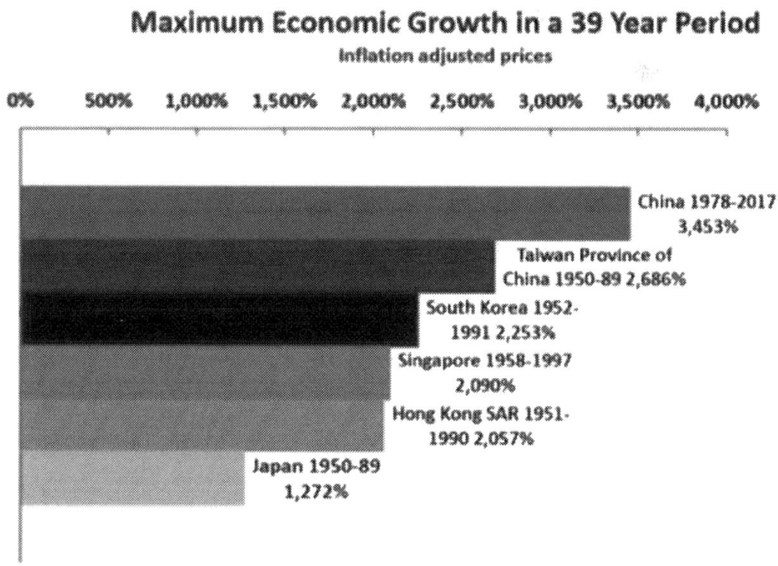

Source: Calculated from the conference board total economy database 2017, World Bank World development indicators, China National Bureau of Statistics.

From 2000 to 2005, China's energy consumption rose by 60%, accounting for almost half of the growth in world energy consumption (Downs, 2006, p. 1). In 2004, China contributed 4.4% of total world GDP, whereas China also consumed 30% of the world's iron ore, 31% of its coal, 27% of its steel, and 25% of its aluminum. Between 2000 and 2003, China's share of the increase in global demand for aluminum, steel, nickel, and copper was, respectively, 76%, 95%, 99%, and 100%. On a global scale, an increase in the rise of personal car ownership alone could mean an extra billion cars on the road worldwide within the next 10 years (NYT, 2010). As a report by Chinese Academy of Social Sciences in 2006 predicted accurately on the mounting worldwide impact of China's resource consumption:

> *China is currently the world's third largest energy producer and the second largest energy consumer. In 2002, China accounted for 10% of total world energy use and is projected to account for 15% of global energy use by 2025. China is the world's largest coal producer accounting for 28% of world coal production and 26% of total consumption. China is the third largest consumer of oil and is estimated to have the world's sixth largest proven reserves of oil. China has roughly 9.4% of the world's installed electricity generation capacity (second only to the United States) and over the next three decades is predicted to be responsible for up to 25% of the increase in global electricity generation. China emitted 10.6% of global carbon emission from fossil fuels in 1990 (second only to the United States) and 14.2% in 2003. This share is projected to rise to 22.2% by 2020.*

This situation has given rise to problems of net energy imports, environmental pollution and ecological destruction, cross-border pollution, and mounting carbon dioxide (CO_2) emissions:

> *The domestic environment has deteriorated rapidly, with some 70% of urban population exposed to air pollution, 70% of seven major water systems heavily polluted, over 400 cities short of water, and 3400 km^2 (equivalent to Japan's Tottori Prefecture) turning into desert every year. Cross-border pollution, notably acid rain and sandstorms, have reached the Korean Peninsula and Japan. Global environmental problems: China is the world's second-largest CO_2 producer after the U.S.*
>
> (Li, 2003, p. 1)

By early 2010, China moved ahead of the United States in being the No. 1 emitter of carbon in the world, although per capita, the United States is still the leading nation.

China's escalating energy consumption is placing heavy pressure on the world's energy prices. Chinese energy demand has more than doubled during the past decade. According to the study of Konan and Jian (2008), China will consume about 41% of global coal consumption and 17% of global energy supply by 2050. Metal prices have increased sharply due to strong demand, particularly from China, which has contributed 50% because of the increase in world consumption of the main metals (aluminum, copper, and steel) in recent years. In energy, China is going global with its investments in "Green Development":

China Going Global Investment Index

Rank	Energy
1	India
2	Iran
3	Pakistan
4	Vietnam
5	Indonesia
6	Philippines
7	Russia
8	Thailand
9	Hong Kong
10	Iraq

Source: The economist intelligence unit.

In health care, China is doing the same around the world:

China Going Global Investment Index

Rank	Healthcare
1	US
2	Japan
3	Switzerland
4	South Korea
5	Israel
6	Germany
7	Sweden
8	Norway
9	Denmark
10	France

Source: The economist intelligence unit.

Due to its rapid growth and rising share in the world economy, China was expected just after the turn of the 21st century to retain its critical role in driving commodity market prices (*World Economic Outlook*, Sep. 2006). It did so, according to almost every study by the end of the first decade of the 21st century (ACORE, 2011; US Congress, 2011). China is willing to offer more than the above world market prices for purchasing raw materials, which attributes great comparative advantages to the developing world but also threatens the West who seek to purchase and control those fossil fuel supplies (Tseng, 2008). Historically, the world has been already burdened by the high energy consumption by the West, particularly by the United States. Today China's growing appetite for international trade drives its mounting demand for resources to sustain its economic growth and to fuel its countless development projects. China has already become the world's largest importer of a range of commodities, from copper to steel and crude oil (Silverstein, 2011). The phenomenal rise of commodities prices worldwide in recent years is claimed to be attributed to China's growing import demand. Some even worried that there would not be enough resources in the world, for example, gas and oil, to satisfy the ever increasing demand driven by China's economic growth. Furthermore, taking China's neighbor, India, into consideration with its population of 1 billion, it will add double pressures on the demand for the same resources.

After a short period of self-sufficiency in energy supply especially in coal and petroleum, China became a net importer of petroleum in 1993, and it took only a few years before China became the second largest oil importer and consumer after the United States in 2009 (USDOE, 2011). China's energy profile used to be heavily weighted toward fossil fuel technologies (petroleum and coal) at a time when reductions are urgently needed to stabilize global climate change. Based on the 2008 statistics from the International Energy Agency, the growth rate of China's energy consumption and its share of the global total final consumption are comparably much higher than the rest of the world (Fig. 1).

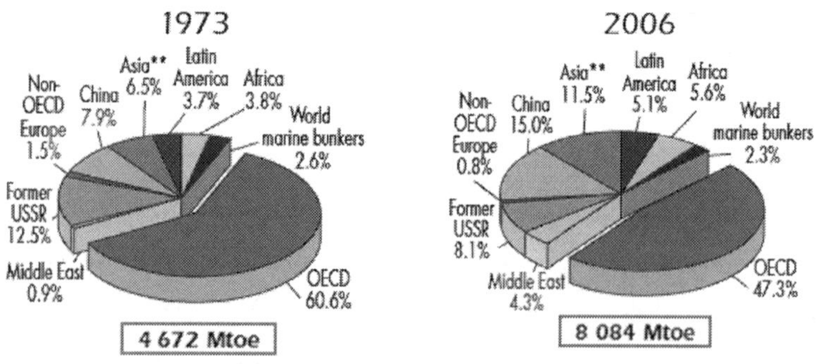

FIGURE 1 Shares of total final consumption 1973 and 2006. *Source: International Energy Agency (2008) Key World Energy Statistics, p. 30.*

It is foreseeable that China's resource imports of oil and gas will continue to increase if its targeted economic growth is to be maintained and it does not move fast enough into The GIR. The implication is therefore clear that not only global commodity price and international geopolitical power relations will be affected but also China's international politics, such as its foreign policy rationale, international aid objective, arms sale consideration, and compulsory expansion of its long-range naval power projection capabilities, will be closely connected energy economic issues. The rise of China, as a key actor in energy consumption, is forcing the current international energy regime to adjust or modify its established rules of the capital market game. This is because the international energy regime "is influenced not only by economic, political, and social factors of resource-rich countries but also by international political factors, particularly change in the international balance of power, adjustment of relationships among countries and changes to international rules" (Xu, 2007, p. 6). However, events in North Africa and the Middle East in early 2011 may have changed all this. Clearly the price of fossil fuels will rise and remain high.

One of the key challenging questions raised here is whether the global ecological system is on the verge of reaching the limit, and whether the expansion of global resource consumption is ecologically possible. The consequences facing China are very severe, and the Chinese growth model could face a fundamental challenge because the peak of resource exhaustion and the imperativeness of ecological sustainability would impose severe limits to its future economic growth, hence fundamental social changes will be inevitable (Li, 2010).

Li and Mingi argue that "World oil production is projected to have peaked in 2008. World natural gas production is projected to peak in 2041. World coal production is projected to peak in 2029. Nuclear energy is projected to grow according to IEA's 'Alternative Policy Scenario'. Long-term potential of the renewable energies is assumed to be 500 EJ (12,000 million tonnes of oil equivalent). The world's total energy supply is projected to peak in 2029" (Li 2010, p. 128).

Furthermore, China's coal production is projected to peak in 2030, oil production to peak in 2016, and natural gas production to peak in 2046. China's long-term potential of nuclear and renewable energies is assumed to be 1000 million tons of oil equivalent. China's energy imports are assumed to keep growing from now to 2020 to sustain rapid economic growth. The Chinese economy is assumed to keep growing at an annual rate of 7.5% from 2010 to 2020. By 2020, China's energy imports are projected to grow to near 700 million tons of oil equivalent, comparable to the current US energy imports. After 2020, China's energy imports are assumed to stay at 8% of the rest of the world's total fossil fuel production. China's total energy supply is projected to peak in 2033 (Li, 2010, p. 130−131).

Clark and Isherwood (2010) found this same pattern of short-term fossil fuel energy supplies in China while doing their study of Inner Mongolia

Autonomous Region (IMAR) for the Asian Development Bank in 2007. IMAR is the second largest coal producing region in China, and in the use of renewable energy (solar and wind in particular) to transition from the environmental problems caused by coal, China needed to provide public policies in its next Five-Year Plan for sustainable development with the financial resources. The 11th Five-Year Plan started that with over half of the 1.5 GW of power from global solar installation estimates being done in China (Chan, 2011). Today, the IMAR is developing such renewable energy resources like wind and solar while controlling its coal production through advanced coal technologies that are "cleaner." Almost every month several megawatts of wind power are being installed and operated in IMAR (Martinot and Li, 2010).

The Rise of China in the Context of Energy Dependency

To keep up the economic growth rate, China has to make the access to adequate energy supplies as a national priority and to a great extent a national *security* priority (Constantin, 2005; Huliq News, 2008 and Li, 2010). China is perhaps one of the few countries that regard energy security as a vital component of their *national interests*. Currently, China is "the world's second largest consumer and third largest producer of primary energy" (Martinot and Li, 2010). There is no sign that China's energy consumption will slow down; on the contrary, it will steadily increase. Thus for energy consumption to keep pace with its targeted economic growth at a moderate rate of 8%–9%, China will have to utilize every fuel source available, including investment on renewable energy and the expansion of nuclear power. It is expected that China's import of energy resources will increase at a steady rate particularly from Russian natural gas and liquefied natural gas shipped through Chinese seaports, which are both difficult options (Clark and Isherwood, 2010).

China's growing interest in resource-rich regions such as Africa, Latin America, Middle East, Central Asia, and Southeast Asia is no doubt linked with its energy security consideration (Brautigm, 2008). How will the rapidly increasing demand for energy, raw materials, and other natural resources shape Chinese policies toward its international relations especially with resource-rich countries?

Can China afford depending on global energy markets, either via exclusive bilateral deals or via direct investment in resource exploration to sustain its economic growth? What strategies will China use to secure its share of the global resource market? To find the answers to these questions, it is of importance to take an energy security approach to explore the geopolitical, economic, energy, and environmental implications behind China's rapidly growing energy challenges and to understand the Chinese anxiety and concern with issues of energy security in attempting to search for new sources of energy supply.

China's economic and foreign policy behavior is increasingly influenced by growing energy, weather, and environmental concerns. Consider this problem that China when compared to other developed nations:

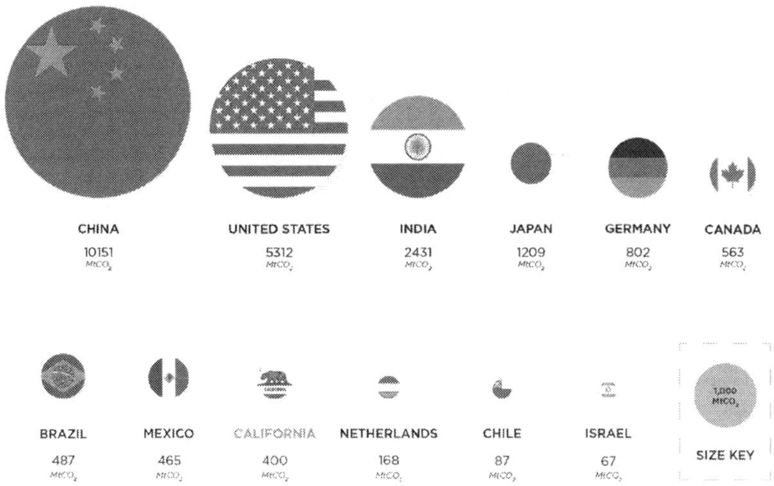

California represents about 1% of global carbon dioxide emissions in 2016*.

These data are from the California Energy Commission report in 2018. China is the world's largest emission polluter; however, not by per capita.

Andrew Chung, principal at Lightspeed Venture Partners in Silicon Valley, notes that the Chinese increase in clean (green) tech financing is reflective of the US drop: "First, a lot of the clean technologies are dependent on policy and government support to scale up. In some other parts of the world, you have more consistency in the way these types of funds are appropriated" (SJBJ, 2011, p. 8). As the world's second largest economy and trading nation in 2010, China's search for energy and its global strategies for energy security have led to heavy debates and in some cases have even resulted in political conflict. China is predicted by some economists and members of the US Congress to be the No. 1 economy and trading nation within the next decade (US Congress, 2011).

The Western nations have been expecting that ideally China's energy vulnerability might drive it toward cooperation with rival oil-consuming nations through participation in multilateral organizations and other forums. Since energy security is no doubt playing a more decisive role in Chinese foreign policy, Beijing's relations with both the existing major energy-

* Note: $MMTCO_2e$ is million metric tons of CO_2 equivalent. US emissions include emissions from California. *Source: California Energy Commission using data from http://www.globalcarbonproject.org/index.htm.*

consuming powers and energy-exporting countries will shape its motivation and justification on energy issues as well as on nonenergy issues.

In recent years, China's "going-out" economic and foreign policy encouraged its national oil companies (NOCs) to try to acquire some Western oil companies, but still secure control over access to some overseas energy supplies, including purchasing equity stakes in foreign oil companies (US Congress, 2011). This strategy has been regarded as "mercantilist" in the West and particularly in the United States where the attempt of a Chinese NOC to buy out the American oil corporation UNOCAL in 2005 triggered political backlash in the US Congress causing the final withdrawal of the Chinese company. The incident indicates the lack of trust of the United States in China's energy diplomacy, because the US politicians felt that the Chinese move was to undermine American energy security. Hence, the US oil and gas company Chevron bought UNOCAL.

In the studies of China's energy security with its economic and foreign policy, a number of geopolitically vital areas cannot be disassociated with China's efforts to maintain both energy security and good international relations within these regions and with the major Western powers. China's energy diplomacy with the Middle East, Russia and Central Asia, the Asia-Pacific, Africa, and Latin America have become a global topic, where Beijing's efforts toward greater energy security through multilateral organizations are discussed. It is still too early to predict whether the world is to witness the evidence supporting the liberal hypothesis that economic interdependence promotes international cooperation or confirming the realist conviction that competition and power accumulation will eventually lead states to conflict and war as history has shown in the past. Energy demand is seemingly accelerating China's "peaceful" rise to global prominence and moderating the conflict aspects of Chinese foreign policy while China establishes hundreds of new solar, wind, and other renewable energy companies. Chan (2011) stated that of the global installation of solar in 2011, over half of the 2.5 GW will be in China.

China has been struggling to develop and promote good relationships with underdeveloped regions that contain potential energy reserves, such as Africa and Latin America, through its unique international aid system linking development aid and trade with energy suppliers. Recently, China has aimed to prepare for technological advances and changes of climatic circumstances that will bring maritime transport in the Arctic waters to make possible the linking of North Atlantic and the North Pacific into closer commercial relations. Some policy makers expect that China will increasingly strengthen its political economy of international relations in the Arctic region and speed up its research through its polar research bases in Antarctica. In addition, China is adopting different policy strategies and objectives to different regions.

Currently, China is one of the key investors in Africa, and its trade and investment relations in Latin America are going to accelerate in the coming years (Hanergy, 2011). China's increasingly dynamic economic relations with these regions through long-term financing of infrastructures, renewable energy technologies, and smart grid systems are seen by some Western critiques as challenging the traditional ties between these regions and their historical colonial ties with the Western powers. Intensification in China-African and China-Latin American trade relations also accelerated the "neocolonialist" argument claiming that China is imposing the regions with a renewed "colonial" relationship. However, despite the criticism on China's energy-oriented policy in its economic and political relations with the two regions, the Chinese style of approach and engagement especially its aid policy and practice has indeed a far-reaching long-term and permanent realignment of power relations in the conventional international aid system and has already changed the system in many ways (Opoku-Mensah, 2010).

Currently, the Iran nuclear issue is testing China's foreign policy orientation in the context of its energy security consideration. China—Iran partnership has grown out of mutual need for products, ranging from arms and technology to consumer goods, and China's soaring need for energy supplies (Dorraj and Currier, 2008, p. 70). Thus, it has been a painful foreign policy decision for China to lend support to the US-led United Nations sanctions against Iran's nuclear program, fearing the grave consequence that this might lead to loss of one of its major energy suppliers.[3] China is being torn between the imperative need for energy on the one hand and the US pressure on its role as "responsible stakeholder" and "strategic reassurance" on the other hand.[4]

From an internal Chinese perspective, energy security has become the essential premise for China to achieve its national goal of quadrupling its GDP in 2020. There is a genuine consensus among Chinese leaders and scholars that energy has become a key strategic issue for China's economic development, social stability, and national security and that the realization of China's key national interests[5] is highly dependent on the access to sufficient energy resources (Liu, 2006; Zhang, 2006). China's "market economy" had locked itself in a "tiger-riding dilemma," i.e., any slowdown in economic growth would put the country in a risky situation, leading to social unrest and political illegitimacy (Li and Clark, 2009). China's government fears that domestic energy shortage and rising energy cost could undermine the country's economic growth and thus seriously jeopardize job creation (Lo, 2011). Beijing

3. According to data released by the General Administration Agency (GAC), Iran supplied 11.3% of China's energy consumption in 2009 (adapted from People's Daily Online, Feb. 10, 2010).

4. "Strategic reassurance," coined by James Steinberg, Deputy Secretary of State, in a conference sponsored by the Center for a New American Security states that "China must reassure the rest of the world that its development and growing global role will not come at the expense of security and well-being of others."

5. China's national interests are defined by the government as including sustained economic growth, the prevention of Taiwanese independence, China's return to as a global power status, and the continuous leadership of the Chinese Communist Party (CCP). Today energy security is defined as a core part of China's national interests.

increasingly stakes its political legitimacy on economic performance and rising standards of living for its people. Consequently, the threat of economic stagnation due to energy shortage represents real risks of social instability, which could in turn threaten the continued political authority of the state and the Communist Party. Energy security, and hence economic stability and sustainable development, are basic strategic political concerns for the leadership.

In fact, some scholars of energy politics point out that state-led pursuit of energy supplies is often seen as the source of international conflicts (US Congress, 211). For example, see Intriligator in this book. However, behind it, other sources of conflict—nationalism, geopolitical competition, competing territorial claims—are most likely to have been at the root cause of those conflicts (Constantin, 2005). One Chinese scholar of strategic studies clearly explains the reason why energy security has become a core component of China's national interest:

> *With external trade accounting for almost 50% of China's economy, China is now highly interdependent with a globalized market. This shift also includes hard social, political and geopolitical choices that deeply impact matters of national security. The more developed China becomes the greater its dependence grows not only on foreign trade but also on the resources to fuel the economy. With these complex and expanding interests, risks to China's well-being has not lessened but has actually increased, making China's national security at once both stronger and more vulnerable.*
>
> (Zhang, 2006)

China's sensitivity on the confluence of geopolitics and resource politics is also derived from the fact that historically China has been a weak sea power. One of China's key weaknesses through centuries of its development and into the modern age is its lack of a strong navy to safeguard its global interest and is perhaps one of the major factors leading to China's massive investment on raising and modernizing its naval capabilities. China therefore has good reasons for acquiring an aircraft carrier to enable it to protect its national interests (Cole, 2006; Nødskov, 2008). China has territorial disputes in the South China Sea over the Spratly Islands with neighboring countries and is also worried about the security of the major maritime transportation routes through which it transports the majority of its foreign trade, as well as its oil imports upon which it is totally dependent. Based on the historical lessons, China has a clear understanding on the linkage between its energy security and international geopolitics, which is spelt out clearly by one scholar:

> *The history of capitalism and its spread globally have shown that it is often accompanied by cruel competition between nation states. Those countries that lose out are not necessarily economically or technologically underdeveloped or those with a low level of culture. Rather, they are most often those nations who forgo the need to apply their national strength to national defense and therefore do not possess sufficient strategic capability.*
>
> Zhang (2006, p. 17)

Today the rise of China is due in large part to its rapid emergence as a major force in world energy markets and energy geopolitics (Chan, 2011; Lo, 2011). Beijing's booming energy consumption and heavy investment for energy security have raised a new range of contentious issues between China and other world powers, which are adding a new layer of issues to the already complex and dynamic relationships. China's economic growth is supported by three primary pillars, namely, (1) export-led growth, (2) real property growth, and (3) government spending; among them export has been the key engine driving its economic growth. The last global financial crisis (2008–09) had indicated a concern for the first pillar because European and especially American consumers can no longer consume at the debt-supported levels as they had in the past (Economist, 2009). One of the perplexing questions is whether the sustainability of China's export-oriented development strategy can be counted on to be sustainable and reliable into the future. Current data suggest that China has weathered this storm and become the new financial center for economic markets. Hence businesses are coming to China not only to invest but also to seek investments. The new Chinese export has become investment "capital" and financing (SJBJ, 2011; Lo, 2011).

New Policy: Change to Economic Growth Strategy by Promoting Sustainable Energy

In Nov. 2005, Chinese Premier Wen Jiabao declaimed at the Plenary of the Chinese Communist Party that "energy use per unit of China's GDP must be reduced by 20% from 2006 to 2010," and this declaration was turned into a policy goal set up by China's current 11th Five-Year Plan (2006–10).[6] In this national policy planning China's energy policies is defined to be "from a growth at any cost model" to "a sustainable, energy-secure growth path." To deal with the rising energy intensity, Chinese government has introduced a number of energy- and emission-saving policies as well as administrative plans and legal frameworks to strengthen energy conservation work. According to HSBC, in late 2010, the next 12th Five-Year Plan (enacted in Mar. 2011) would focus on 3 key issues:

1. "Achieving more balanced and sustainable growth is the key,"
2. Requiring "real reforms of income distribution, industrial regulations and fiscal system," and

6. Five-Year Plan, short for *The Five-Year Plan for National Economic and Social Development*, and even shorter for the plans are 11-5 year Plan, is a national goal-setting policy paper. At the macro level, it disposes national key construction projects, administers the distribution of productive forces and individual sector's contributions to the national economy, and maps the general direction of future development including specific policies and targets. The current Five-Year Plan for 2006–10 is also called the *11th Five-Year Development Guidelines*. The 12th Five-Year Plan or 12-5 year Plan came out in Mar. 2011.

3. Taking "steps towards financial reforms ... (that will) unleash the power of consumers and inland regions."

As Lo (2011) reports the Central Government and his corporation in Shanghai, China, have aggressively begun focusing on just these 3 three target areas plus more. Lo adds 3 other key elements to the 12th Five-Year Plan, however: (4) "strengthen environmental protection," (5) "enhance innovation," and (6) "improve living standards" by changing the focus of Chinese "export-oriented" economic model to a "domestic demand–oriented" economic model (Lo, 2011, pp. 7–9) Some concrete policy measures were implemented in line with these macro policy goals. In Dec. 2007, China's Information Office of the State Council issued the country's first ever white paper on its energy conditions and policies –

> *China's Energy Conditions and Policies. China's National Energy Administration (NEA) was set up in 2008 to coordinate energy issues concerning various ministries, commissions, and state-owned energy companies. In order to promote the development of emerging energy industries and meet the carbon emissions reduction targets of 2020, the NEA has compiled a development plan for emerging energy industries from 2011 to 2020 that will require direct investments totaling 5 trillion yuan according to China Daily.*[7]

Consider the growth of renewable power systems in China. The Chinese government is to launch a series of policies to support new energy development through China's 12th Five-Year Plan, or 12th Five-Year Plan for National Economic and Social Development, from 2011–15 (Lo, 2011) which is focusing on new energy, including wind, solar, and nuclear power, and the plan is under final review. The new energy policy will increase China's proportion of nonfossil energy in overall energy consumption from 12% to 13% by 2015, according to China's Energy Research Institute. This development trend is noticed by an international consulting organizations with the following data.

Energy security and environmental problems in China should be resolved not only through self-reliance efforts but also through international economic cooperation. To diversify access to energy supplies and reduce dependency on certain exporters, China is taking many political and economic measures and providing economic aid to strengthen its cooperative relations with resource-rich countries (Tseng, 2008; Ziegler, 2006 NOTE: cite more recent data). However, the dynamic debates on energy security are still going on in China (Downs, 2004). Certain indications can be drawn from the debates that many of China's analysts and policy makers are not fully convinced of the benefits of reliance on world energy markets. The political consensus today is to move

7. *People's Daily* online, Jul. 22, 2010, available from http://english.peopledaily.com.cn/90001/90778/90862/7076933.html.

toward "green" (renewable energy power generation) and integrated energy infrastructure systems that are sustainable.

In addition, China has been alert at the soaring demand in global energy in recent years and at the possibilities of long-term global energy shortage, called "peak oil," which now includes "gas" (US Congress, 2011). Hence, China's energy security will be one of the most important parts of its broader foreign policy in the years to come. The world will soon focus on China's new economic and energy policies, its energy market reform, and its new strategies in meeting the political challenges of rising energy costs and environmental pollution (Clark and Li, 2003; 2013). Much global attention will be given on China's move toward technology development and innovation in generating clean coal and natural gas power along with new institutional developments. Chinese energy policies and each of China's steps and practices bear significant implication on greenhouse gas emissions and climate change. The Silk Belt and Road by President Xi has already seen results as areas around the world document.

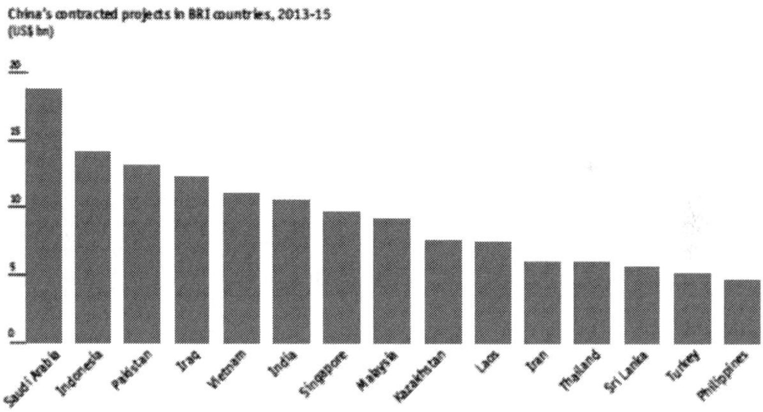

Source: Global Investment Tracker; The Economist Intelligence Unit.

Already being burdened with serious environmental problems and energy shortage, the ongoing global economic downturn presents China with a historic opportunity to rethink its growth strategy to move toward a more stable and sustainable path. Today, a promising optimism is that China seems to be firmly committed to the creation of a largely self-sustaining innovation system as part of a knowledge-based economy of the future. China is sparing no effort to meet its 11th Five-Year Plan energy conservation goals, in which China will cut its per unit of GDP energy consumption by 20% from 2005 levels by the end of 2010 (Fig. 2).

China's policy determination in clean and renewable energy can be clearly seen from its ambitious plan published in 2007, "Middle and Long-term Development Plan of Renewable Energies," which was approved by the

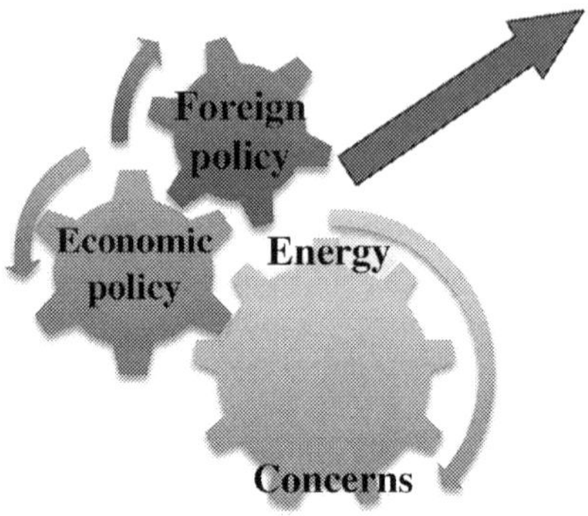

China	$34.6 billion
United States	$18.6 billion
United Kingdom	$11.2 billion
Rest of EU-27	$10.8 billion
Spain	$10.4 billion
Brazil	$7.4 billion
Germany	$4.3 billion
Canada	$3.3 billion
Italy	$2.6 billion
India	$2.3 billion

FIGURE 2 China's ongoing and future policies connected with energy concerns. *Sources: Xing Li, Aalborg University, Denmark and Pew Charitable Trusts, 2010, p. 7.*

People's Congress in 2008 as *The Renewable Energy Law*. The new policy is determined at moving the country toward renewable energy to reduce energy consumption and cut the surging carbon dioxide emissions. The policy expects to derive 10% of China's energy supply from renewable sources by 2010 and 15% by 2020. To meet the 2020 goal the total expected investment will be 2 trillion Chinese yuan (US $133.3 billion). If successful, by the end of 2010, China would emit 600 million tons less carbon dioxide a year, and by 2020 the annual reduction in carbon dioxide emissions would reach 1.2 billion tons. The target of this policy plan reflects another of China's policy concerns in coping with the environmental and economic challenges of climate change. The linkage of energy policy and climate change policy can be read from the policy document, *China's National Climate Change Programme 2007*, prepared by one of China's key government institutions, the Development and Reform Commission. Some examples of China's successes can be seen in communities that are becoming sustainable (Wang and Li, 2009; Kwan, 2009).

Through the legal framework stipulated in the new laws, the Chinese government has set efficiency goals and imposed taxes and regulations designed to curb demand and reduce emissions of greenhouse gases. In addition, the government energy and environment institutions are imposed with defined guidelines and responsibilities. The new policy toward alternative energy is supported by financial incentives including direct subsidies and innovative policy measures, tax-related incentives, custom duties, and pricing incentives. Some concrete policy incentives are (1) connecting "intermittent"[8] sources of electricity like wind or solar to the national grid; (2) connecting utilities mandated to open transmission lines to renewable generators, with ratepayers bearing part of the extra costs; (3) feed-in tariffs guaranteeing

8. It refers to energy-generation installations that are not state owned.

renewable energy producers a steady, high price for electricity so as to enable them to compete with coal producers; and (4) tax breaks, preferential loans, and other financial incentives encouraging investors to support renewable ventures (China FAQs, 2010).

China, however, has focused its efforts in the 11th Five-Year Plan to grow the renewable energy sector primarily for export. As is pointed out by Lo (2011) for the 12th Five-Year Plan and by the report that SGCC wrote on the 11th Five-Year Plan, "The strategy of building a world-leading strong and smart grid with ultra-high voltage grid as its backbone and subordinated at various voltage levels featured as being IT-based, automated, interactive, based on independent innovation … Since 2009, SGCC has started 228 demonstration projects of 21 categories in 26 provinces and municipalities." (SGCC, 2010, p. 1).

It is likely that China will meet and even exceed its renewable energy development targets for 2020 by applying other alternative energies including hydro, wind, biomass, and solar photovoltaic (PV) power. It is expected that more than one-third of China's households could be using solar hot water by 2020 if current targets and policies are continued (Martinot and Li, 2010). China expects the policy objectives to be reached through the integration of a number of relationships: the responsibility of the state and the obligation of the public, institutional promotion and market mechanism, current demand and long-term development, and domestic practice and international experience.

In recent years, China has won the global recognition for its achievement in the development and application of alternative energy. China overtook the United States for the first time in 2009 in the race to invest in wind, solar, and other sources of clean energy. American clean energy investments were $18.6 billion last year, which were a little more than half the Chinese total of $34.6 billion. Just a few years ago, China's investments in clean energy totaled just $2.5 billion (*Los Angeles Times*, Mar. 25, 2010). In recent years, it is increasingly recognized that China's "green leap forward" policy has made it become the world's largest makers of wind turbines and solar panels surpassing Western competitors in the race for alternative energy. As one of the key US newspapers points out:

> *China vaulted past competitors in Denmark, Germany, Spain and the United States last year to become the world's largest maker of wind turbines, …. China has also leapfrogged the West in the last two years to emerge as the world's largest manufacturer of solar panels. And the country is pushing equally hard to build nuclear reactors and the most efficient types of coal power plants. These efforts to dominate renewable energy technologies raise the prospect that the West may someday trade its dependence on oil from the Mideast for a reliance on solar panels, wind turbines and other gear manufactured in China.*
>
> *New York Times*, Jan. 30, 2010

Clean renewable energy strategy emphasizes that a sustainable growth path based on equity is leading the transition to knowledge and information economy. When referring to China's alternative renewable energy policy, some studies have shown that China is facing both opportunities and challenges. The potential opportunities are plenty, such as solar energy, wind energy, biomass energy, small hydropower, geothermal energy, and ocean energy, whereas the challenges are apparent as well, such as the lack of coordination and policy consistency, weakness and incompleteness in incentive system, lack of innovation in regional policy, immature financial system for renewable energy projects, and the limited investment in research and development of renewable energy (Zhang et al., 2007). There is still a long way to go before China's renewable energy market becomes mature and socially and culturally embedded.

CONCLUSION: CHALLENGES AND OPTIMISM AHEAD

This chapter aims at providing a framework of critically understanding China's transformation from a self-reliance development path to a "market-driven" dependent growth strategy to now a "social capitalist" economic system. The chapter's emphasis is on the economic and ecological consequence of China's insatiable demand for energy driven by the growth-based industrialization policy in the past decades. This chapter argues that since the beginning of the 21st century, energy has become a key concern on the agenda of China's economic and foreign policy-making calculations as it moves rapidly into The GIR. Among China's core national interests—securing energy resources, generating national renewable companies and systems, and gaining market access and political recognition—energy and economic security is the top priority for developing a sustainable nation. It is expected with the rapid economic development and the improvement of people's living standard, energy demand in China will unavoidably continue to increase, which will be inseparable with its environmental problems, such as the emission of sulfur dioxide, carbon dioxide, and particulates, among other issues of pollution and waste.

Above all, the Western definition of "market-driven" economies in energy is questionable in China such that different definitions and meanings are needed for "market" and therefore "capitalism." And that is what China has done. It has redefined capitalism so that it has a societal focus, direction, and set of policy along with financial strategies. For example, the rapidly emerging renewable energy industry in China has created a new market finance mechanism for long-term debt, which involves the Chinese business financing the entire sale, installation, and operations along with maintenance of the renewable energy technologies and products. In short, China may have discovered a new form of The World Bank.

Chinese policy makers understand the fact that due to a growing need for and even competition over energy resources and maritime transportation security, global resource-based competition and geoterritorial claims on sea areas and shelves will become harsh and could lead to armed confrontations. To understand the implication of the underlying dynamics of international political economy in resource-rich regions, it is of great importance to understand the source of international economic competition and the political conflict for access to energy and natural resources to understand the interactions between individual national interests. The international geopolitics and geoeconomics in the acquisition and distribution of states' wealth and power are manifested in their respective country's economic and foreign policies.

China's soaring demand for energy in connection with its export-oriented economy poses a variety of new challenges for its economic and foreign policy. Hence the country will be more and more dependent on the purchase of natural resources abroad for sustaining its economic development. Any crisis to China's access to overseas resource and maritime shipping routes will have a negative impact on China's growth and trade-dependent economy. China will endeavor to protect the strategic areas concerning its national interest. It has no choice. In recent years, China's energy diplomacy in the context of the political economy of global energy developments has drawn the attention of the West, especially in connection with the sensitive regions, such as the Middle East, Central Asia, Latin America, and Africa. As one Chinese scholar bluntly states, "The determining factor shaping the rise and fall of a country ultimately is not just the size of its total economic volume but also the strategic ability of the country; that is, the ability to use national forces to achieve political goals" (Zhang, 2006, p. 22).

However, despite the above global reality described by this realist perception, China's deep sense of its energy insecurity and vulnerability is changing its development policy toward clean and renewable energy. China is accelerating R&D on renewable energy supply and advanced energy conservation-based techniques and products; it is making necessary structural changes in industrial and agricultural sectors moving to non—energy-intensive industries. Furthermore, China is trying to rely primarily on domestic resources while strengthening mutually beneficial international energy cooperation. The optimism that China is presenting to the world is not groundless. China is not only one of the world's leading producers of renewable energy but also overtaking more developed countries in exploiting valuable economic opportunities, creating green-collar jobs, and leading development of critical low-carbon technologies.

Such optimism in China's own "green revolution" is also confirmed by the front page of a recent report by Climate Group (2009), "As one of the world's major economic powers, China will have to be at the forefront of this journey.

This report shows that it can be." Nevertheless, China still has a long way to meet its policy objectives on energy and environmental sustainability. Due to its size and population the consequences of failure in China's case are much more serious than many other counties. China should not be left struggling alone on the road to optimism; and the whole world must pay more attention to China. World peace and a sustainable planet depend on global harmony and collaboration beyond conventional competition over supply and demand.

Jiaxing, China: Leading Smart, Green, and Healthy City/Region

Wang Weiyi[a]

Jiaxing University, Zhejiang Province, People's Republic of China

OVERVIEW OF JIAXING

Jiaxing, a prosperous and beautiful city, also known as Hexing, or Jiahe in the ancient time, has a 7000-year history of civilization and 1780-year history of city formation.

Located in the north of Zhejiang Province of the southern Yangtze River Delta Region, Jiaxing borders the East China Sea in the east, Taihu Lake in the north, and Qiantang River in the south. The Grand Canal runs across the city. Neighboring cities within 100 km are Hangzhou, Ningbo, Shanghai, and Suzhou. Jiaxing covers an area of 3915 km^2 with a household population of 4.614 million. Under its jurisdiction are two districts Nanhu and Xiuzhou; 3 county-level cities Pinghu, Haining, and Tongxiang; and 2 counties Jiashan and Haiyan. Jiaxing has a coastline 121 km long.

Owing to the unique geographical location, convenient transportation, rich innovation, and human resources, Jiaxing enjoys a fast and well-developed economy. According to the newly published data in Feb., 2018 by Jiaxing

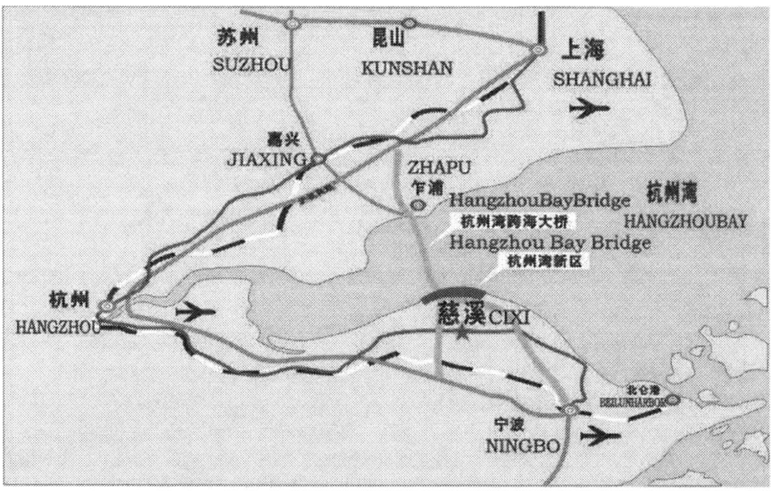

a. Ms. Weiyi WANG, Master of Arts, Associate Professor, College of Foreign Studies, Jiaxing University, Zhejiang Province, P.R. China.

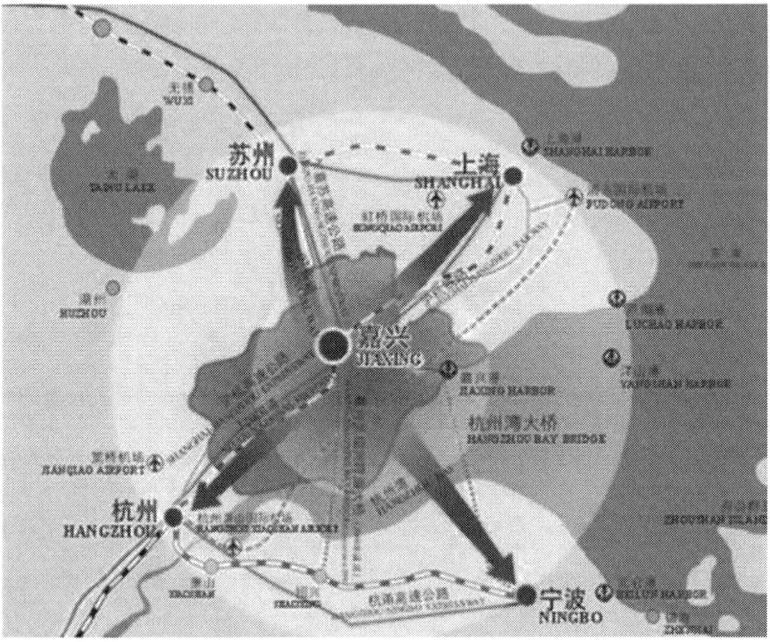

From Travel China Guide (2017).

Statistics Bureau, GDP in 2017 reaches RMB 433.5 billion(≈667 billion USD) with per capita: USD 13,917. The household population is 4.656 million and the city is near where the Silk Road ends and begins in China.

Jiaxing is a prefecture-level city in northern Zhejiang Province, China. Lying on the Grand Canal of China, Jiaxing borders Hangzhou to the southwest, Huzhou to the west, Shanghai (62.5 miles) to the northeast, and the province of Jiangsu to the north (Wikipedia, 2018).

In recent years, encouraged by the national 5-sphere integrated plan[9] and the 4-pronged comprehensive strategy[10,11] and by sticking to opening up and innovation-driven development, Jiaxing has seized the chance of "Internet +"

9. Source: Xi Jinping's Report at 19th CPC Congress, Oct. 18, 2017.
10. Five-sphere integrated plan refers to "economic, political, social, cultural and ecological civilization development" (Source: Xi Jinping's Report at 19th CPC National Congress, Oct. 18, 2017).
11. Four-pronged comprehensive strategy refers to "building a moderately prosperous society, deepening reform, implementing the law-based governance, and exercising strict governance over the Party in all respects" (Source: Xi Jinping's Report at 19th CPC National Congress, Oct. 18, 2017).

to speed up the transformation and upgrading in terms of agriculture, industries, modern service industry, environment and city management, etc.

Honor List achieved in recent years:

No.	Honor
1.	National Civilized City
2.	National Garden City
3.	National Hygienic City
4.	National Greening Model City
5.	China Good Public Security City
6.	China Historic and Cultural City
7.	China Excellent Tourism City
8.	National Innovation Pilot City
9.	National Demonstration City for IPR Protection
10.	National Demonstration City for Science & Technology Progress
11.	National Pilot City for Information Consumption
12.	China-EU Green & Smart City Cooperation Pilot City
13.	National Pilot City for Public Information Utility Development
14.	National Model City for New Smart City Development
15.	Demonstration City of "Broadband China"
16.	National Pilot City of Information Economy
17.	Wuzhen National Pilot Zone for Internet Innovation & Development
18.	National Demonstration City of Innovation-driven Development

GENERAL FRAMEWORK OF SMART, GREEN, AND HEALTHY CITY IN JIAXING[12]

Guidelines, Basic Principles, and Goals

As we know, city is the main carrier of human creation and continuation of civilization as well as a center of economic, political, cultural, and social activities. With the fast development of information technology (IT), Smart City is now recognized as one of the effective solutions to some problems of city development. However, for the lack of integrated systemic construction, there occur information islands and barriers. Therefore it is of utmost importance to make a systemic planning with information dominantly driven by mechanisms and technologies so that the construction of a smart city can deeply integrate IT applications with city modernization, realizing the coordinated development of urban and rural areas.

Guidelines: guided by the new vision of "Innovative, Coordinative, Green and Open Development that is for everyone"[12] and based on "Jiaxing 13th Five-Year Plan for National Economy and Social Development," Jiaxing will be built into a smart city by means of institutional mechanism reform, innovative moves, scientific and technological advances, and industrial development, so

12. General framework of SGHC in Jiaxing is reported by Jiaxing SGHC Office in May, 2017.

that Jiaxing will turn out to be an internationalized, modern networking and green city with urban and rural coordinated and balanced development.

Basic Principles

- To let the people enjoy a safe, effective, convenient, green, and fully preferential city life;
- To promote rational and free flows and optimization of all factors of production and resources, shorten the digital gap between urban and rural areas, and realize coordinated development;
- To form a new model of governance with the integration of the government and people, achieving the goal of the refine and precisive governance;
- To promote the integration of modern IT with various industries to deepen and widen its applications for accelerating the transformation and upgrading of the traditional industries;

Goals

- To greatly upgrade the information infrastructure;
- To observably improve the fully preferential public service;
- To significantly strengthen the city governance;
- To remarkably promote the innovation-driven development of information economy;
- To evidently elevate the standards of city life such that it is more ecological and livable;
- To significantly ensure comprehensive cyber security.
- To commit to an approach of "Government Guidance and Market Operation" by encouraging social capital to participate in investment and operation and creating new business models via market intensification.

GENERAL FRAMEWORK OF JIAXING SMART CITY

This framework consists of 5 architectures including business, systems, data, technical standards, and deployment.

Business Architecture

Business architecture focuses on citizens, enterprises, and governments, developing the civic services, city governance, industrial economy, and ecological living environment. The architecture is shown in Table 1.

System Architecture

System architecture is based on the principle of "Sharing, Integration, and Open." "Sharing" refers to building the information infrastructure

TABLE 1 Business Architecture

Civic Services	City Governance	Industrial Economy	Ecological Livability
Administrative affairs Health Education Culture Tourism Social security Community	City operation Administrative management Comprehensive management Emergency management Macroeconomy Market regulation Military & civilian integration	Internet economy Industrial & information integration Industrial park Electronic business Modern logistics Modern agriculture Financial payment	Water treatment Transportation Environmental management Energy utilization Spongy city Precision meteorology Town with its own characteristic

Source: "General Framework of SGHC in Jiaxing" p. 8.

characterized by "widespread perception and safe interconnection" to support the instant use of the censoring equipments and computers. "Integration" aims to make the city data focused and integrated by monitoring the status and resources of the information infrastructure and setting up the information sharing and support system together with various open apps. "Open" intends to build an open resource model to create all kinds of characteristic apps to expedite the development modes of emerging economy and improve the modernization of city governance. The architecture is seen in Table 2.

Data Architecture

Data architecture is composed of city public data, business operation data, perceptive data, crowd sourcing data, and comprehensive management data. The architecture is indicated in Table 3.

Technical Standards Architecture

Technical standards are in accordance with the national standards, including general standards; city infrastructure standards; IT and support standards; safety and security standards; construction, operation, and maintenance standards; and application and service standards. The architecture is shown in Table 4.

Deployment Architecture

The deployment architecture adopts an approach of 3-level platforms. The system will be open and provide access services for the citizens by uploading and downloading data from the nation and province, sharing the data with

TABLE 2 System Architecture

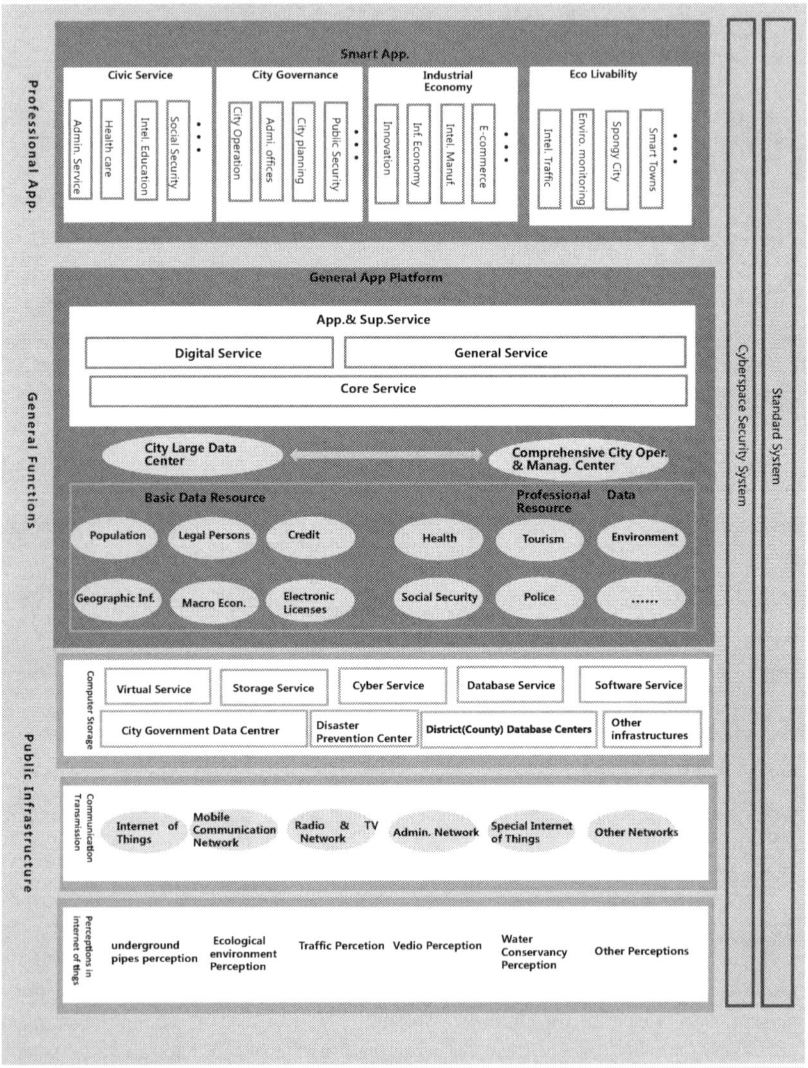

Source: "General Framework of SGHC in Jiaxing" p. 9.

TABLE 3 Data Architecture

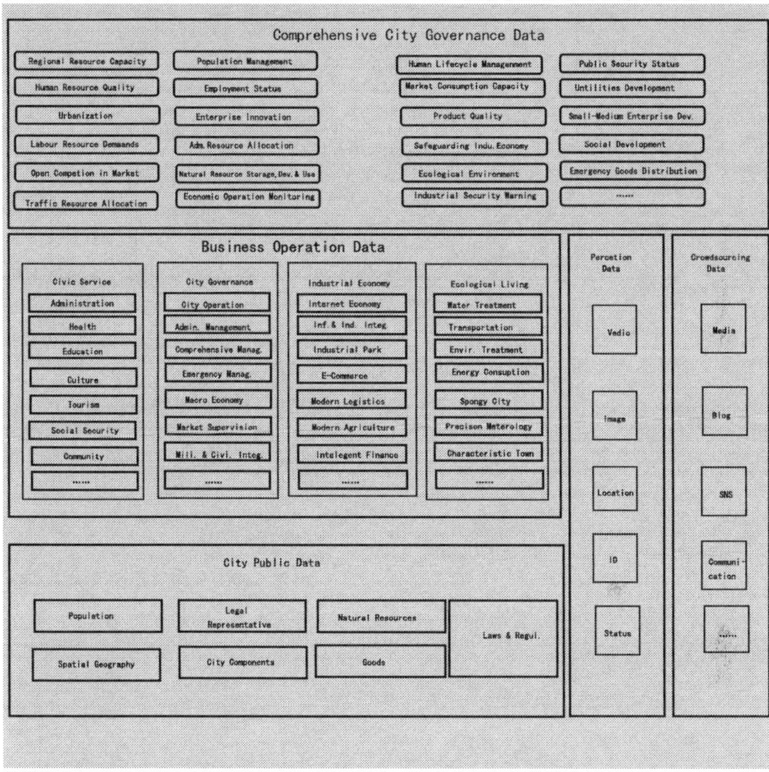

Source: "General Framework of SGHC in Jiaxing" p.10.

TABLE 4 Technical Standards Architecture

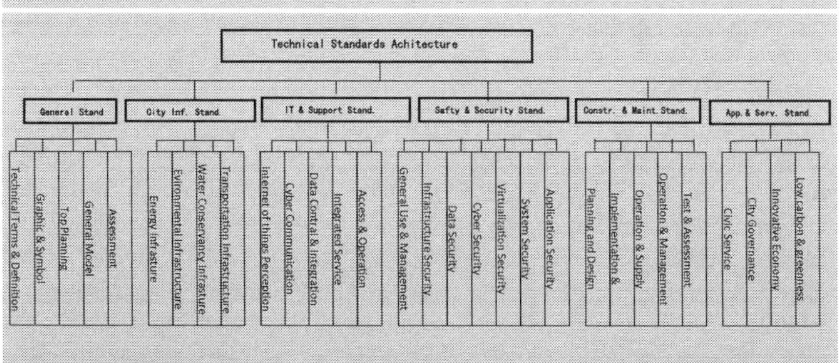

Source: "General Framework of SGHC in Jiaxing" p. 11.

TABLE 5 Deployment Architecture

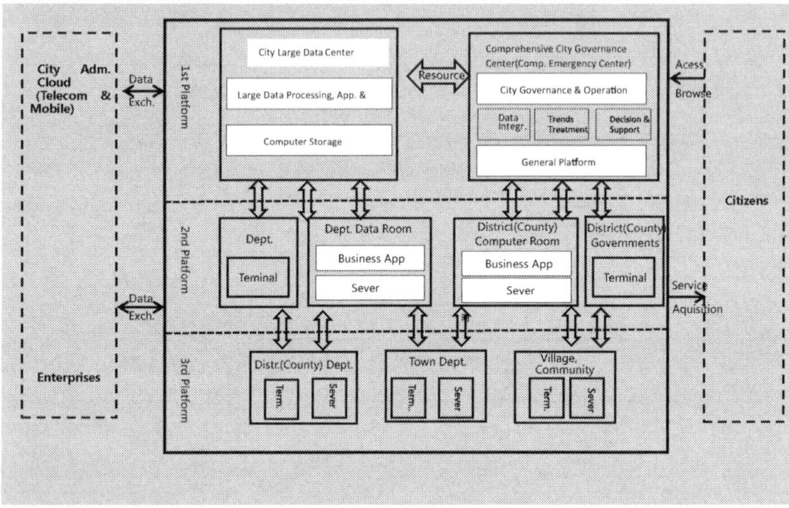

Source: "General Framework of SGHC in Jiaxing" p. 12.

neighboring cities, and sharing the data with Baidu, Alibaba, Tencent (BAT), and the like. The architecture is shown in Table 5.

Level 1 (City Level) Platform

Level 1 platform means that the city will establish a municipal comprehensive operation and management center and a large data center.

The large data center is the "heart" of the smart city, which carries the overall data and resources of the city. Six data bases will be set up and completed including household population, legal representatives, spatial geographic information and geological resources, macroeconomy, electronic licenses and certificates, and public credit registry, covering such important professional data as the city governance, civic services, infrastructure, industrial economy, ecological environment, etc. integrating the data resources from government affairs, enterprises and institutions, and internet. The citizens will also be provided with convenient services by introducing the data from BAT as the third parties. The electronic government data share-exchange platform will be further improved for the data support for interdepartment coordination. An open, integrated, and innovative platform for the city data will be built to have a standard data service interface. A data cleaning and processing platform will also be established to provide high-quality and effective data for the government departments' application of large data analysis. Large data analysis and application in city management, civic services, industrial economy, and ecological environment will be further developed to render services for urban decision making, enterprises, and citizens.

The municipal comprehensive operation and management center (or municipal comprehensive emergency command center) is the "brain" of the smart city. Based on effective data from the large data center, a "city operation panorama" will be produced to present the operation tendency, analysis, assessment, predication, and assistant decision for city management, civic services, infrastructure, industrial economy, and ecological environment. The municipal comprehensive emergency command center will help to connect the information systems among the relevant departments, proficiently preventing or responding to all emergencies. By integrating and improving the monitoring system, the internet public opinions can be constantly monitored, analyzed, and handled. Meanwhile, the channel of collecting public opinions from citizens and netizens will be further improved by connecting with "Mayor Hotline," "Civic Service Hotline," and cyber media communication platform.

Level 2 Platform (Municipal Departments, Districts, County Level)

Level 2 platform is composed of the platforms and professional databases from the county-level cities, districts, counties, and municipal departments. This platform provides the basic support for the needs from the departments and local areas by deploying the applications from longitudinal connection and lateral matching, undertaking the data aggregation, and forming the comprehensive tendency of different fields and regions. In principle, all county-level cities, districts, counties, and municipal departments are not allowed to set up their own computer centers except for the requirement from the superior counterparts. All units should establish their own comprehensive business platform and database. This system can share the data among its sections. The professional database can be linked to the municipal large data center, business systems, and comprehensive operation and management center.

Level 3 Platform (Town, Village, Community Level)

Level 3 platform consists of the departments of county-level cities (districts and counties), town governments, subdistrict offices, communities, trading areas, and industrial parks by collecting and uploading data, receiving and carrying the instructions from the higher authorities, and handling various tasks.

Main Tasks

Based on the "General Framework of SGHC in Jiaxing" and the requirements of the "National Pilot Zone for Internet Innovation," the "National Pilot Zone for Comprehensive Integration with Shanghai," the "National Pilot Zone for Urban and Rural Integration," and the "National Model City of Watery Region in the South of China," the Jiaxing government has decided to implement 8 main tasks in the construction of Jiaxing Smart City.

Strengthening Urban and Rural Integration to Set an Example in Civic Service

By establishing the system of "Internet + Administrative Affairs," the citizens and enterprises can enjoy public services by means of "One Registration" application, "One Window" reception, and "One Net" handling. In health service, citizens' health information be shared and used among the medical institutions. The medical settlement and referral can be completed easily. In education, Smart Campus will be encouraged for improving an educational service with high and equal quality. In culture, "Internet +" model can help to set up "Digital Library," "Digital Museum," "Digital Cultural Center," "Digital Art Gallery," "Digital Exhibition Hall of Intangible Cultural Heritage," etc. In tourism, "Internet + Tourism" can upgrade the public services and comprehensive management. Likewise, "Internet + Social Security" can provide "one-stop" service for both urban and rural areas. The Smart Pay based on "Internet +" will be further implemented. In the construction of Smart Community, intelligent monitoring, migrants' management, parking services, etc. will also be focused on.

Focusing on Precision to Build a Model in City Governance

A unified administration system of the whole city will be accelerated to make the system linked to the data of Zhejiang Provincial Administration Service Network. By integrating the information and factors of the urban and rural planning, major projects, land resources, environmental protection, and transportation, "One Platform with Multi Disciplines" will be established for upgrading the intelligent capacity. A Social Credit System will be set up to make the overall relevant information be covered and applied. The Comprehensive Emergency Command Platform will improve the efficient coordination of all walks. The Unified Production Safety Supervision Platform will monitor the key industries in full coverage. The Unified Comprehensive Market Supervision System will improve the food safety monitoring and prewarning system. The National Defense Emergency Command Center will be docked with the information from public security, civil affairs, and medical institutions to provide data for war preparation and economic services. The comprehensive perceptive network for underground pipelines can help to collect and renew the data of water supply, drainage, electric power, gas, radio and TV cables, and telecoms.

Strengthening Integrative Innovation to Accelerate the Industrial Transformation and Upgrading

By speeding up the construction of internet economy platform, a large number of industrial parks and bases with their own characteristics will be built. In IT industry, smart terminals will be further enhanced. The internet of things will be further refined. Software and information services will be optimized. The application software and business intelligence software will be greatly developed. In intelligent manufacturing, human beings in key posts will be replaced by robots. Digital workshops will be encouraged for trials, and the new model of C2M based on internet and personalized customization will be pushed on. In intelligent business, e-commerce will be promoted in the traditional industries to provide one-stop cross-border e-commerce services for the enterprises. In intelligent logistics, the comprehensive information platform for logistics of sea, land, and air will be established to develop the intelligent storage system. In smart trading areas, the comprehensive information service platform will be built for parking guidance, customer flow guidance, and mobile pay. In intelligent industrial parks, the innovative intelligent service platform including finance, comprehensive commerce, and virtual resource renting will be set up. In agriculture, the traceable information supervision system will be built to improve the quality of agricultural products. The internet of things in special agricultural parks will be greatly piloted.

Improving Green Intelligence to Build a Garden City

Based on the comprehensive operation and management center platform, municipal departments such as land and resources, planning, and public security can implement the joint law enforcement by collecting, transmitting, analyzing, assessing, and handling the information data from soil, air, water, and natural disasters. The convenient intelligent traffic service management and coordination platform will be accelerated to make the relevant departments join in an efficient traffic command system. The environmental business support and decision service system will be pushed on by setting up the intelligent environmental protection system coordinated with environmental protection, prewarning, and emergency. The comprehensive water treatment platform will improve the management of water network and river basins. The Spongy City will develop the module of internet of things to monitor the whole process from "source, supply, drainage, sewage, and disasters." The regulation and operation platform of PV power generation will promote the model of distributed PV power generation to pilot the new energy trading service.

Enhancing the Overall Planning to Build Intensive Infrastructure

As a demonstration city of National Broadband China," "Broadband Jiaxing" will optimize and upgrade the internet infrastructure by popularizing and applying "4G + technology," trying to be a pilot city of commercial use of "5G," accelerating the deployment of new internet (IPv6) and applying for a particular channel of internet access. "Sensing Jiaxing" will utilize the internet of things and intelligent equipments to improve the networking integration of public security video surveillance, strengthen the intelligent sensing and monitoring system of all component parts in the city, and bring the sensor network and sensing devices into a unified management. Based on "Tianditu (Map World),"[12] the space–time information cloud platform will be developed to form one trend map of the whole city. The common software will be intensified and encapsulated to help the governments and enterprises to develop and apply the information and resources conveniently.

Focusing on Sharing and Opening to Promote Digital Application and Innovation

Jiaxing Large Data Center will help to integrate, mine, and analyze the cross-department and cross-sectoral data to release its value to the most extent. The city data resource will be further developed to build the basic database to totally cover the population, legal representatives, macroeconomy, natural resources (spatial geographic information), electronic licenses, social credit, etc. so that a professional database will gradually be established in the fields of city governance, civic service, infrastructure, industrial economy, ecological environment, administrative affairs, enterprises, and the internet of things. The city unified data exchange sharing platform will provide the catalogs of the government information and resources and the standard data service interfaces classified by the importance and sensitivity so that the city departments can utilize the analysis and application based on the large data to improve their quality and efficiency. Governments, enterprises, commercial sectors, etc. will be encouraged to make innovations.

Consolidating Security Basis to Build a Clear Network Space

The mechanism of information security and dynamic risk assessment will be established to enhance safety protection by compiling the catalogs of the key infrastructure and checking its products and service access, so that the network spatial governance will be improved. Such governance mechanisms as real-name registration, spam message management, cyber fraud monitoring, and fake websites discovering and blocking will be set up. The comprehensive network governance platform will also be built to strengthen the supervision of public opinions, the capability to deal with the internet incidents, and the tip-offs of the cyber criminals and harmful information. The authentication system of unified identity in network space will be established by consolidating data security and safety risk assessment, and providing classified data, encrypted data retrieval, and intelligent protection of sensitive and private data. The relevant regulations concerning the construction, implementation, and supervision will be further enhanced.

Building Smart Towns to Shorten the Digital Gap Between Urban and Rural Areas

As the component parts of a smart city, smart towns needs to be built as well. The city government will elevate and integrate the infrastructure in communication, network, and the internet of things of small towns, providing convenient civic services for the towns. The city governance and service resources will be extended to the small towns so that the town residents will enjoy the fast, convenient, and standardized service within the town. By developing "Internet +" economy and based on the town characteristics, the traditional industries will be transformed by integrating with internet so that new industrial models and business formats will be brought up. The demonstration platform for characteristic towns will be set up to present the achievements of local industries, culture, economy, and tourism.

CASES IN JIAXING

Case 1. "Internet + Jiaxing": A City With Smart Transformation

(Source: Xin Jia Xing, vol. 377, Nov. 11, 2016, pp. 38–40)

Located in Tongxiang City under the jurisdiction of Jiaxing, Zhejiang Province, Wuzhen is well known for being the permanent venue of the World Internet Conference (WIC), which has been held annually since 2014. Organized by Zhejiang government and the Cyberspace Administration of China, the conference is a platform for discussing important issues regarding internet development and security, showcasing the latest developments by major players in the technology and innovation sectors from home and abroad. In 2016, Chinese internet giant Baidu, Inc., flaunted its new driverless cars during the conference period and allowed the delegates to experience them. Recently, the participants gave thumbs-up in Alibaba's Tmall unmanned supermarket after receiving discount for smiling at the Light of Internet Expo at the fourth WIC in Wuzhen, Dec. 3–5, 2017.

Israeli scientist Dan Shechtman, the winner of the 2011 Nobel Prize in Chemistry, experiences "smiling to get discounts" function in Alibaba's Tmall unmanned supermarket. *Source: Xin Jia Xing Vol. 377, Nov. 11, 2016.*

Apart from being the hosting site of this event, which has been attended by President Xi Jinping, other world leaders, renowned scientists, and entrepreneurs, Wuzhen also has a reputation for being one of the most picturesque water towns in China listed in The United Nations Educational, Scientific and Cultural Organization's World Cultural Heritage. In 2009, the Pacific Asia Travel Association named Wuzhen as its first ecotourism location in the Chinese mainland.

Since 2014, Wuzhen has hosted 4 sessions of WIC. There are many surprises that the people have never seen before. The ancient water town connecting with internet has become a smart town. What's more, Jiaxing government has taken it as the center of a circle to strongly promote the smart transformation all over the city, witnessing the marvelous changes brought about by the power of internet revolution.

From Wuzhen "Butterfly Change" to Jiaxing Transformation

Every time when the participants arrived, they surprisingly found that Wuzhen had taken on a new look. The driverless cars developed by Baidu, Inc., and automakers of Chery, BYD, and Shou Qi Group made debut on a road 5 km long and the highest speed reaching 60 km/h. Mr. Wang Jin, senior vice president and general manager of the company's Autonomous Driving Unit introduced that Baidu, Google, and Uber are the main developers of driverless car technology in the world. The cameras on Baidu's driverless vehicles have an accuracy of 90.13% for judging objects, the highest level in the world. The accuracy for judgment of pedestrians reached 95%, and for traffic lights, 99.9%. However, the accuracy still needs to be improved, Wang added. Wang said that Baidu's goal is to realize the commercial potential of driverless cars on a small scale in 3 years and expects to produce them on a large scale in 5 years.

Driverless Car on the Road in Wuzhen. *Source: Wuzheng Town Government.*

The 5G vehicular networking technology invested by NOKIA and the intelligent parking system developed by CETHIK are also displayed during the conference period. Therefore high-tech driverless cars, intelligent auto scheduling, and parking systems, which used to be seen in science fiction movies, have become a reality in Wuzhen.

In the Hall of Zhejiang Large Data Exchange Center, the data transactions are rolling on the screen in real time. Mr. Lai Lei, deputy general manager of the center, said that the data transaction can promote the transformation and upgrading of the traditional industries by integrating the data with government and enterprises to improve the public services such as medical care, social security, transportation, environment, employment, etc. ATIP (ARTDUODUO Intellectual Property) Trading Center shows that its transaction volume within 22 days has exceeded 1 billion RMB (≈US $153 million).

In addition, many makerspaces such as PING AN Maker and Maker Institute CAS, Phoenix Tea, and Tencent WeStart settled in Wuzhen, which has turned the small town into a testing ground for network technology, changing the residents' way of life.

Wuzhen is only the epitome of "Internet + Jiaxing." There are 10 other small towns characterized by Internet +. For example, in Xiuzhou District, an intelligent logistics town was established. Besides, in the high-tech industrial park, the farmers can enjoy the clean green energy and gain benefits by installing the distributed PV power station on their house roofs. In Jiashan

County, Xitang Ancient Town has become a smart town integrated with internet. An intelligent manufacturing town named after "Silicon Valley" aims to build a low-carbon and high-intelligence industrial park with the beautiful scenery of the south of the Yangtze River. In Wuzhen Internet Hospital, patients can see medical specialists from over 1500 hospitals via the Internet Doctor Alliance, which can provide more than 170,000 doctors and received 0.16 billion patients last year. Now the "Internet +" model has been widely applied in administration, civic services, medical care, city governance, culture, education, and electric power in Jiaxing. The citizens can enjoy a convenient and comfortable life via internet.

Intelligent Manufacturing Town in Jiashan County. *Source: Jiashan Government.*

From "Innovation-Driven Development" to "Driving Innovation"

President Xi Jinping once pointed out that innovation is a new golden key that can bring about new development and achievements in international competition. Encouraged by his remark, Jiaxing has seized the chance of the permanent site of WIC to undergo systemic changes. Due to the higher population and less land and natural resources, it is difficult for Jiaxing's traditional industries to survive. However, information economy has offered a good opportunity to evade the inferiority in factors of production, labor, space, and environmental carrying capacity.

An innovation-driven development plan has been carried out. At the strategic level, "Internet + economy" is adopted as the first economy. The city government implements the strategy of "building an SGH city" by making new policies and replanning the information infrastructure and industrial layout. An innovation-driven action route has been clearly presented. At the national and provincial levels, Jiaxing has been listed in the Pilot City of National Information Economy. Wuzhen is recognized as the Comprehensive Pilot Zone of National Network Innovation-Driven Development.

Owing to the disruptive technologies and thinking dimension, internet can propel the city development forward in all aspects and become the engine of smart transformation. Take Jiaxing Science City (JSC) as an example. Ten years ago, it was a rural area. Now it has been developed into a land of large information and innovation industries. More than 500 enterprises have been incubated with an yield of 10 million RMB (US $1.5 million) per mu (1 mu = 667 m^2) and tax revenue of 0.5 million RMB (US $77,000) per mu. JSC is almost equivalent to the Zhongguancun Science Park in Beijing. In Xi Tang Ancient Town in Jiashan County, an investment project of 5.98 billion RMB (US $0.92 billion) in optical communication with the whole industrial chain has been put in practice and brought another 15 support items to join in. This global mega optical communication industry innovates the new model of industrial cluster.

According to the statistics by Jiaxing Information and Economic Commission, 400 large projects in information economy has been started in Jiaxing since the 2nd WIC (2015). Among them, 98 projects with an investment of over 30 million RMB (US $4.61 million) are related to internet. The total investment reaches 70 billion RMB (US $10.7 billion). The output of electronic information industry and retail sales of e-commerce amount to hundreds of billions of RMB (6.5 RMB ≈ 1 USD). The new technologies, new models, new industries, and new economy are emerging every day in the terms of large data, virtual reality (VR), 5G car networking, intelligent manufacturing, energy networking, online finance and smart city, etc. Jiaxing has become a hot land of internet entrepreneurship and innovation both at home and abroad.

There are great changes in the transformation of such traditional industries as leather, clothing, bags, and home textiles. Take China Sweater Town located in Puyuan, Tongxiang City, as another example. More than 10,000 shops have joined in e-commerce. The retail sales amounts to 6 billion RMB. The makers have deeply integrated the sweater industry with internet through the chain of designing, spinning, weaving, printing and dyeing, garment manufacturing, inspecting and testing, and logistic distribution. Network application platforms like "China MES Union(CMESU)," "zhaohuoma. Com," "py3D. cn," and "QILI Software" are springing up.

China Sweater Town in Puyuan, Tongxiang City. *Source: Tongxiang Government.*

Smart City Needs Wisdom

In 2015, Jiaxing was approved to be the pilot city of China Smart City Building. It is good news but a great challenge as well. How to build the smart city needs wisdom. Early in 2016, Jiaxing government made a general framework of SGHC by planning "One Core and One City with Multiple Functions." According to the plan, by the end of 2020, the indexes in information, an integration of information and industry, and information economy will be listed before No. 3 in Zhejiang Province; the output value of information industry will cross 200 billion RMB; the network retail volume will reach 160 billion RMB; e-commerce trading volume will amount to 400 billion RMB; 5 leading enterprises in network economy with an annual output value of 10 billion RMB will be incubated; the online services involved in administration, tourism, medical care, education, traffic, logistics, etc. will be developed with the popularity rate of 90%; and the volume, efficiency, and competitivity will be at the top of the list.

Wuzhen is a tiny place on the earth but a best access for Jiaxing to connect with world internet. As WIC continues to be held in Wuzhen annually, you will see what happens in Jiaxing in reality.

Case 2: A "Smart" Day in Wuzhen

(Source: Xinhua, Dec. 2, 2017)

Source: Wuzhen Town Government.

After checking in a hotel in Wuzhen in east China's Zhejiang province, Mr. Fu Ruyi ordered a set meals by scanning the QR code on his bedside table. It was very late on Thursday night, but the food was delivered to his room within 30 min.

Fu entered one of Wuzhen's scenic spots by "swiping his face" via a facial recognition system on the morning of Dec. 1, which streamlines the ticket-checking process to last no more than 3 s.

At the pier, Fu, the president of an information technology company, scanned a QR code and all the sculling boats on the river appeared on his phone screen. He hailed one and left the boatman a message.

Sitting as comfortably as he could on the boat, he tapped open the "Wuzhen Summit" app, which he used to browse the latest news and the schedule of the WIC, scheduled from Sunday to Tuesday in the picturesque river town.

The introduction page for a session on Monday caught his attention. He was invited to attend the session. He made a digital note of the venue, timetable, and event participants.

App users who were not invited to the conference or could not visit Wuzhen are able to browse the exhibition halls and interact with exhibitors via the app.

"There have always been new things to amaze me every time I have visited Wuzhen for the past four years," Fu said as he stepped ashore from the boat. He went directly to a special conference "volunteer."

"How do I go to the Waterside Resort" he asked. "Please walk along the road and turn left at the first intersection. The hotel is on your right about 100 meters down the road," a robot answered in a pleasant voice.

Based on facial and phonetic recognition, the robot helpers are able to answer questions about conference schedules, locations of venues, weather, and even the nearest toilets.

There are several dozen robots on duty at hotels, exhibition halls, venues, and streets.

Fu continued exploring. At the street corner, he came across another helper, a sturdy garbage collection robot. He said "Hey" to it and the robot moved in his direction. On its way to Fu, the robot opened its lid, saying, "Please give me your garbage. Thank you"! Fu threw the empty coffee cup into the robot.

Garbage Collection Robot displayed in WIC. *Source: Wuzhen Town Government.*

Fu arrived at his destination, the Waterside Resort, where he had a business meeting with a manager of a VR rehabilitation training company.

Fu said VR has proven beneficial in training patients who have had a stroke or spinal injuries by making the dull process more interesting, like playing video games at home. On his way back to the hotel, Fu passed a cashier-free minimarket, where he bought drinks, snacks, and a magazine using mobile payment.

His smart phone was running out of battery, and he needed to recharge it before he was cut off from all the services and information, almost the whole world, it seemed. As he picked up the pace, he noticed a box full of shared power banks at the reception desk of a restaurant. He suddenly relaxed and walked in.

Case 3: Make Efforts to Build a Strong Internet Economy and Construct New Kinetic Energy for Development

(Reported by Jiaxing Municipal Commission of Economy and Information, Source: Xin Jia Xing, vol. 374, Aug. 2016, pp. 20–21).

Bird's-eye view of Jiaxing City Proper. *Source: Jiaxing Government.*

Making Policies to Set Internet Economy as First Goal

To strengthen the internet economy, Jiaxing government has made a strategic decision to take internet economy as a priority and internet industry as the "No.1 industry." A series of policies and plans have been made including "Opinions on Building a Strong Internet Economy City," "The Development Goals of Internet Economy," "Internet + Action Route," "Implementing Opinions on e-Commerce," and "Policies and Opinions on Accelerating the Development of Information Economy." The indexes of internet economy will be added to the target responsibility assessment.

Meeting with a foreign delegation in IT industry. *Source: Jiaxing Economic and Information Commission.*

Promoting the Construction of Important Platforms in the Whole Region

As a permanent site of WIC, Wuzhen will be further developed as the core area of internet experience and international cooperation by introducing Shanghai-Hangzhou quantum trunk line of Shenzhou Quantum Communication and "Wuzhen Internet Hospital" invested by We Doctor Group Limited and Media Large Data Exchange Center by Zhejiang Daily. JSC will be the main platform for information industry agglomeration and integrated services by forming the industrial cluster of internet intelligent terminals with the settlement of Wingtech Communication Co. Ltd. and Techain Electronics Technology Co. Ltd. In Xiuzhou High-Tech Industrial Park, a number of intelligent furniture manufacturers such as Softide Group Co. Ltd., Zhejiang Dingmei Intelligent Decoration Co. Ltd., and their accessories companies have gathered together. The leading enterprises like Royal Dutch Philips Electronics Ltd. and DBG are investing in Jiaxing Economic Development Zone. Nanhu District and Xiuzhou District are listed in Pilot Zones of Zhejiang Information Industrial Development and the Characteristic Bases of Zhejiang Software and Informative Service Industry. A large number of small towns with internet characteristics are booming up. Wuzhen is now a world-famous internet town. Xiuzhou Photovoltaic Town is elevated to the model of the provincial level.

Introducing the VIP Investment Projects and Cultivating the Whole Industry

In 2016, there were 358 investment projects in information economy. Among them, 117 projects were involved in information industry with an investment amount of 7.3 billion RMB. A total of 240 projects focused on "Internet + Manufacturing" with an amount of 2.1 billion RMB.

In addition, Zhejiang Large Data Exchange Center, 5G Car Networking, Shenzhou Quantum Communication, and Wuzhen Internet Hospital successively settled in Tongxiang City. In Jiashan County, the whole industrial chain invested by SEFC Optical Fiber and Cable Co. Ltd. was built. The computer precision electronic components company invested by CASETEK Holdings Ltd. was also established. The logistic warehouse invested by "Cainiao" (Green Hand), Alibaba, was completed in Jiaxing Economic Development Zone.

Signing Ceremony on Photovoltaic High-tech Industrial Park in Xiuzhou in 2016. *Source: Xiuzhou District Government.*

ACHIEVEMENTS

Information Industry Leads in All Industries

According to statistics, in the first half of the year of 2016, the total output value of electronic information manufacturing reaches 48 billion RMB by a growth rate of 16.1%, higher than that of other industries by 11%. The total profit amounts to 2.34 billion RMB by a growth rate of 41%. Moreover, the industrial agglomeration is more significant. The intelligent terminal, optical communication, and electronic components are clustered in Nanhu District. PV application and smart household products are manufactured in Xiuzhou District. PV and communication products are produced in Jiashan County. Pinghu City is famous for micromotor manufacturing. Haining City is well known for PV and magnetic materials. Tongxiang City focuses on the new internet industry. All districts (cities and counties) under the jurisdiction of Jiaxing have their own characteristic industrial clusters.

The Integrated Development Leads in the Social Transformation

The building of SGHC is accelerated in all aspects including intelligent traffic, power grid, security, administration, health care, civic service, education, culture and environment, etc. Manufacturing industry integrated with internet is progressing at a fast rate. Haining City, Tongxiang City, and Nanhu District have been recognized as the national models. Pinghu City and Xiuzhou District are listed in the national pilot zones. Three enterprises are confirmed as the ISO pilot enterprises of China "Information and Industrialization Integration." The current index of information development of Jiaxing is 0.953. The "Integration of Information and Industrial Development" index is 72.50. The application rate of digital technology for product design increases by 70%. The application rate of production control system is 80%. The Enterprise Resource Planning rate arrives at 70%, and the flow control rate is 65%.

Communication Infrastructure Leads in the Third-Class Cities in China

The communication network infrastructure in Jiaxing has been built from optical network to wireless network with full coverage in three aspects: households equipped with 100M optical broadband, the city equipped with 4G network, and key public sites freely covered by WIFI.

Entrepreneurial Innovation Exchange and Cooperation Is Gathering in Jiaxing

The projects invested by Tencent, Baidu, CETHIK, Zhejiang Daily Media Group Co. Ltd., and We Doctor Group Ltd. are settled in Wuzhen. Jiaxing

Economic Development Zone (JEDZ) has reached agreements with Ciwen Media for building the headquarters, Ping An e-Commerce for financial industry, and Foxconn for logistics. Nanhu District has signed the contracts in internet economy with Letv Finance and Pingpong Cross-border e-commerce by investing a volume of 2 billion RMB. VR (Virtual Reality)/AR (Augmented Reality)[13], the first game industrial park in Jiaxing, is located in Xiuzhou District. Several world famous network companies are planning to invest in Pinghu City. Nano Internet OLED has finished its first-phase construction in Haiyan County with an investment of 20 million RMB.

Inauguration Ceremony & Propulsion Conference in 2017. *Source: Jiaxing Science City.*

Case 4: An Exploration and Practice on the Project of Intelligent Medical Care for Citizens

(Reported by Jiaxing Municipal Commission of Health and Family Planning, Source: Xin Jia Xing, vol. Dec. 2017, p.27−28)

MAIN PRACTICES

Open "Healthy Jiaxing" Palmar App By Setting Up "One-Stop" Health Information Platform

Jiaxing Municipal Commission of Health and Family Planning has made efforts to set up the "Healthy Jiaxing" Palmar application system including APP

13. VR(Virtual Reality)/AR(Augmented Reality, is a new field of research developed on the basis of vitual reality technology. It is also an important development direction of human-machine interface technology).

and WeChat public number by integrating the medical care resources and data of the hospitals at all levels. The function is divided into "Smart Health Care," "Public Health," and "Health Inquiry." The integrated regional medical care resources provide more holistic and convenient medical services such as making a registration or appointment, seeing a doctor, checking the bill and medical reports, etc. Compared with the traditional way of making registration and appointment in the hospital, the Palmar App via mobiles can significantly shorten the time for the patients to queue up to register and wait for treatment. "Healthy Jiaxing" App is now generally adopted in all hospitals at municipal level and the medical institutions in Jiashan County and Xiuzhou District.

A Community Hospital can check the CT results via "Cloud Clinic." *Source: Jiaxing Health and Family Planning Commission.*

Encourage the Interdiagnosis Settlement to Build "Zero Queuing Payment" in the Treatment Process

In 2016, Jiaxing Municipal Commission of Health and Family Planning focused on the project of easy payment by means of the hospital information construction. Jiaxing Traditional Chinese Medicine Hospital was listed for a trial. Later in Jun., all municipal hospitals implemented the easy payment system. The settlement of inspection and dispensation can be completed by the doctors in the clinic room and the patients need not repeatedly queue at the cashier windows. Besides, the patients can pay online via "Citizen Card + Citizen Card Online Account," "Citizen Card + Alipay or WeChat," and "Banking-Medical Service System."

Start Using Self-Help Equipments and Reach the Goal of "One-Stop Self-Help Service"

All the municipal hospitals in Jiaxing have been equipped with self-help devices. Both out- and inpatients can use the devices to acquire self-help service at every link of medical treatment. Up to now, 212 outpatient self-help devices and 150 inpatient self-help devices are in action.

Self-help Service Devices in Jiaxing No. 1 Hospital. *Source: Jiaxing No. 1 Hospital.*

Build the Electrocardiogram System and Share the Data in Real Time

Based on the Jiaxing Municipal Intelligent Medical Care Platform, Jiaxing's No. 1 and No. 2 hospitals have set up an electrocardiogram consultation center connecting with the community clinics, local hospitals, and ambulances. The data can be transmitted to the center simultaneously so that they can be shared in real time. The regional electrocardiogram system can also make the high-end medical care resources cover all the community clinics and local hospitals and solve the problems of seeing a doctor or paying higher expenses. The real-time transmission via information technologies can make the emergency patients enjoy the consultation in ambulances, greatly shortening the rescue time and increasing the efficiency. The regional electrocardiogram consultation centers in both hospitals have passed the authentication issued by National Chest Pain Center.

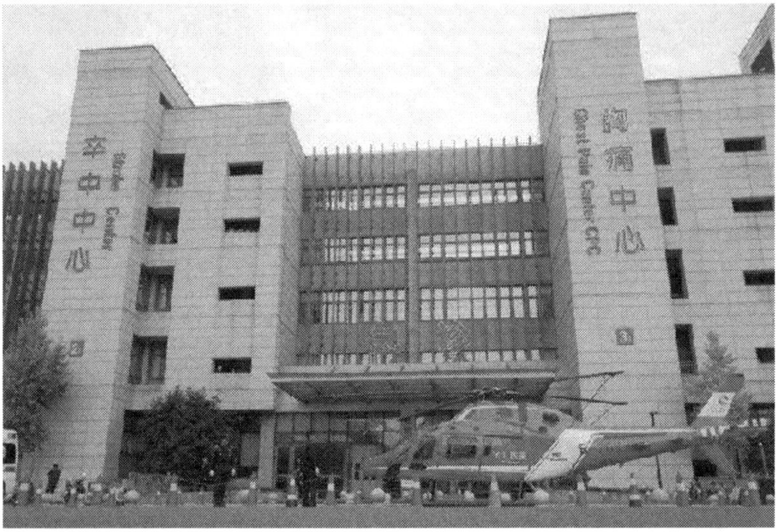

Chest Pain Center CPC in Jiaxing No. 1 Hospital. *Source: Jiaxing No. 1 Hospital.*

ACHIEVEMENTS

Upgrade the Modern Medical Care Service

With the deep integration of the information technologies and medical care service, there appear a variety of approaches such as "Healthy Jiaxing Palmar App.," interdiagnosis settlement, and self-help devices, which can cover the whole region, all the citizens, and the whole process of medical treatment. This simple, quick, and efficient intelligent medical service system has greatly improved the modernization and capacity of medical care service. Now fewer outpatients are queuing in the peak time. The hospitals are in better order, and the doctors and nurses can distribute their time more reasonably and efficiently. According to the survey conducted by Zhejiang University, the satisfaction rate in registration, treatment, settlement, and payment is above 96%. The respondents admit that the treatment process has been optimized and the medical service improved.

Promote the Patients' Sense of Gain

The intelligent medical care service system makes the patients reduce the queuing time and roundtrips to the cashier and hospital and save the waiting time of treatment. As a result, the patients have felt much more convenient and the relationship between doctors and patients has been improved as well. Take the intelligent settlement as an example. By the end of Nov. 2016, the total amount of intelligent settlement done by all the hospitals was 847,240. If a patient saves 10 min to line up for payment, the total amount of social time cost can be reduced by 141,206 h. As the survey shows, 9.85% of the respondents thinks that the "Intelligent Medical Care Service" can save 0–9 min, 48.36% of the respondents think that it can save 10–30 min, 31.73% think it can save 30–60 min, 5.69% think it can save 60–90 min, and the rest think that it can save above 90 min.

Improve the Modernization of Medical Care Management

By means of large data, cloud computing, internet of things, and mobile internet, the Project of Intelligent Medical Care for Citizens has improved not only the capability of medical care service but also the modernization of medical care management. First, the supervision has been transformed from the fragments to the whole process. The supervision departments can comprehensively supervise the whole running process of all medical institutions. Second, the medical care service can be transformed from the extensive type to the intensive. Based on large data analysis, the difference of medical service can be detected and classified for management. Through the analysis of the click frequency on "Healthy Jiaxing Palmar App,"- the patients' demand situation can be watched. Thereby precise and individual services can be provided. Third, the government authorities can make policies from experienced decision to scientific decision. The large medical database has provided the large capacity and complete scale of data for medical institutions, which can break the boundaries and make more scientific decisions.

Get the High Praise of the Whole Society

The Project of "Intelligent Medical Care for Citizens" is coconducted by Jiaxing Municipal Commission of Health and Family Planning, the relevant departments, and medical institutions and enterprises. It aims to solve the problems of difficulty, complexity, and expense in medical treatment. As it is carried on, the patients feel much more convenient and cheaper than ever. The project has won the high praise of the whole society. The recent poll shows that the satisfaction rate arrives at 99.78%.

Case 5: Administrative Reform—At Most One Visit

(Reported by the Reform Office of Nanhu District, Source: Xin Jia Xing, vol. 388, Oct. 2017, pp. 31–33)

Background

As a Pilot Zone of National Administrative Reform, Nanhu District in Jiaxing began piloting a system aimed at streamlining and bringing the administrative services online, as well as making sure that if you have to visit a government department, you only do it once. This scheme, loosely dubbed "At Most One Visit," has seen the integration of several key governments and placed them all under one roof at new administrative service centers.

In Jul., 2017, the Nanhu District Administrative Examination and Approval Bureau was set up. Under the bureau there are 8 functional sections including Network Information Office, Effectiveness Supervision Section, Reception Center, Investment Projects Section, Registration Section, Section of Infrastructure, Transportation and Agricultural Affairs, and Civic Affairs Section. These sections are taking the responsibilities of former 16 departments, covering 174 items of administrative licenses and 28 items of civic services. Another team from former 6 municipal departments joined in for undertaking 71 items. There are 110 staff members in total instead of the former 260 taking the responsibilities; 16 seals have been encased. Only one seal for examination and approval is applied.

Nanhu District Administrative Examination and Approval Bureau Inaugurated in Jul., 2017. *Source: Nanhu District Government.*

Main Practices

I. *Upgrading* "One Seal for Approval" *and initiating* "At Most One Visit": To achieve the goal of one seal for all examinations and approvals, Nanhu district government gives full play to institutional advantages and creates the new model of examination and approval as follows:

1. *Implementing the model of* "2 + X." According to the nature of transfer matter, the specific functions are classified and integrated. Workflow is optimized. By taking "Investment Project" and "Business Registration" as the main threads and "Trade Licenses" as a subordinate line, a new model of "one certificate" has been set up.
2. *Building the model of* "Three Separations" (i.e., *Acceptance, Reconnaissance, and Approval*). By optimizing the check points, breaking up the examination and approval process, training the staff on "Almighty Post," and opening up an interactive module, all sections can be interconnected. Through the cooperation online and offline, "One Window for Acceptance," "Joint Reconnaissance," and "One-Stop Approval" are carried out as a result.

II. Carding "Two Lists" and making clear the number of "At Most One Visit."

The "Two Lists" refers to the list of items for "At Most One Visit" and another list for "At Most One Window." The details are as follows:

1. *Publicizing the list of items for* "At Most One Visit." Nanhu district government has sorted out 679 items for "At Most One Visit," which accounts for 94% of all items. Among them, Nanhu District Examination and Approval Bureau has classified 200 items for "At Most One Visit" and 12 items for "Zero Visit." Then the list is publicized in the media.
2. *Specifying the list of* "At Most One Window." To upgrade "At Most One Visit" to "At Most One Window," Nanhu District Examination and Approval Bureau has made efforts to deepen the mechanism of "One Window for Acceptance" by taking advantages of "Integrated Handling" and requiring the staff to give the applicants "Full Guidance at One Visit." For an application matter with multiple items and several trips to the check windows, the staff member at the acceptance window will be required to take it all by helping the applicant to plan the route of handling, making a list of submission materials, and distributing all the forms at one time. All items for "At Most One Window" are publicized on the website and other media. Up to now, 347 single items and 18 multiple items have been listed for publicity.

III. Taking "Three-One"Approach of Examination and Approval and Improving the service quality of "At Most One Visit"

1. "One Network" *management.* To build "One Network" management, Nanhu district government focuses on providing individualized services by upgrading "Zhejiang Administrative Services Network," developing the function of real-time push of "Information Sharing Platform for Examination, Approval and Supervision," and exploring the mechanism of integrated review and coordinated operation. A total of 22 staff members from 16 relevant departments have been

appointed to timely download the review results and upload the current status of the supervision process to Nanhu District Examination and Approval Bureau. Besides, the district government has also strengthened the construction of "Smart Hall" with the functional divisions such as the advisory counter, self-help zone, self-help filling forms area and baby room, etc. and equipped with mobile library, sample forms automatic query, silent kiosks, SMS reminding, real-time data analysis, and the like. Hence a comprehensive administration platform with integrated examination and intelligent information has been established. By introducing OSM (Onsite Management), the space, room, equipments, and goods in the hall can be all under the unified management.

Nanhu District Administrative Examination and Approval Center. *Source: Nanhu District Government.*

2. "An Integration of Examination and Approval." To implement the fast track of examination and approval for the investment projects, Nanhu district government has issued a new policy, promising that the time of examination and approval for the investment projects can be shortened from 100 to 50 days, and that the file transfer can be cut down from 13 to 7 relevant departments. Thanks to "Internet + Administration," Nanhu district has become the first government in Zhejiang Province to get through the "Information Wall" among the departments and break the "Information Island," making the investment project examination and approval enjoy "Zero Visit." In trade registration, Nanhu district government has set up the first system of enterprise full electronic registration and issued a policy to clearly define the staff members' appointment, authority, and handling time. Efficiency is

greatly improved. In the first half of 2017, the total amount of registered capital has increased by 97.4%, comparing with the same period of last year.

3. "One-Stop Service": Several fast tracks have been opened up. First, the consultation system is enhanced by keeping reservation, sorting out and publicizing the list of relevant items, and exploring the virtual examination and approval module. Second, "green window" (or "straight channel") is opened for the key investment projects. Third, "faster track" is developed by laying down the special network connecting with towns and villages so that the trade registration can be accomplished online. In addition, the government has signed the agreements with Industrial and Commercial Bank of China, Agricultural Bank of China, and Bank of Communications of China. The investors and businessmen nearby can submit their application documents in the local service outlets of these banks.

APPLYING "FOUR RUN-UPS" TO ACCELERATE THE PACE OF "AT MOST ONE VISIT"

1. *The first* "Run-up" is to let the data run and render the intelligent service. Based on the "Internet +" module, Nanhu district government has built a virtual examination and approval hall where the registration, examination, and approval can be done online beforehand. The application, examination and approval, licensing, and filing can be carried out via the online administrative service platform so that the applicants may reduce their roundtrips and institutional transaction cost.

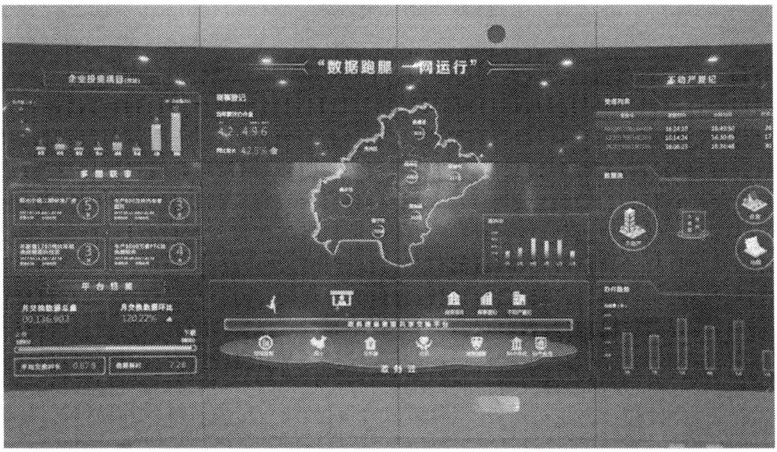

The Screen shows "Let Data Run in One Network." *Source: Nanhu District Government.*

2. *The second* "Run-up" is to let the commissioners work and render the precision service. A total of 10 professionals have been chosen as the commissioners to render their door-to-door service, appointment service, and preparation service. In addition, a duty system is established. The commissioners will be regularly dispatched to the key investment projects or the concentrated areas such as JSC and Nanhu Fund Town for one-to-one service.

Jiaxing Science City (JSC). *Source: JSC.*

3. *The third* "Run-up" is to let the agents render the free service. These agents are chosen from the staff members of the relevant departments, institutions, and communities. After a period of professional training, they are sent to various places for helping the applicants.
4. *The fourth* "Run-up" is to let the express delivery work for individuals. The government has contracted with the local postal departments for delivering all the documents in time and encourage the applicants to take express delivery. The postal charge will be paid by the government. So far, 15,000 licenses have been delivered both in and out of the city.

Features and Highlights

1. One certificate for all licenses

   Nanhu district has made a breakthrough in licensing reform and become the first government in China practicing "One Certificate for All Licenses." The government took the industries of entertainment, printing, and catering as priority and began to reform the licensing procedures. First of all, the government redefined all the licensing items as "one matter," which used to be involved in several departments, and then integrated the resources of the relevant departments by taking the approaches of "one list of application," "one window for acceptance," "one section for handling," "one certificate for all licenses," and "one file for all the items." For instance, before the reform, if the applicant wanted to run the entertainment business, he used to go through a variety of procedures involved in registration, business license, hygiene license, environmental impact assessment, etc. After that, these procedures and licenses can be handled together by one integrated section named "Social Affairs." The examination and approval links are deducted by 50%. The number of the approvers is reduced from 10 to 2 persons. The duration is shortened from 150 to 5 working days.

One Certificate for All Licenses. *Source: Nanhu District Government.*

2. A streamline service for investment projects

   To shorten the time of examination and approval for investment projects, a streamline service is rendered to the project from its registration to construction licensing with the integration of the departments such as development and reform, economy and information, environment, water conservancy, housing and urban construction. Other relevant departments like planning and fire also join in the streamline service. On the one hand, the payment for planning license and environmental impact assessment can be handled before the construction licensing. On the other hand, all the documents and examination and approval results can be shared online to reduce the duplicate submission of the materials. Now the duration is cut down from 50 to 30 working days.
3. One window for real estate transaction, tax, and registration.

   To practice "At Most One Visit" for real estate transaction, tax, and registration, Nanhu District Examination Approval Bureau invites Nanhu

Real Estate Registration Center and local Taxation Bureau to open a window for one-stop service. In particular, senior citizens above the age of 70, the disabled, and pregnant women can enjoy faster track service, which only takes an hour for them to submit the documents and get the license.

4. One network for zero visit

   Thanks to the deep integration of "Internet + Administration," Nanhu district government has developed the whole process of electronic registration platform. By introducing the electronic seal and signature and face recognition system, the restrictions of region, space, and time for investment projects have been broken through. Application, examination and approval, licensing, and filing can all be done online. Thus the model of "Internet + Examination and Approval" is established with the characteristics of "Zero Paper," "Zero USB," "Zero Interview," and "Zero Charges." The duration from preparation to licensing is deducted from 15 working days to 1 day. Now 12 items like enterprise investment project records, accounting qualification examination and approval, prior approval of enterprise name, etc. can be accomplished online.

5. An integration of self-help application

   An integrated intelligent machine for self-help application is developed with 55 items involved. Applicants can query and print the list of the relevant items, scan and upload their documents, and check the status. WeChat and mobile apps can be linked for making appointments. This model of "Unmanned Reception" supported by IT can significantly shorten the queuing time to bring "At Most One Visit" and even "Zero Visit" into reality.

Integrated Intelligent Self-help Machine. *Source: Nanhu District Government.*

6. One trip of on-site reconnaissance

   Based on the separation model of "reception, examination and approval, and reconnaissance," the key points, workflow, and time limit related to the onsite reconnaissance is scientifically sorted out and integrated, and then the relevant personnel will make a unified form and a standard list of joint reconnaissance. After that, they will go together for conducting the reconnaissance.
7. Implementing "Smart e-Police Platform"

   Following the tendency of "Internet +," 11 police stations in Nanhu District have innovated and developed "Smart e-Police," making the migrants and overseas personnel register their accommodation via "Zero Visit." Besides, the public number of "Cloud Intelligent Police Work" supported by apps is also put into action. The enterprises and citizens can make appointments and receive the related information online.

Case 6: Improve the Public Security Management by Constructing "Intelligent Security Community"

(Reported by the People's Government of Jiashan County, Xin Jia Xing, vol. 388, Oct. 2017, p. 28)

In 2005, when President Xi Jinping was the party secretary in Zhejiang Province, he paid great attention to the scientific development of Jiashan County. He hoped that Jiashan County would make great achievements not only in economy but also in society. As IT and large databases develop fast and deeply, Jiashan has launched the campaign of "Intelligent Security Community" and made a significant progress. The details are as follows.

Starting Planning "Intelligent Security Community" and Putting It as a Priority in the Government Projects

Jiashan government has applied the intelligence concept to the security of communities and put it as a priority in the county development projects. A leading group is organized with the members of the deputy party secretary, town (or subdistrict) leaders, community (or village) heads, and residents, and then the construction standards are determined. In 2014, 11 communities were piloted for building the "Intelligent Security Community." Among them, the gated communities were equipped with the integrated security devices and the open communities were installed with the module of "Virtual Gates + Cameras." Besides, the mechanisms of safeguard and sustainability were established including project bidding, construction process, financial aids, the responsibilities, reward, and punishments, etc.

Source: Jiashan Public Security Bureau.

Transforming "Personnel Management" to "Intelligent Management" by Introducing Cutting-Edge Technologies and Building Intelligent Platforms

To improve the capability of the community security and management service with intelligent means, virtual section and data intelligent collection were conducted. First of all, at the entrances of the communities, 5 systems were installed with plate grabbing, facial recognition, radio-frequency identification (RFID), machine-aided cognition (MAC), and ID information collection. A virtual section is formed for the purpose that "the image is left when a person passes," "the plate number is left when a vehicle passes," and "information is left when a person or an object passes," Second, to make full use of the intelligent data, a management and analysis platform for "Intelligent Security Community" is also set up. Based on the data collected by means of facial recognition, plate cognition, RFID, mobile phones, key position inspection, community cameras, household recording machines, and app transmission, "one-stop" check and comparison can be realized like Baidu Search due to data interdependency. Third, Data Perception is applied via the internet of things. Through the two-level platforms of "Intelligent Community" and "Community Police Station," the intelligent devices can detect the situation of the community and criminal activities in real time and make a forecast and early warning. For instance, when an electric bicycle installed with RFID passes the community entrance, the detecting device will make the recognition automatically. If the identification is not matchable, the device will make a warning.

In Transverse Direction: Plate Cognition, Face Recognition, Mobile MAC, RFID, Visitor's Registration; in Longitudinal Direction: Data Collection and Check, Data Distribution, Data Control, Mining and Analysis. *Source: Jiashan County Government.*

BUILDING "INTELLIGENT SECURITY COMMUNITY" INTO A NEW MODEL OF INNOVATIVE SOCIAL GOVERNANCE

Adhering to the idea of "Integrated and Open Governance," Jiashan government has organized relevant departments and social groups to participate in the construction of "Intelligent Security Community." First, they can monitor the suspects round the clock with intelligent devices and data. By means of facial recognition and other data collecting devices, the suspect's daily activities can be detected. So far 29 escaped criminals and drug traffickers have been arrested. Second, "Intelligent Security Community" can help support case investigations in the last place where the suspects might hide. Through the data collection, mining, and analysis via the community platforms, relevant information can be transmitted to the police station in real time, and then the policemen can be dispatched for control or management immediately. Since 2015, 230 cases have been dealt with and 6 stolen vehicles have been found. Third, "Intelligent Security Community" can integrate its intelligent service with the daily life of the residents. Through the mobile apps, they can receive various service information including weather forecast, security hints, daily necessities, and the like. Now the 2.0 module of "Intelligent Security Community" has been developed with the subprojects of remote monitoring system, visitor cloud system, family security system, family monitoring system, and electric appliances and fire-alarm system.

REFERENCES

ACORE, March 2011. American Council on Renewable Energy. US-China Quarterly Review.

APCO, 2010. China's 12th Five-Year Plan: How it Actually Works and What's in Store for the Next Five Years. APCO Worldwide.

Brautigam, D., April 2008. China's Africa Aid. The German Marshall Fund of the United States.

Chan, M.L., May 2009. Building Strong Grid Policy, State Grid Corporation. International Conference on UHV.

Chan, M.L., 2010. Green White Paper Series. Technical Advisor. JUCCCE, Shanghai, China.

Chan, S., March 2011. President of SunTech, "Global Solar Industry Prospects in 2011". SolarTech Conference, Santa Clara, California. www.SunTechSolar.com.

Chase-Dunn, C., 1982. Socialist States in the World-system. Sage, Beverly Hills.

Chase-Dunn, C., 1989. Globalization: a world-systems perspective. Journal of World-systems Research 5, 165–185.

China FAQs, 2010. Renewable Energy in China: An Overview. World Resources Institute.

Clark II, W.W., Cooke, G., 2011. Global Energy Innovation. Praeger Press.

Clark II, W.W., Isherwood W. (Lead Co-Author and Co-Editor). Utilities Policy Journal. Special Issue on China: environmental and energy sustainable development. Winter 2010a.

Clark II, W.W., Isherwood W. Utilities Policy Journal. Special Issue on Inner Mongolia, China: environmental and energy sustainable development. Winter 2010b.

Clark II, W.W., Li, X., 2003. Social capitalism: transfer of technology for developing nations. International Journal of Technology Transfer 3, 1–11.

Clark, W.W., Li, X., 2013. The Political-Economics of the Green Industrial Revolution: Renewable Energy as the key to National Sustainable Communities, Chapter #22. UK, Oxford.

Cole, B.D., 2006. Chinese Naval Modernization and Energy Security, a Paper Prepared for the Institute for National Strategic Studies. Washington, DC.

Constantin, C., 2005. China's Conception of Energy Security: Sources and International Impacts. Working Paper No. 43. University of British Columbia.

Dorraj, M., Currier, C.L., 2008. Lubricated with oil: Iran-China relations in a changing world. Middle East Policy XV, 66–80.

Downs, E.S., 2000. China's Quest for Energy Security. RAND, Santa Monica, CA.

Downs, E.S., 2004. The Chinese energy security debate. The China Quarterly 177, 21–41.

Downs, E.S., 2006. China - Executive Summary. The Brookings Foreign Policy Studies Energy Security Series The Brookings Institution.

Economist, July 16, 2009. Special Issue and Articles on "Collapse of Modern Economic Theory".

Eisenberg, L., 2009. Los Angeles Community College District (LACCD)" and "Appendix A. Sustainable Communities. Springer Press, pp. 29–44.

Hanergy Corporation. www.hanergy.prc.

Jiahua, P., et al., 2006. Understanding China's Energy Policy. A Background Paper Prepared for Stern Review on the Economics of Climate Change. The Chinese Academy of Social Sciences (CASS), Beijing.

Kaplinsky, R., 2006. Revisiting the revisited terms of trade: will China make a difference? World Development 34, 981–995.

Konan, D.E., Zhang, J., 2008. China's quest for energy resources on global markets. Pacific Focus 23, 382–399.

Kwan, C.L., 2009. Rizhao: China's Green Beacon for Sustainable Chinese Cities. Sustainable Communities. New York: Springer Press, pp. 215–222.

Li, M., 2010. "Peak energy and the limits to economic growth: China and the world". In: Xing, L. (Ed.), The Rise of China and the Capitalist World Order. Ashgate, Farnham, England, pp. 117–134.

Li, X., Clark, W.W., 2009. Crises, Opportunities and Alternatives Globalization and the Next Economy: A Theoretical and Critical Review. In: Xing, L. (Ed.), Globalization and Transnational Capitalism: Crises, Challenges and Alternatives, Chapter 4. Aalborg University Press, Aalborg-Denmark, pp. 83–107.

Li, X., Shaw, T.M., 2018. The Political Economy of Chinese State Capitalism. Chapter in book for.

Li, Z., March 2003. Energy and environmental problems behind China's high economic growth – a comprehensive study of medium- and long-term problems, measures and international cooperation. IEEJ. http://eneken.ieej.or.jp/en/data/pdf/188.pdf.

Liu, X., September 2006. China's Energy Security and its Grand Strategy. Policy Analysis Brief, The Stanley Foundation.

Lo, V., April 25, 2011. China's 12th Five-Year Plan. Speech by Chairman. Shui On Land and President. In: Yangtze Council, Asian Society Meeting, Los Angeles, CA.

Martinot, E., Li, J., 2010. Powering China's Development: The Role of Renewable Energy. Worldwatch Institute. Worldwatch Annual Reports.

Huliq News, 2008. China's Energy Demand Increases Global Pressure to Seek Out New Sources. Available at: http://www.huliq.com/597/73705/chinas-energy-demand-increases-global-pressure-seek-out-new-sources.

NYT. New York Times, January 30, 2010. China Leading Global Race to Make Clean Energy.

Opoku-Mensah, P., 2010. "China and the international aid system: transformation or Cooptation?". In: Xing, L. (Ed.), The Rise of China and the Capitalist World Order. Ashgate, Farnham, England, pp. 71–85.

Pew Charitable Trusts, 2010. Who's Winning the Clean Energy Race? G-20 Clean Energy Fact Book. The Pew Charitable Trusts, Washington.

Scientific American, April 5, 2011. Clean Tech Rising. www.scientificamerican.com/technology.

SGCC, August 18, 2010. State Grid Road Map of China. State Grid Corporation of China. Framework and Roadmap for Strong and Smart Grid Standards.

Silverstein, K., 2011. Energy avenues open with China. EnergyBiz. http://www.energybiz.com/article/11/01/energy-avenues-china-open.

SJBJ. San Jose Business Journal, April 1, 2011. Clean Energy Financing Jumps to Record $243B. The News, Data from Pew Chartable Trusts, p. 8.

Thøgersen, S., Østergaard, C.S., 2010. "Chinese globalization: state strategies and their social anchoring. In: Xing, L. (Ed.), The Rise of China and the Capitalist World Order. Ashgate, Farnham, England, pp. 161–186.

Time, February 28, 2011. World: Now No. 2, Could China Become No.1?, p. 17.

Tseng, Yu-H., 2008. Chinese foreign policy and oil security. Internationals Asian Forum. International Quarterly for Asian Studies 39, 343–362.

US Congress, April 5, 2011. China's Energy and Climate Initiatives: Successes, Challenges, and Implications for US Policies. From Energy Strategy Institute (EESI) and World Resources Institute (WRI).

US DOE. US Department of Energy. 2011. Annual Global Reports. www.usdoe.usa.org.

Wang, W., Li, X., 2009. Ecological Construction and Sustainable Development in China: The Case of Jiaxing Municipality. Sustainable Communities. New York: Springer Press, pp. 223–241.

Zhang, W., Summer, 2006. Sea Power and China's Strategic Choices. China Security, pp. 17–31.

World Bank World Development Indicators, 2017. China National Indicators of Statistics, Calculated from the Conference Board Total Economy Database. www.WorldWatchReports.org.

Xu, Q., 2007. China's Energy Diplomacy and its Implications for Global Energy Security. FES Briefing Paper 13. Friedrich-Ebert-Stiftung, Berlin, Germany.

Zhen, Y.Z., Hu, J., Zuo, J., 2009. Performance of wind power industry development in China: a Diamond Model Study. Elsevier Press Renewable Energy Journal 34, 2883–2891.

Zhang, P., et al., 2007. Opportunities and challenges for renewable energy policy in China. Renewable and Sustainable Energy Reviews 13, 439–449.

Ziegler, C.E., 2006. The energy factor in China's foreign policy. Journal of Chinese Political Science 11, 1–23.

FURTHER READING

Clark II, W.W., October 2014. "Asia leapfrogs ahead: into the green industrial revolution" BOAO Review. Journal of Economic Commentaries for the Asian-Pacific Region 20–24.

Clark II, W.W., September – October 2016. To change ahead, GIR needs engines of real economics. APO News, Green Economy, 3 p.

Clark II, W.W., 2017. Sustainable Communities and Cities Design Handbook, second ed. Elsevier Press.

Clark II, W.W., Cooke, G., 2014. The Green Industrial Revolution, 2015 and earlier shorter edition in Mandarin. Elsevier Press.

Clark, W.W., Fast, M, 2018. Qualitative Economics, Second ed. Springer Press.

Clark II, W.W., Isherwood, W., December 2007. Report on energy strategies for Inner Mongolia Autonomous Region. Asian Development Bank. https://doi.org/10.1016/j.jup.2007.07.003 in Woodrow W. Clark II, and William Isherwood, (Authors and Co-Editors). Utility Policy Journal. Special Issue on line as China: environmental and energy sustainable development. Winter 2010.

Hongbin, Q., Sun, J.W., October 2010. Economics: From Quantity to Quality of Growth, China's Next Five Year Plan: HSBC Global Research (revised March 2011). www.research.hsbc.com.

IEA, 2010. International Energy Agency. Key World Energy Statistics. IEA, France.

Peyrouse, S., 2007. The Economic Aspects of the Chinese-Central Asia Rapprochement, Silk Road Paper, Central Asia-Caucasus Institute & Silk Road Studies Program - A Joint Transatlantic Research and Policy Center. Johns Hopkins University.

Shoumatoff, A., May 2008. The Arctic Oil Rush. VANITY FAIR. Available at: http://www.vanityfair.com/politics/features/2008/05/arctic_oil200805.

Spears, J., 2009. China and the Arctic: The Awakening Snow Dragon. ChinaBrief, IX, pp. 10–13.

Sundra Solar Corporation, Beijing. 2011. www.sundrasolar.com.

Tankersley, J., Lee, D., March 25, 2010. China takes Lead in Clean-power Investments: US Falls to No. 2 in Funding for Such Alternative Sources as Wind and Sun. Los Angeles Times.

The Climate Group, 2009. China's Clean Revolution ii. The Climate Group, London.

The Development and Reform Commission, PRC, 2007. China's National Climate Change Programme 2007. Beijing, China.

UNCTAD, 2005. Review of Maritime Transport 2005. UNCTAD.

US DOE. US Department of Energy. US-China Clean Energy Research Center. http://www.whitehouse.gov/blog/2010/09/03/us-and-china-advancing-clean-energy-research-through cooperation.

Valsson, T., 2007. How the World Will Change with Global Warming? University of Iceland Press, Reykjavik.

Chapter 6

Interactionism in Everyday Community Life

Woodrow W. Clark, II[1], Michael Fast[a,2]

[1]*Research Professor in Economics, Pepperdine University Graziadio School of Business (PGSB), West Los Angeles, CA Campus, Los Angeles, California;* [2]*Department of Business Economics, AAlborg University, AAlborg, Denmark*

Chapter Outline

a. Professor Michael Fast, PhD.

Climate Preservation In Urban Communities Case Studies. https://doi.org/10.1016/B978-0-12-815920-0.00006-X

INTRODUCTION

Some practical empirical cases are discussed that apply qualitative economics to the scientific perspective of interactionism in everyday life from George Herbert Mead (1934) and Herbert Blumer (1969), who in the early 20th century started a discussion in this area that both led to and created the discipline of "sociology." Subsequently, people have referred to their work as the basis for inter(cross)-disciplinary studies, combing the social "sciences," which include the actual sciences: physics, chemistry, biology, etc.

For the purposes of this chapter and with the theoretical considerations outlined earlier as a form of a new "praxis" (Brown, 1978), which has cases that are far from the typical analyses seen in conventional business economic and social science publications, what is demonstrated here is qualitatively economic (Clark and Fast, 2008) and uses interactionism and life-world perspective traditions. From these theoretical conceptions, empirical data provide clear examples of this perspective to understand businesses activities, actions, and events of people and groups. Energy is focused upon, given both backgrounds and publications in this area (Clark, 2009 2010, 2017a,b).

The act of exchange between humans is the central theme as well in life-world traditions that need intersubjectivity as another way to understand companies (Fast, 1992). Qualitative economics approaches the exchange act, which is a core concept of neoclassical economics, very differently. First, exchange needs to be defined from a qualitative economics perspective as the exchange interaction between two or more human. Following Blumer and Schutz, among others, exchange is not just monetary but also physical and symbolic through actions and gestures, as discussed in relation to intentionality and actions to be taken.

Given the discussion on exchange, the basic issue is about subsets of interactions. Exchange, as in the buying and selling of goods and services, is the foundation of modern economics. Yet, exchange is part of all human interactionism. When a person interacts with another for any reason, that person is part of an act, event, or series of activities to exchange something. Linguists have long noted this in their analyses of discourse and verbal communications. Again, such studies are components and parts of interactionism, which forms the core of qualitative economics.

ACTUAL CASES OF INTERACTIONISM

The first case moves ahead from new ideas that came from "dead economists" (Buchholz, 1989), such as the commercialization of a new fuel cell technology that could exceed environmental and political demands. The technology was done in a research laboratory during the 1990s and was commercialized into a business application in 2005 for stationary power applications. Commercialization for mobile purposes occurs next, allowing vehicles to exceed legislative regulations for zero emission regulations in California and other nation states.

Such a dramatic change in curbing emissions from power and fuel generation reflects Schumpeter's theories and perspective of "creative destruction." Fuel cells provide that opportunity compared with conventional battery technologies, which have dominated energy storage with few significant advances or changes for over 150 years. Moreover, fuel cells provide a commercial opportunity to mitigate environmental and climate degradation. Well-documented evidence from a variety of economists indicates that research and development are the heart of the capitalist engine through "organized inventive efforts undertaken by university-trained scientists and engineers, working in special facilities, tied to particular business firms, and focused on advancing their product and process technologies" (Nelson, 1990, p. 199).

With a thorough examination not only of the creativity process and technological invention itself but also of commercial and business applications, one can see how qualitative economics describes, understands, and explains business actions and events. Perhaps the place to start with this analysis is back to the very core of economic capitalism theory, with Adam Smith. There is no need to repeat the earlier critique of his perspective of economics. Instead, in *The Wealth of Nations* (1937 edition) Smith highlight, one point of interest for an interactionism perspective within qualitative economics when he placed the act of exchange at the center of economic life or as a universal characteristic of humans, as "a certain propensity in human nature … the propensity to truck, barter, and exchange one thing for another" (Smith 1937, p. 13).

However, interactionism is more than the exchange of business activities. It is far more complex in that it presumes the ability of the actors to think and reflect upon their actions. In Mead and Blumer, the interactionist perspective involves the actors literally talking to themselves. Or, as Blumer calls it, it is the "generalized other," which provides the act of exchange with a far more comprehensive and significant definition. This basic act is what distinguishes humans from other forms of life. People interact and this is best understood through the science of linguists.

The problem is that economics is not a science. Perkins (1996) describes the science of "Physics (which) is often described as the fundamental science, as it seeks to understand the 'rules' or 'laws' by which the universe operates." In short, "science must be qualitative and quantitative." Today, economics is mostly quantitative and concerned only with using the neoclassical economic theory of Adam Smith.

Anyone familiar with business activities, events, buying, and selling on whatever level knows, for example, that business is not done or consummated at one point in time. It develops over time and is the culmination of multiple business interactions among many actors after considerable reflection and thought. For some companies, and in more and more countries, business is concluded after extensive legal scrutiny as well. Often these interactions (especially financial and legal aspects) result from the need for basic definition of words, concepts, and legal terms. A case in point is fuel cell technology, which was commercialized when Clark was at the Lawrence Livermore National Laboratory (LLNL). To bring fuel cells into the "market" required an area of focus: then and now it was the vehicle business (see the subsequent discussion). Keep in mind that in the 1990s (and even today) there were not many hydrogen fuel cell vehicles (HFCV).

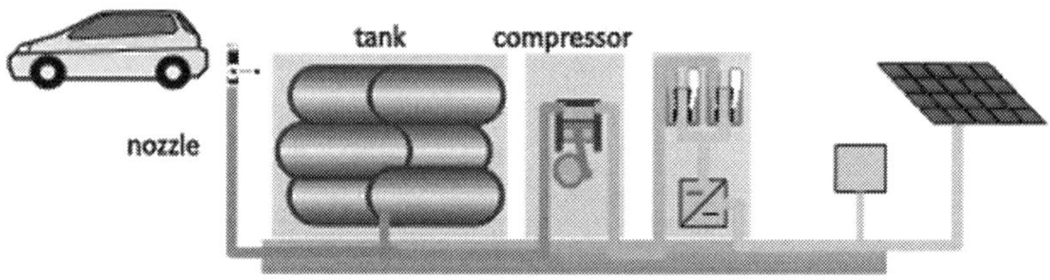

The perspectives of exchange and interactionism are thus an engagement between actors over time. People have established relationships from different areas of expertise to produce HFCV and also to refuel them. They have engaged with each other for some purpose.

The key here is that renewable sources for hydrogen, not natural gas, are critical. Honda has recognized this with its refueling station in southern California using solar power electrozied into hydrogen for Honda cars that are currently available.

As seen here, HFCV refuel at a station that obtain hydrogen from solar panels behind the car. This raises another issue regarding having refueling stations. Clark wrote California Governor Schwarzenegger's Executive Order in February 2004 for the state to have hydrogen refueling stations. That process was slow and full of mistakes and biased opinions regarding how to build these stations. The result was that the electric car, especially Elon Musk's

Tesla, came to the state of California and is now available around the world. Musk developed the idea of the Tesla car due to his experience founding Solar City company that he started in Northern California, where Tesla is now located. In early 2018, Musk launched another company SpaceX that will be sending people into space for both work and vacations.

Other current scientific cases are the result of interactionism and the 21st century through the formation of smart, green, and healthy (SGH) communities, which will be the focus of later chapters. With Clark's four earned graduate degrees and after serving as an advisor to the State of California, he founded his third company, Clark Strategic Partners, which provided expertise and science on how to make SGH communities.

New technologies should be seen as consisting of technique, organization, and knowledge for new companies (NewCos) and the growth of other companies. The turn of the 21st century demonstrated this clearly with both Apple Corporation and Microsoft as well as Google, Facebook, and others, especially in Silicon Valley in northern California. Next, we consider the knowledge basis for a new advanced fuel cell technology.

Related to HFCV are zinc–air fuel cells (ZAFC), which are not for transportation but for energy storage. ZAFC are also an interesting case study; they were developed at US national laboratories (including LLNL) into an environmentally sound technology to be commercialized. Firms want to commercialize the technology to meet and even exceed the zero-emission vehicle (ZEV) requirements (ZEV) in California, as well as other American states and a growing number of foreign countries. The ZAFC is a fuel cell invented, patented, and commercialized for mobile (eg, vehicles) and stationary power sources (eg, utilities). ZAFC and the zinc recovery unit (ZRU) are the subjects of the case study in this chapter. ZAFC technology is capable of powering a standard car for 250–300 miles between recovery stops, which take 10 min from the ZRU.

The science laboratory solution to ZEV regulations was in energy storage devices, such as a flywheel or ZAFC. Therefore, a team created the ZAFC to use the demonstrated high energy density capability of zinc–air (>540 kJ/kg [>150 Wh/kg] at up to 100 W/kg. The high energy density is partly the result of the high operating voltage of the cell (1.2 V) and the use of atmospheric oxygen as a reactant. Not only is the energy density insufficient, the available batteries require long recharge times (4–8 h) to ensure maximum life and acceptably low life cycle cost. Although rapid battery recharge is possible, in most cases it is undesirable: To recharge a bus battery in 15 min requires the design of the battery to reject waste heat associated with 0.5 MW.

The refuellable fuel cell has no expensive components other than the air electrode (with a quoted large-volume cost of $\$120/m^2$ [Alupower, Ltd.]). The high surface power (5 kW/m^2) gives the fuel cell a low producer's cost proportional to power: approximately $50/kW, half of which is attributable to the air electrode. Zinc recovery equipment is also inherently inexpensive. Zinc is electrodeposited onto cathodic surfaces of a series of bipolar-connected stainless-steel sheets, whereas the opposite surface of each sheet is catalyzed for oxygen evolution. Other components such as a 10-gallons/min slurry pump and metering devices are also inexpensive.

A different case in point is the Los Angeles Community College District (LACCD), which has nine campuses in the County of Los Angeles and which is the largest college campus system in the world. The nine campuses had been mostly built in the 1960s but need updating and with the LACCD board and senior staff to make them Smart Green Campuses (SGC). In the first decade of the 21st century, the plan was to create Leadership in Energy Environmental Design. As a result of California State Proposition 39, people in communities could vote for funds to be taken from them for their kindergarten through grade 12 schools and colleges. The public voted two times with over two-thirds of the people for Los Angeles County to finance over $5.3 billion to redo the college campuses. Because of Clark's experiences with California state government, LLNL, academics, and business involving green energy and economics, he was appointed the energy manager for the nine LACCD campuses.

The plans for each of the colleges (each campus has about 20–30 buildings plus parking lots and areas for equipment and even local police. The college bond funds were limited to remodeling and updating the campuses; no funds were for staff or faculty.

The red areas below are for solar panels on roofs and over parking lots. See photos.

(A) **(B)**

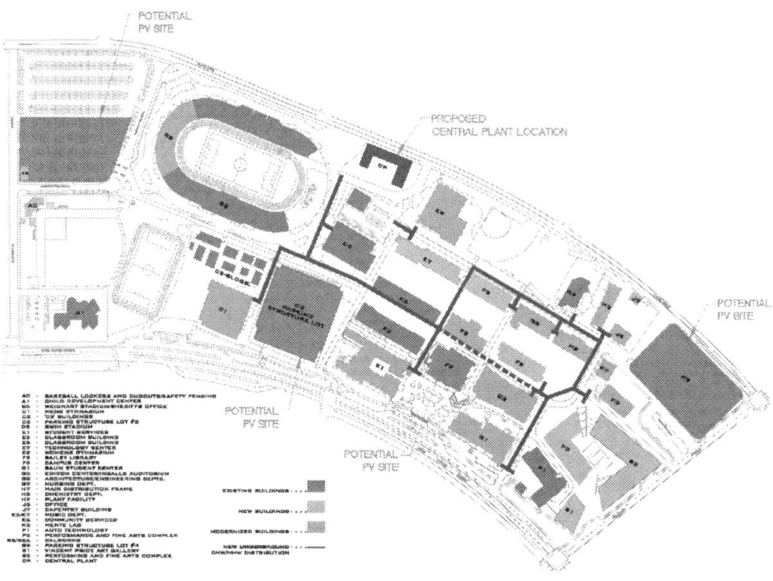

(A) Solar Panels over Campus Parking. (B) Area Check from Southern California Edison.

The local central power plant company, Southern California Edison (SCE), in whose district this campus was located, gave the LACCD $2.8 million to support the project. Pictured here was the chair of SCE (in the middle), who gave the check to one of the LACCD employees and Dr. Clark for the solar panel project.

MANIPULATION OF CALIFORNIA INTO AN ENERGY CRISIS

Situational Analysis

What is troubling about these analyses, aside from their influence on the policy makers and the general public, is that they reflect the objectivist philosophical tradition that promotes one paradigmatic view of the world. This particular viewpoint, while internally contradictory and inconsistent (e.g., western economies such as France and Germany have a considerable amount of government involvement in their "markets"), relies on statistical methods and probably theory to justify their philosophical (and politicized) position. It is exactly the distinction that Reinert was conveying in his analyses of Anglo-Saxon neoclassical versus German-American Renaissance historical roots in economics (Reinert, 1996, 1997, 1998).

Quantitative models and simulations have not been successful. Economics is a field dominated by statistical manipulations of numbers. It is not a field that uses mathematical models or more subjectivistic traditions. As McNeill and Freiberger (1993) note in economics, "the mathematical and linguistic realms stand quite apart" (McNeill and Freiberger, 1993, p. 96). Because economics is seen as concerning money, and money is viewed as numerical, mathematics brings powerful tools to the field of economics.

Yet the precision of math leads to overly crisp estimates and idealized models that seem to describe a society of robots. "Hence, economics also employs verbal concepts like recession. Language handles real-life questions better and treats details more subtly, but it also narrows the scope of models and shortens chains of reasoning" (ibid., p. 96). The analogy to language is a central theme of Clark and Fast (2008), which explores linguistic theories as applied to describing, understanding, and predicting business and economic everyday life.

Energy is a good example of how neoclassical economics works, but it does not work in the economics sector and others. The basic parameters are that economic theory presumes a balance or equilibrium between price and quantity. In energy, as the California energy crisis reached its peak, two American Nobel Laureates along with 10 well-known economists issued a manifesto in the spring of 2001. In general, the manifesto argues (and this was often repeated throughout 2001) that the energy sector under deregulation must be allowed to function on its own. That market power has its own natural way of correcting itself because of the demand and supply of energy. In other words, the argument goes that the market system will balance itself.

Not long after the manifesto, several individual scholars modified their stance and noted that the California market was flawed and "dysfunctional" because it was not a true deregulated market. This was even though several of them had helped to craft it and vigorously supported the California form of energy deregulated markets. Space does not allow a full review of the printed opinions in the press or professional journals from this group or others, but similar ongoing discussion regarding the same perspectives existed well into 2002.

A number of economists argued (see The Economist, January–March, 2002; Fortune January, 2002) that the collapse of ENRON proves how neoclassic economics works. As the argument goes, a company cannot perform in the market, and thus fails. Other companies come in and take its place. The free market economy moves on. The problem with that argument is that it ignores what ENRON did in the first place: set up and influence the deregulation market and then manipulate it and use or profit from investors, so that in the end, the collapse of the company is based on the value of its shareholders and creditors who lost their investment.

The result was that deregulation in California just before the turn of the 21st century, through two Republican governors who served two terms of 4 years each, had set a "market" model based on neoclassical economics in California which other states began to follow. In fact, by 2001, over 22 states enacted deregulation, as the map shows.

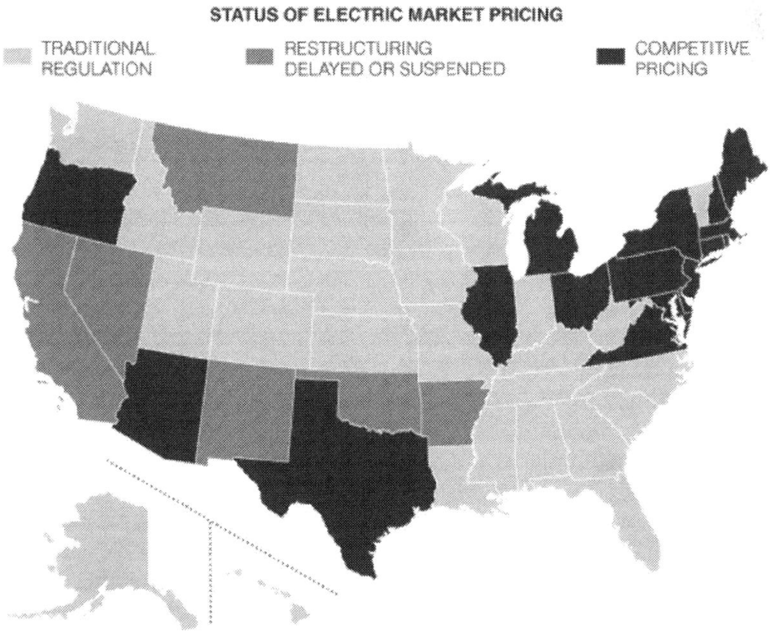

The neoclassical approach to economics, which has become dominant not just in the energy sector, rests entirely on the basis of exchange of goods and services. Reinert (1994, 1995, 1996a,b, 1998) wrote series of articles related to these issues. In a series of articles, the author dismantles the neoclassical (Anglo-Saxon) view of economics by contrasting it with the German tradition, in which manufacturing or creating products is the key to economic analyses and everyday life. We concur with that perspective and will subsequently present the subjectivist paradigm, which provides an entirely different approach to economics through the use of interactionism in everyday life. Therefore, economics is *about* everyday life and is *in* everyday life.

The ENRON Case: Private Instead of Public Monopolies

The date is fall 2002. All of the facts about the ENRON bankruptcy will not be known for many years owing to extensive legal and accounting litigation. Even so, a picture has unfolded about ENRON that symbolizes the problems with deregulation and privatization.

Whereas the *Economist* (various articles in 2002) and *Fortune* (January 16, 2002) focused only on the accounting errors of Andersen, the basic fact remains that ENRON executives instructed and managed Andersen's activities. In other words, the collapse of ENRON is not only an accounting firm gone wrong but also the core issue of unfettered free market economics in the energy sector.

To be perfectly clear, ENRON is (was) a "market" energy company that represented the mythic of neoclassical economics at its best: a "market force (firm)" unencumbered in its business activities as a result of its political influence and strong impact in forming the deregulated market, manipulating the prices (owing to supply), and finally directing the spurious accounting of its cash funds. Hence, Governor Davis asked the US Federal government to investigate ENRON's "role in the California Energy Crisis" (January 31, 2002).

As a point of departure for the ENRON case, consider an outline prepared by the staff for the governor of California (Staff, 02) as they prepare to address the impact of the ENRON bankruptcy on the state. For a good updated overview of the ENRON case, consult McGraw Hill PowerWeb (2002). As the *Economist* noted a year into the energy crisis:

"The end was not unexpected, but it was still spectacular. On December 2nd ENRON, once America's seventh-biggest company, filed for Chapter 11 bankruptcy. Only days earlier, its bonds had been downgraded to junk and Dynegy, a smaller energy-trading rival, had pulled out of a planned takeover. ENRON is the largest company ever to go bankrupt. Disentangling the resultant mess (and lawsuits) will keep legions of lawyers employed for years to come." (Economist, December 6, 2001)

The monetary impact in California in terms of long contracts was relatively low. For example, about \$84 million in contracts was obligated by the University of California system alone. Nevertheless, the prices paid for the supply of energy to the state were controlled by "market" companies or "merchant firms" such as ENRON so that about \$9 billion in additional costs for power were paid by the state to ensure power in winter 2000–01.

In 2000 and the first half of 2001, ENRON, among other market force suppliers, reaped enormous profits from California's deregulated energy market. Governor Davis stated in a letter to the Federal Energy Regulatory Commission (FERC) Chairman, "Since the summer of 2000, I have on numerous occasions demanded that the FERC investigate allegations of market manipulation by power generators and marketers, including ENRON. In my testimony before the FERC on November 9, 2000, I urged the FERC to investigate and remedy California's dysfunctional electricity market. Last June I called upon the FERC to investigate and refund the detailed overpayment by California consumers and businesses" (Davis, January 31, 2002).

As the charts show, the prices paid for power over this short time were by hundreds of percents higher than normal or historical prices during this period.

FERC did finally take action in the late spring of 2001, as Governor Davis acknowledged in a press release and noted the actions taken by the state and its citizens to curb the immediate energy crisis:

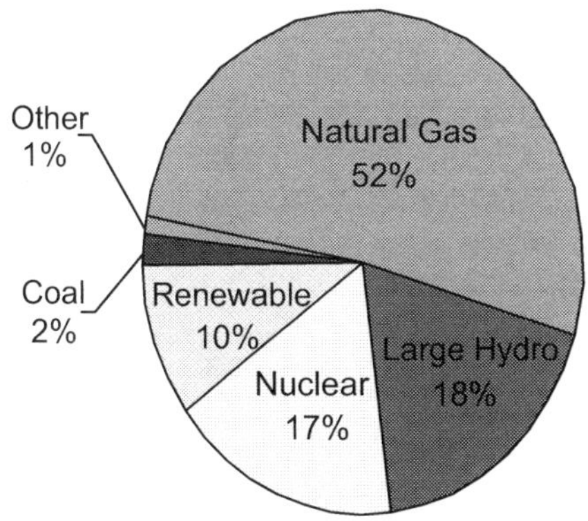

CHART 1 Share of Current In-State Power Generation Fuel. *Source: Rand Institute, February 2001.*

"I appreciate the action taken by you and your fellow FERC commissioners to finally reinstate price caps last June. Your action, combined with California's successful conservation efforts, our construction of 11 new power plants, and long-term contracts, caused energy prices to decline." (Governor Davis, press release, Jan. 31, 2002).

Subsequent reports and studies have verified the state of California's analysis regarding how a far more devastating energy crisis was averted over the summer of 2001 with these and other policies (California Energy Commission, Feb. 2002). While acknowledging the role of more rain in the winter of 2001 (and hence supplying more water for hydroelectric power), the fact was that the summer of 2001 was one of the 25 hottest in Californian history. The California government programs and citizens conserved over 12% throughout the summer of 2001. The market suppliers were forced to keep prices reasonable through price caps at the state (public utility) and federal (FERC) levels. Long-term contracts further ensured a constant supply of power rather than relying on market manipulation in the spot market.

Yet public records confirm that on Apr. 17, 2001, for example, then ENRON Chief Executive Officer Kenneth Lay made eight recommendations to Vice President Cheney regarding federal energy policy. One of those recommendations was continued opposition to price caps. A memo containing those recommendations was released the week of Mar. 20, 2002. On Apr. 18, 2001, the very next day, Vice President Cheney told the *Los Angeles Times* that the White House emphatically opposed price caps.

When the White House released the final report of its energy task force, seven of ENRON's eight recommendations had been fully adopted by the report. However, this was normal. Further documentation reveals that ENRON staff and senior officers as well as members of its board had tried to influence the White House during the Clinton administration and systematically pursued its self-interests in states such as California that were considering deregulation. As Kuttner (2001 and then Mar. 25, 2002) notes, "Perhaps the most damaging effect of all is the eclipse of an opposition politics. Embrace of deregulation by "pro-business" Democrats is more than a mistaken philosophical conversion."

The White House's energy policy, as enforced by FERC, particularly its opposition to price caps, helped to sustain the outrageous prices of the wholesale energy market. "This prolonged California's energy crisis and continued the massive transfer of wealth from our state to out-of-state energy companies such as ENRON" (Staff, 2002). It is still unclear how much direct influence ENRON had over the White House.

"What is clear, however, is that the White House's energy policy was nearly a mirror image of ENRON's energy policy. It is also clear that policy was detrimental to California and the wholesale energy market throughout the West. If ENRON was able to use its market power and public influence to unfairly manipulate the California energy market, then FERC must take action" (Staff, 2002).

Here is a checklist (Kuttner, 2002) of four icons that collapsed with the fall of ENRON:

A pension double standard. ENRON's retirement plan was heavily invested in its own stock. Executives cashed out over a billion dollars while ordinary employees were locked in. Janice Farmer, who retired with $700,000 in ENRON, told a Senate hearing last week that she was left with a $63 monthly Social Security check.

Bogus accounting. Since the Great Depression, the one form of regulation that even Wall Street has supported is the regulation of stock trades and corporate accounting … (when) once asked an ultra-Chicago economist if there was any regulatory agency that he endorsed. "The SEC [US Securities and Exchange Commission]," he said instantly, explaining that capitalism itself depends on honest information. ENRON's entire game was to make its business plan so complex that neither investors nor regulators nor even its own auditors could penetrate it. While its core energy business made money (at the expense of consumers), it had speculative off-the-books subsidiaries. These borrowed heavily to make risky investments and eventually took the whole company down.

The business press. ENRON's breathless cheerleaders included not only its own insiders and stock touts but also a business press that pronounced it the epitome of the new economy. It surely was that, epitomizing all the smoke and mirrors. In the wake of its collapse, ENRON's former sycophants have turned on it, with *Forbes* and *Fortune* running scathing denunciations. *Businessweek* asked giddy ENRON boosters what they thought now. (Gary Hamel, chairman of Strategos, before the collapse, said: "ENRON isn't in the business of eking the last penny out of a dying business but of continuously creating radical new business concepts with huge upside." Afterward, he said: "Do I feel like an idiot? No, but if I misread this company in some way, I was one of a hell of a lot of people.") Well, yes. The one holdout among the business press is, of course, *The Wall Street Journal,* still trying to blame the debacle on regulators.

Deregulation generally. ENRON's collapse debunks the conceit that a market economy can be efficiently self-policing. ENRON fleeced consumers by manipulating prices of electricity and gas; it fleeced investors and its own employees. Tycoons do this because they can. ENRON should signal a whole new era of reregulation of everything from electricity to pensions to accounting standards. It is another warning that Social Security pensioners cannot trust Wall Street.

The Wake-Up Call to Economics

The *Economist* (December 6, 2001), and economists in general, continue to return to the same theme as the one from over 2 decades ago, when the ENRON bankruptcy began:

> *The company's opaque accounting makes it hard even now to understand why it got into trouble, and whether the cause was bad luck or worse (see our special report, http://www.economist.com/printedition/displaystory.cfm?story_id=896844) The close links between ENRON's chairman, Kenneth Lay, and George Bush will keep the affair in the political limelight. And ENRON's staff, whose retirement fund was, at the company's urging, mostly invested in ENRON shares that they, unlike e the company's bosses, were then unable to sell, deserve public sympathy. But if America's capital markets are to stay the cynosure of the world, some quick lessons need to be drawn. (Economist December 6, 2001)*

The "bottom line" to the neoclassical economic community who unflaggingly support market power is that "proper" deregulation or privatization will sort out the ENRON and other companies. Some will survive and others will disappear. Business as usual is the economic creed. Or, as the *Economist* concludes in one article, "In the drama of capitalism, bankruptcy plays an essential part—until the next boom" (Economist December 6, 2001). However, in the ENRON case, the culprit is the accounting firm that ventured from its traditional and honorable function in the audits to the consulting role now seen in all accountancy firms. As the *Economist* wrongly put it:

> *The most important concern auditing. ENRON has restated its profits for the past five years, chopping $600m off its earlier numbers. The company's auditor was Andersen, now a target of many lawsuits. Last year ENRON paid Andersen a fat audit fee of $25m; it also paid the firm $27m for consulting services. (Economist December 6, 2001)*

The *Economist* goes on to cite other wrongdoing by Andersen. That was Dec. 2001, just after the ENRON bankruptcy. Since then, a number of firms have confessed to questionable accounting practices and a number of accounting firms have owned to their use of "liberal accounting methods." Certainly the *Economist* is correct in saying that "After ENRON, the SEC should do what its former chairman, Arthur Levitt, has long urged: ban accounting firms from doing consulting work for their audit clients. Accounting rules also need updating" (Economist December 6, 2001). Nevertheless, what is lost in this spurious debate is the core issue of deregulation itself.

"Last are regulatory lessons," according to the *Economist* (ibid, 2001):

> *There is a risk of turning any bankruptcy into an excuse for massive new regulation. Some have argued that energy is too important to be left to markets of the sort that ENRON pioneered; or that, since it was engaged in financial*

> *speculation, ENRON should have been regulated like a bank. Neither conclusion is justified. Energy deregulation has brought huge benefits in lower prices and more secure supplies: energy trading will continue to grow regardless of ENRON's collapse. Nor would it be wise to subject all companies with financial arms to stifling bank regulation. ENRON's energy exchange was, however, explicitly exempted from oversight by financial regulators: that should be changed. (Economist December 6, 2001)*

There has yet to be any evidence supporting these claims from the *Economist* or economist proponents of deregulation anywhere in the world. In the end, the best lessons of all will come from the mere fact of ENRON's bankruptcy. The *Economist* argues that:

> *The first is the regulation of auditors. For years the profession has insisted that self-regulation and peer review are the right way to maintain standards. Yet ENRON has shown that this is no longer enough. Second is the urgent need to eliminate conflicts of interest in accounting firms. Andersen collected audit fees of $25m from ENRON, its second-biggest client, last year, but it earned even more for consulting and other work. Lastly come America's accounting standards. GAAP standards used to be thought the most rigorous in the world. Yet under British standards [sic], ENRON would not have been able to overstate its profits by so much. And, once again, although ENRON may have been egregious, it is not a lone offender. The ENRON scandal shows that America can no longer take the pre-eminence of its accounting for granted. That is a far bigger concern than any number of congressional investigations. (Economist, January 19, 2002)*

The *Economist* (January 19, 2002) and *Fortune (January 16, 2002)*, other international business journals, and most economists worldwide come to same conclusion: ENRON was a case of bad accounting (or at least mixing accounting and consulting) rather than a mistake to deregulate public good markets such as energy. The culprit here is the mythology of neoclassic economics that fosters and encourages market forces to manipulate.

The Economics and Commercialization of an Advanced Storage Technology to Prevent Climate Change: The Fuel Cell

This case focuses on the commercialization of a new advanced technology from LLNL in California. The concern is with the fundamental business notion of how innovation is brought into the marketplace (Bailey and Wolfram, 2012). Hence, the empirical study here violates many of the commonly accepted economic perspectives in business economics. The interactionism perspective is far more robust for understanding economic phenomena and is radically different from the subjectivism paradigm. It also meets a need to describe, understand, explain, and perhaps predict events.

This case concerns an advanced innovative technology (ZAFC) for the automotive industry in the long term, but in the short term (within 4–5 years) for stationary power. In the energy and environmental industries, for a variety of reasons, the data were collected over a 2-year period. One of the authors (Cooper and Clark, 1996) was a participant–observer with the commercialization of the ZAFC. However, there was another compelling reason for this case study.

Technology(s) as the Key Element in the Creation of New Business Ventures

Today, electric vehicle propulsion is severely limited by the low energy density of available batteries, for which lead/acid is the only mass-produced technology with an acceptable cost. Battery exchange requires spare sets of batteries (which more than doubles battery investment), special vehicle designs, and expensive off-loading equipment. All-electric vehicles powered by secondary batteries cannot simply be restored to full energy.

Understanding that the ZAFC as a "fuel cell" rather than a battery is an extremely important scientific and business concept, the *Economist* (Oct. 1997) declared that the "fuel cell" had finally arrived ready for the marketplace. That article further linked both the vehicular and power utility applications of fuel cells to energy generation and storage. Subsequent research indicates that communities can be powered from renewable systems when advanced storage devices are introduced as part of the energy grid (Isherwood et al., 2000). By the end of the 1990s, the US government seriously began to recognize the importance of new advanced technologies that were environmentally benign, complemented foreign policy, and created new industries (Hydrogen and Fuel Cell Letter, 1998).

Three years previously within the LLNL, and only after intensive discussion, the laboratory staff had concluded that they had a fuel cell instead of a battery. In fact, many of the publications about the technology had initially labeled it as a battery. The basic distinction is that a fuel cell requires refueling (in this case, of the zinc) of the cells so that the chemical reaction (in this case, with air) and another substance (electrolyte) will process a chemical reaction generating electricity. For batteries, although the process is similar, the typical battery must be recharged, usually with some sort of external unit. Recharging, unlike the refueling process, which takes 8–10 min, takes 4–6 h depending on the extent of the battery discharge.

After the successful demonstration of ZAFC on a standard-sized bus from Santa Barbara, California, commercial viability (defined as a "successful laboratory and demonstration on vehicles of ZAFC for performance, reliability, safety and cost-benefit") of the ZAFC into a demonstrable prototype would require additional funding for 18–24 months. There were no US federal government resources. Only internal laboratory funds were used. Other economic models had to be found.

The issue of how to commercialize ZAFC for vehicles is an economic and business problem normally outside the realm of a research laboratory. Hence, the case is an example of an organization creating a new social construction or reality and operating within its perimeters. Therefore, a new business-economic model had to be developed and demonstrated for the ZAFC technology within a national laboratory. The interactionism perspective is appropriate because it agrees with a dynamic and creative approach to a business problem (technology commercialization). Nonetheless, by 2002, the company created in the United States along with an Australian partner firm had collapsed into bankruptcy. The individual actors would blame each other and even mention nation or cultural barriers to the new venture. Nevertheless, the reality was clearly something else. As discussed subsequently, the interactionism perspective was able to provide a perspective and approach to understanding the problems for an advanced technology commercialization that provide a concrete, positive, and productive direction for other such ventures.

Since the turn of this century, much has happened in the commercialization of fuel cells. Aside from national and state centers being formed to promote the research and development of fuel cells, new companies and established firms have been rapidly commercializing a variety of fuel cells. Two points need to be made. First is that the commercialization of fuel cells, like all other new technologies, has historically involved and been promoted by the public sector. Interaction among scientists and business people with the government as a source of grants, incentives, and tax credits has been critical. These relationships have even turned into formal organizations and associations for both networking and strategic planning. It is worth mentioning two in California that focus on the mobile or vehicle application of fuel cells (California Fuel Cell Partnership) and the power plant uses (California Stationary Fuel Cell Collaborative). Both organizations have been models for similar ones in other states, nations, and regions.

The second point is even more significant: the role of government in setting standards, protocols, or regulations that send public sector signals to the private sector that it must create new products that are environmentally friendly or at least nonpolluting. New alliances and forms are created, known as "civic markets" (Clark and Lund, 2001; Clark and Bradshaw, 2004), that bring together both the public and private sectors to address, define, and solve societal problems. The auto industry is certainly the largest and most affected industrial sector. Yet, so are other industries that pollute the environment, such as energy, water, transportation in general, and waste sectors. The public message is that related health and societal impacts on communities and regions can be measured. These externalities must be reversed through government and business actions to commercialize new technologies.

As noted in Chapter 1, the impact of all electric cars has been felt. Aside from the scandal over electric cars in the 1990s when they were brought into California from the major American car companies, as people in California drove more because of their freeways, once the cars arrived there, GM and others took the cars back and destroyed them. That was a major setback for those concerned about increasing environmental and pollution problems.

Yet in the second decade of the 21st century, electric cars came back via Elon Musk and his Tesla cars, named after Nikola Tesla over a 100 years earlier. Musk received significant funding from the US government for his car, which led to corporations financing him. Earlier in his career, he had helped start PayPal and Solar City, both companies that were focused on resolving environmental problems. Now other major car companies are manufacturing and distributing all-electric cars in California and around the world. Many have moved the marketing of their hydrogen fuel cell cars out of California.

However, the problem is still where and how to charge the electric cars. As noted earlier, that is where Hangery Company in China has a significant technology role with their all solar-powered cars (http://www.hanergy.com/showCar/carshow.html). Clark was at the kick-off event on Jul. 2, 2016, and summarized the impact of a totally green technology for the cars that Hanergy created: http://v.qq.com/x/page/o0313vci9m6.html.

Case 3: Economics of Entrepreneurship, Small and Medium-Sized Enterprises, and New Ventures

A variety of economic theories and models identify the deployment of innovations in the firm. The conventional liberal approach developed by Michael Porter at Harvard University uses a structural analysis of the firm to understand its competitive advantages and plot strategies against other firms (Porter, 1980, 1990). To accomplish an analysis of the firm, Porter created what he calls the "five-forces" framework: entry barriers, substitutes, buyers' and suppliers' bargaining power, and intraindustrial rivalry. Each force needs to be adequately understood so that the firm becomes competitive.

The government interactionism perspective is reflected in the "push economic model" (Clark and Lund, 2001; Clark et al., 1998). "The technology pipeline" demonstrates how new ideas and research move from the concept stage logically into the marketplace as "stars" or successful industries of tomorrow. All the researcher or inventor needs to do is follow the logical process outlined by business schools: business plan, venture capital, and industrial alliances. Basically, this approach argues that research and development (R&D) needs to be pushed out from universities and laboratories into the private sector.

In basic economic terms, the federal government approach is a supply-side "technology push," which for business applications is unsatisfactory. Economists will talk about the demand-side "technology pull" approach, which is "market" dominated. This implies that the customer knows what it wants better than do the laboratory scientists. The need to think, consider, and explore new cognitive ideas to start a business is critical (Weick, 1979): a new model of "technology interaction" in which the research scientist within the laboratory must be in constant interaction with the private sector to create new technologies for a 3- to 5-year time frame.

On the other hand, the neoclassical economic model argues that the private sector itself (often described as the "market") must see the demand and therefore needs for new technologies come first from the market or industry. In other words, the private sector must "pull" technologies that it defines and sees as important from university and laboratory R&D. The market or industry therefore must define where public R&D funds should be expended. The market defines the most significant research areas for publicly funded R&D. In the EU, "science parks" were instituted that connected university faculty and students to facilities that allowed them to build new business. They were not like an "incubator" because these science parks provided both private and public money to start NewCos (Clark, 2003a,b).

Both liberal and neoclassical economic models ignore the give-and-take or interactionism in conducting everyday business activities. In short, businesses are interactive and iterative and not a static or even linear process. As seen subsequently, studies of R&D characterize the demands of industry as being short-term and often too narrow to satisfy markets and shareholder pressures. Universities are oriented to the long-term in their research programs, in part because of the time needed for graduate students to finish their dissertations (the time frame is about 5–8 years, often the time needed for students to conduct research, write the thesis, and publish it). However, laboratories have a 2- to 5-year time frame, which places them closer to the commercialization time frame of industry, which is within months of funding and certainly not more than a year or two before R&D must see products in the marketplace. These time frames are important because industry or the market rarely funds R&D that takes over 2–3 years before it is commercialized. Competition and certainly shareholders demand a far quicker return on R&D investment than the more long-term laboratory and university R&D cycle.

Entrepreneurs enter the R&D scenario as teams or groups that provide funding for demonstrations and final commercialization of a new technology. Schumpeter (1934) viewed the entrepreneur as an individual who is the creator of new business within a society.

While entrepreneurial ventures create new businesses within a community, the new venture or entrepreneurial enterprise is rarely an individual business activity but much more of a team and collaborative effort. Much of the literature attempts to refine Schumpeter's individualistic definition (Freeman, 1994; B. Berger 1991 among others).

The development of a new company requires many actors who constantly communicate, exchange knowledge, and debate issues as they start the new venture. This far more international and global form of interacting is due in part to the Internet, but also primarily to long-time personal relationships from graduate school, international meetings, and conferences (Fast, 1993). Usually the new company will consist of a technical person, a marketing or business person, and a financial person or backer. As Freeman noted in 1989 and again in 1994, there is a distinction between the "business and technical innovator" and it should rarely be the same person.

The "American entrepreneurial spirit" fosters innovation through research and development. The creation and growth of many new firms is directly linked to new discoveries that are financed when they are more "mature" by venture capitalists for entrepreneurs who attempt to commercialize the technologies into product(s) and must get over the "Valley of Death" (lack of funding and cash flow). Organizing and management of the resulting new company(s) generally only benefit directly from either seed or venture capitalist forms of high-risk finance, experienced management, access to funds and markets, and a network of business contacts (Clark et al., 1998).

The outcome is even more significant: new industries and new capital are recycled into more companies. Rarely, however, are the funds placed into the second "D" (demonstration) and certainly not into the R&D itself. The role of government (and perhaps getting equity for the investment) must continue to interact with new applied technologies, especially in the areas of energy and environmental concern.

Methodological Considerations in Theory Building

The scientific interactionism perspective in business economics is an excellent example of qualitative methods and action research. One of the authors (Clark) was a participant–observer in these case studies. He was the manager of strategic planning at the national laboratory (fuel cells), the senior policy advisor to Governor Davis (California energy crisis), and both the founder of a new company and professor of Enterpreneurship. Hence, Clark was able to have access to key decision makers, data, and technical staff for all of the cases. The participant–observer role also allowed Clark to participate in meetings and interview key staff members. The results of this constant interaction were then implemented into action plans for further business development.

Qualitative methods are common within anthropology, linguistics, and some areas of sociology and political science. Some of these fields use methods from the conventional functionalist paradigm, such as survey, questionnaires, and statistical methods. Instead, as outlined earlier, participant–observation is a preferred research method, in addition to interviews and action research, which are other primary modes of data gathering. When necessary, we provide documents and reports such as emails, court records, and legal documents (Clark and Fast, 2006). Qualitative methods for gathering data are a significant part of the interactionism perspective.

In anthropology, the product for most research is normally a written ethnography of a culture, group, or class of people. Participant—observation, interviews, and sometimes action research are common research methods. In more recent years, many anthropologists employ local people, use survey techniques, and take videotapes as part of their research. Nevertheless, basic theories in the field of anthropology follow the functionalist paradigm in a number of ways: (1) they provide a static rather than dynamic view of the subjects; (2) they find data that fit into some preconceived notions of the people in groups, cultures, subcultures, etc.; and (3) they fail to understand the interaction between people in everyday life.

Yet, the qualitative methods of anthropologists are useful for understanding the dynamics of interaction within any human organization. When applied to businesses, the outcomes can be dramatic, revealing, and insightful. Suchman (1978, 1987 and 1994) did just that with her groundbreaking work at Xerox, when she applied qualitative methods to understand the interface (as business engineers would call it) between copy machines and humans. Since then, her anthropologically trained team has explored other avenues of research about machine and human interactions. Nardi (1993, 1996a,b) at Apple completed an edited volume with articles on the human—computer interaction or interface through the use of qualitative methods while constructing new theoretical perspectives.

An anthropologist conducting qualitative research is directly relevant to this study. In the course of conducting research, everyday discussion and dialogue are often the prime sources for data. Linguistics, a field related to anthropology, depends heavily on qualitative methods because the researchers will solicit words and sentences from native speakers of the language. A native speaker would be asked for sentence examples and then for interpretations and meanings. Alternative sentences would be suggested and critiqued by other native speakers. Often such sessions are in groups or a classroom setting in which the participant can hear, ask, and diagnosis the words and sentences.

In the case that follows, the exercise of clarifying words, phrases, and meanings was initiated by senior staff members, not for linguistic research reasons but to understand the exact meanings of ideas among technical experts. In other words, although technical people all understood each other on one level, they did not share specific definitions and meaning on another level. Therefore, several work sessions met in which intensive discussion over terms and meanings were clarified and sorted out. In other words, the scientists themselves participated in the research to assist their own technical needs.

Such meetings served to clarify ideas and terms and became a consistent and frequent interactive data collection method. In Blumer's terms, the scientists needed to make indications to themselves within the context of their social world so that they could communicate among themselves as well as to the outside world of other scientists and more particularly to the business community. The need for common definitions and understandings allows for more complex levels of technical and business activities. In short, deriving common meanings, and hence understandings, creates the social context in which to conduct technical research, and by extension, commercial business.

The following pages present an empirical case in which Clark was a participant—observer over 2 years (1995—97). The research, and hence the applied actions, continue. By definition, interaction provides a dynamic and constantly changing framework, which in this case led to an entrepreneurial business development for an advanced technology from a large research laboratory in the United States.

The concern of this chapter is an understanding of the application of the interactionist perspective to a new technology being commercialized into a business venture. The case documents one federally funded technology, the ZAFC, along with environmental concerns regarding its use and by-products from an industrial ecology framework (Cooper and Clark, 1996). A significant part of action research methods within an interactionism perspective is to argue for a sustainable development or environmental considerations in the creation of any new technology and related business ventures.

Graedel and Allenby (1995) discuss industrial ecology as part of an overall sustainable development consideration. They note that "the evolution of technology and economic systems such that human economic activity mimics a mature biological system from the standpoint of being self-contained in it material and resource use" (Graedel and Allenby, 1995 p. 24). In other words, the global environment is interlinked with business and industrial activities. Everything that human beings do affects the environment.

For Graedel and Allenby, the globe is a closed system operating as if it is a biological "life cycle." In short, humans interact with themselves and their environment. When anyone develops a new business, they must take that perspective into account as well. The ZAFC does just that, because once it is mined, zinc is converted into electricity; it becomes a slurry in the fuel cell process and then is reconstituted into pellets to be fed back into the fuel cell (Cooper et al., 1995; Cooper and Clark, 1996).

PRACTICAL APPLICATIONS IN ENTREPRENEURSHIP AND BUSINESS DEVELOPMENT

The Organization and the Actors

The US government was the primary funding agency for the LLNL and other US Department of Energy R&D labs since the 1950s. That ended in 2008 because President George W. Bush did not want the government to fund the laboratories. Instead, private companies needed to do that. In this case study, LLNL has a contract with a major American university system for the operation of the laboratory. The US government had funded over 735 R&D laboratories through various departments and agencies. The National Aeronautics and Space Administration (NASA), for example, had 12 laboratories. The USA Defense Department through its various military branches has over 20 laboratories. The US Department of Energy funds more than a dozen laboratories for over US $16 billion annually. Many of the federal laboratories are renowned in their fields, which also made the United States a world leader in new technologies and business. However, this is not true now.

Today, LLNL has over 8300 full-time employees, around 1200 of whom are doctoral-level scientists and another 2000 are engineers at the college and graduate levels. The annual budget remains at about $1 billion. The facilities cover one square mile, with over $5 billion in equipment and buildings representing one of the world's most renowned research and development organizations. The laboratory is among the largest (in terms of annual revenues) research laboratories in the United States and probably the world.

In late 1995, two divisions collaborated with others seeking research projects under the overall title of "industrial ecology." Although ZAFC technology (Clark et al., 1996; Cooper et al., 1995) had been created under the chemistry and materials division, it clearly fit into a broader multidisciplinary effort within the laboratory for industrial ecology-oriented technologies. Fuel cell research and development are administratively within the energy program division because it supplied the research funds to chemistry.

Almost all federal laboratory technologies available for commercial application have been historically taken "off-the-shelf" and "pushed" (some call it marketed) into applied research areas or commercial use. The current laboratory technologies are described in books, magazines, or newsletters, such as the biannual collection of technologies from the laboratory. For example, NASA publishes *Tech Briefs* through a private contractor, which is a monthly magazine featuring new NASA technologies. In most cases, a one-page description outlines each technology available for transfer to businesses, listing the laboratory contact person with telephone, fax, and email addresses.

The key interactive people at LLNL in the ZAFC case are five people within LLNL: the lead scientist and inventor of the technology (electrochemist); leader of transportation programs (civil engineer); his marketing assistant (social science undergraduate degree); the lead technology manager in related technologies (mechanical engineer); and the Clark as participant—observer (qualitative economist), who assists in business planning and strategic alliances. The interaction within this group was frequent and sometimes intense. Several situations arose in which the group had to (or was forced) to define some scientific concepts terms because of technical confusion among the various disciplines. Intense all-day meetings, briefing sessions, and casual meetings had to be held frequently as the research changed the definition of the ZAFC and the effort to seek commercialization funds changed business strategies. All were interactive processes demanding constant dialogue and discussion.

ORGANIZATIONAL INTERACTIONISM DEVELOPMENT FOR UNCERTAINTY AND CHANGE

Operational Concepts

Consider a theoretical refinement of the interactionist perspective drawn from transformational linguistics (Chomsky, 1975, 1988, 1980). Later economic ideas from Chomsky (2012) are explored from the "plutonomy and precariat" perspective in greater detail. At this point, however, the concern is with a few theoretical concepts to help order empirical results, as with the fuel cell technology case.

Gullestrup (1994) presents a multilevel construct in which social reality is viewed as having two levels that began earlier (1972) and then moved ahead in 1997. Hofstede (1980, 1991) presents many levels of culture within any society. The difference with linguistics in terms of creating levels is that the linguistic theory argues for a surface and deep "structure" in language. Or, as in this business case, an organization and its actors are attempting to bring a new technology to the marketplace and the interaction among the various actors needs to be understood in terms of concepts, meaning, and behaviors. In an earlier series of theoretical papers, there are presentations on some preliminary details about this operational approach to levels of interaction among organizations and actors in everyday business life.

Structures in language theory are not rigid, as in the conventional functionalist paradigm in social science. Instead, the notion of structure in linguistics is fluid, flexible, and multilevel or nonlinear. Structures are not stagnant in either time or space, but dynamic. In language, structures simply provide a set of universal rules (grammar) and database (lexicon) upon which to draw. Language skills and their application to everyday life must be flexible and fluid by definition. This dynamic and nonlinear nature of linguistic structures is the opposite of the conventional functionalist paradigm.

For the interactionist application of linguistic transformational theory, a surface structure allows the action researcher to describe and understand organizations and the actors within them. A far more detailed discussion of deep and surface structures is given in a later chapter. Clark and Sorensen (2001) develop some of the ideas in relationship to entrepreneurship. Later, Clark and Bradshaw (2004) present some theoretical constructs in attempting to describe and analyze how the California energy crisis can be explained.

For now, consider surface structures as part of any research investigation, which continuously analyzes actors and groups within organizations of any size. The surface structures are seen publicly in terms of events and situations that occur within or between organization(s) among the actors. These events are seen in terms of business activities such as sales, promotions, manufacturing, and distribution. This continuous disharmonic dynamic through constant interaction of the actors defines and redefines surface structures.

Deep structures are the basic building blocks of surface actions and interactions because they define and refine concepts, ideas, and symbolic objects in terms of their meaning. In other words, deep structures are the "meaning" attached to actions. Deep structures cannot be seen as surface structures because they are the mental representations or definitions of ideas that people bring to situations. It is the deep structural level in which people communicate (or do not communicate). Here, people interact because they define and thus understand one another.

One other theoretical operation from linguistics is useful within the interactionism perspective: transformations. In the study of language, deep and surface structures are interactive in that one level influences the other. A linguist would create rules and representations (Clark, 1996) to describe and understand everyday language use. That is a major intent of this book with the advanced technology case of fuel cells. The notion of "representations" is useful and appropriate for the case. Next, representations are discussed and examples are given from the case.

Basically, a representation is similar to what Mead and Blumer call "notions" to oneself. In other words, when people interact, they are constantly creating, thinking, and reflecting on their actions as well as those of others. These notions are representations in that humans will define and act upon how they think about their action or that of others. The ability of humans to make such representations allows them to change, adjust, and advance ideas. In short, for the case here, representations provide agreement among the actors and therefore further collective group actions.

Surface Structure Representations

For the fuel cell case and within the context of the surface structure analysis, two primary organizations are interacting in the initial stages: a large power energy company in Sydney, Australia and a large American national laboratory. The laboratory is responsible for technology development whereas the power company needed a new technology to meet the legislative demands for ZEVs in Australia (similar to the earlier ZEV requirements in California) and as a major sponsor to create a "Green Olympics" in 2000.

The laboratory provided the ZAFC technology by attempting to commercialize it. The role of the American government in supporting the initial research and development of the technologies, although helpful, was not enough to make the end product usable for commercial application to vehicles. As Clark and Paolucci (1997) document, the auto maker and battery consortia in the United States were admittedly opposed to fuel cell R&D from federal funding sources. Because these consortia either controlled or advised the federal funding sources, few R&D funds were directed toward fuel cell R&D.

The constant interaction between the technical and business staff of both the power utility company and the laboratory proved to be essential in building trust and establishing an iterative dialogue. In fact, the key power utility executive saw the fuel cell as a large entrepreneurial venture. Thus, he left the power utility company to form with partners, a new company to commercialize the fuel cell. The global surface structure event was the formation of a new consortium with the two initial partners (the power utility and the laboratory), and then when the new entrepreneurial company was formed, a broader and more dynamic new surface structure business network consisting of firms in the United States, Australia, and eventually Europe. The common goal by early 1997 was to commercialize and operate ZEVs for the Australian Olympics in 2000.

Within each organization, there are surface structures that provide both constraints and opportunities for the actors. In fact, because of the need to commercialize the fuel cell, but with limited financial resources, the actors with each larger organization had to reconstitute a group to work on the fuel cell. This group then had to endure extraordinary surface structure organizational changes in 1996, such as that the power utility company was merged, a national Australian election that totally changed the government, and privatization along with deregulation that came into the marketplace. For the laboratory, the conservative American congress was continuing to cut R&D budgets. The laboratory itself had to become more oriented toward defense and nuclear weapon "stewardship" (e.g., dismantle and stockpile), which restricted all internal R&D funds.

Overall, the US government retreated from seeking industrial partnerships and commercialization of its technologies. For example, industrial ecology became an unacceptable mission for many government laboratories. Technology transfer was drastically cut and even eliminated in most US government agencies and departments. A pall was felt and a dark cloud was cast over applied and demonstration research such as the fuel cell. The result was that the group of actors at the laboratory focused their efforts on commercialization of the fuel cell because of individuals' commitment to the concept and belief in fuel cell technology for a cleaner environment. The tasks required to do so, however, were not really part of either the actors' job description or their defined daily routine. Certainly, one of the critical decision points for the Australian executive to leave the power utility company there came as a consequence of the organizing turmoil both there and in the United States.

In the early days of establishing the collaboration between the laboratory and the power utility company, however, the team coordinated and collaborated on new levels by achieving results internally and with the new global consortium. Only later when the new entrepreneurial firm was formed did the relationships develop into much deeper and broader areas, which included and then excluded some of the earlier actors in the commercialization process. For the international interaction to occur between a large corporation and research laboratory, the interactionism perspective specifies that a number of representations had to be made between the actors and their respective organizations.

Particularistic legal constraints and covenants are defined as rules and apply to both organizations owing to government regulations, foreign ownership rules, and the nature of business. They will not be considered here (Clark, 1996). The interactionism perspective provides a framework in which actors within groups and organizations can be described and understood in situations (e.g., meetings, discussions, or work sessions), activities (e.g., daily work as with computers, conference calls, telephone use, and research) and events (e.g., conferences, travel, and larger-level or management meetings). Each representation is an example of the discussion from Blumer and Mead in which A → B → A (A interacts with B) and then interacts back again to A after some thought and planning: A ← B ← A).

Deep Structure Representations

Beneath the surface structures with their set of rules are the deep structures with their own set of rules. Think of the deep structure as the level of interaction that defines business actions within situations. Everyday business life is an endless series of surface structure business situations with shared concepts and meanings to conduct business formed in the deep structure interaction between actors. The derivation of these meanings comes from the often nonbusiness interactions in situational events that are part of everyday life.

In short, deep structures provide the meaning behind the observable business behavior of actors and organizations seen at the surface structure. The linguistic theory of generative grammatical is equivalent to the dictionary or lexicon of a grammar for providing the definition of ideas and concepts in the deep structures depending on the everyday interactions. The interactionism perspective is particularly significant because there is an extreme need to define these ideas and concepts in the international business arena, even when business is conducted in the same language. As in the fuel cell case, English may be spoken by Australians, Americans, Danish, Germans, and Swedes, but it is subject to misunderstandings and therefore there is a further need for constant interactions, actions to make decisions, and definitions.

The outcome of defined concepts and even the meaning of specific English words (e.g., "commercially viable" had different meanings between the Americans and Australians, which led to misunderstandings) are the representations in the deep structure. Owing to the interaction within the deep structure and the interactive transformations with the surface structure, terms, concepts, and hence common goals become clearer. With common "lexical" definitions, the actors and their organizations can move forward into continuous business activities.

For fuel cell technologies, the senior laboratory staff had to convene an all-day meeting in early Jan. 1996 (over 2 years after the project began) to define technical terms across the academic multidisciplines involved in the commercialization process. The result was a "glossary of terms" that could be used in any interactions within the laboratory and worldwide with other partners (see Firth, 1990 as an example of discourse analyses). These basic deep structure definitions of terms allow accurate communication and understanding for surface structures interactions in everyday discussions among the various laboratory and international actors and other organizations. Understanding the business process for commercializing this technology is the basis for an interactionism perspective. When actors have defined representations of basic concepts and ideas, they can conduct business and create new mutually beneficial economic opportunities.

Organizations as Collectivity in Competency Building

The interactionism perspective has a multidimensional theoretical range in that it describes and incorporates concepts from industrial ecology into everyday life. For example, industrial ecology is applied to specific products within a company. In the fuel cell case, that life-cycle, recycling, and refueling component of the fuel cell is readily apparent.

With the case of fuel cell vehicles seen as part of the transportation system in a much larger industrial ecology context than the individual firm. In other words, the use of zinc as a fuel cell element eliminates environmental pollution by vehicles, but more importantly, it applies an environmentally friendly element to energy power and storage. By mining zinc and then recycling it through a recovery unit (making pellets) after use in the fuel cell on board a vehicle, the environment can be protected at the source of mining the raw material as well as for the end user within the transportation system (no air pollution and 100% recycling of the zinc itself).

Consider another organizational surface structure level for which the fuel cell has significant implications for basic infrastructure environmental issues. Either as a hybrid electric vehicle or potentially as a self-sustaining electric vehicle, the fuel cell has a significant impact on vehicular environmental performance and hence the transportation infrastructure. With other fuel cell–powered vehicles, the internal combustion engine (ICE) and fossil fuels are relics of the past, antiquated industrial technology from the 20th century.

Imagine the new transportation infrastructures of the future: a world of silent vehicles, clearly a problem for pedestrians, runners, and cyclists (especially with earpieces for listening devices), blind and disabled, accustomed to hearing vehicles approaching. A number of new technologies will be required to assist the transportation infrastructure. Intelligent roadways will need to have sensors for traffic of all sorts. Gas stations will become refueling stations. Traffic control will use other markers and signals. Endless new technologies and businesses would be needed, as the entire infrastructure would convert from the ICE to ZEV vehicles.

Three issues emerged from an interactionism perspective as being a problem with ZAFC: (1) a general collective resistant among American companies to change or venture into new technological areas; (2) the lack of capital within small or midsized firms for starting new internal ventures; and (3) the collective belief that current infrastructure and vehicle configurations were established and enough to satisfy the customers.

In other words, the collective and hence conventional wisdom within the industry was resistant to change and innovation. For a new innovation such as the fuel cell, either a redefinition of the collective organizations had to be made or resources needed to be sought outside them. The latter appeared to be the only alternative.

Parallel to this period for fuel cell commercialization, the interactionist model worked on another level and signaled what appears to be the successful path. One basic and growing principle of American industry is that an increasing amount of business comes from international and global markets. In other words, whereas American creativity and innovation are world renowned, American industry is not.

Although the commercial outcomes of the interaction have yet to be demonstrated, it appears that the fuel cell will be funded for its final development by an international multinational consortium. European bus makers saw the technology as a near-term (by the year 2000 through the EU Commission) commercialization of fuel cells for buses which would lead to worldwide markets for other cities and events. By 2004, 33 fuel cell–powered buses produced from Daimler were placed in 10 European cities. Clearly, the global strategy has yet to be grasped and appreciated by American auto makers.

Networks and Change

A significant component of the interactionism perspective at the microeconomic level is the use of "networks" or relationships between people and organizations. Therefore there is a need for personal connections, partnerships, and relationships between technical staff and separately between business executives are collaborations in which often intense exchanges of information are commonplace. Some networks (Håkansson, 1994; Håkansson and Snehota, 1994) form in many different ways, but primarily link businesses with compatible strengths (and in some cases weaknesses) to achieve common goals. Other networks form between government and private industry (Sørensen and Nedergaard, 1993). Networks can form on horizontal and vertical planes, depending on the nature of the interactions among the actors, organizations, and situations involved (Clark, 1995a,b, 1996). Jones et al. (1998) consider networks as being distinctively from "markets" or the conventional approach to economic and business development.

With the interactionism view, the understanding of networks as interpersonal linkages between the interactions of people in those networks takes on a new meaning and dimension. In short, networks cut across and even violate company or firm organizational structures.

Whereas government and business networks are familiar in other industrialized countries, they are new concepts for American industry and government. Nevertheless, the interactive network between an American government laboratory and an Australian government public utility and a variety of international private sector corporations is a significant achievement. The basic explanation of why it is working can be seen within the interactionism perspective of networks.

The network approach as a theory often has its basic assumptions in the functionalistic paradigm and draws on system theory and structural functionalism (especially on Blau, 1960; Pfeffer and Salanzik, 1978). For example, in an earlier article, Jones et al. (1996) saw networks as defined by the functional roles in an industry. Their approach changed significantly by 1997, when a general theory of network governance was proposed.

But the idea of networks is an important issue in the interactionism perspective because the result or manifestations of interactive representations are interpersonal relationships linked together into personal networks that are formed over time. When one examines the relationships and representations formed by and among researchers, networks are apparent. One technological colleague calls another to verify engineering research data. A researcher finds a new piece of information and communicates with others about the findings. The same set of interpersonal networks exists in business. Furthermore, interactive networks are being formed between the public and private sectors. New forms of partnerships between government and private industry are especially becoming more common in the American economic system.

Such concepts of networks and relationships are not new to other industrialized countries. Swedish scholars have researched networks for 2 decades (Håkansson, 1994), which concern "core corporations" and their relationship to smaller (SME) companies. Although the worldwide recession dramatically changed these networks (Håkansson and Snehota, 1994), the basic form of the Swedish economy, the largest and most significant in Scandinavia, remains the same: close networks of partnerships between large corporations and private industry.

Denmark, on the other hand, developed an entirely different network economic structure (Sørensen and Nedergaard, 1993). Private industry and government have a strong and interrelated relationship. However, there are few large "core corporations" in Denmark. For the Danes, a large corporation has 1300–2000 employees. As in the United States, official statistics classify a firm as being large when it has 500 or more employees. There are only a few companies of that size in Denmark, whereas several have many times that number of employees outside the country. On the other hand, Sweden has many corporations that employ over 10,000 people each within the country.

In other words, the concept of networks in Denmark means the personal relationship between and among SMEs rather than large corporations and SMEs. The attempt to structure networks, as many do from within the functionalist paradigm, both misrepresents the phenomenon of how networks themselves came into existence and grossly exaggerates the need of scholars to orchestrate or organize them overly.

In much of the literature, when they are defined, networks can be almost anything: associations of similar small companies that collectively have more purchasing power; regional groups of vertically organized companies; companies and local educational institutions; or even governmental units with common goals and purposes.

In fact, successful networks (those that accomplish something over a period) are formed to achieve the collective or common goals of the members. The successes of networks in Denmark over the past years have often been recognized as contributing to the country's economic success during the worldwide recession of the early 1990s. While this assertion is debatable, more recent events have proven that the economic successes in Denmark (turn of the 21st Century and 2008 with the Global Economic Collapse) are a role model for other nations.

Networks are centered on basic shared philosophical concepts. Networks must use or leverage resources within a triangular relationship among government, private industry, and support institutions such as education, training, and technical assistance centers. Whereas a network can certainly change and even disappear or purposely disband, the initial formation of a network needs facilitation and continued support from a dispassionate outside party.

The key to networks within the network theory is "trust." This can be only achieved after people work together on a common problem and find that others are able to keep secrets, share valuable information, and exchange new ideas. The interactionist perspective best describes, understands, and explains how networks are created and operate. In themselves, networks are neither theory nor scientific. Instead, the understanding of networks allows both the scholar and the business person to pursue shared goals. If networks are institutionalized as formal permanent structures, they will implode from their own administrative weight. The very notion or idea of a network is something that exists at moment or situation in time to accomplish some task(s). People know one another and form the network to solve the problem at hand.

The fuel cell case demonstrated a form of network building in three ways.

First, as an action researcher, the participant—observer (Clark) knew from experience how international companies might act best in development. He had understood some surface and deep structure rules used in international business activities. Critical to any interaction with a company, especially in Germany, France, Scandinavia, and most English-speaking countries, is to establish linkage, rapport, and trust. That was done in the fuel cell case by coordinating the interaction between the laboratory technical staff and the key international actors through conferences and meetings outside the United States. The international firms then sent staff to visit the laboratory.

Second, to protect priority information as well as further establish trust in a legal manner, nondisclosure agreements were executed and signed. Although to some these legal agreements may appear to violate "trust" in the conventional personal use of the term, for international business purposes, such agreements are (1) a signal of trust building, (2) satisfaction to company legal departments, and (3) necessary to have specific technological discussions.

Finally, constant communication via telephone calls, fax, and email are essential. When working in the international business world, such communications can be difficult because of different time zones, language problems, and costs. Email allows constant communication and overcomes these problems; nevertheless, it, too, has limitations. The best communication method is personal contact. This was purposefully done with visits among the various consortium companies.

The LLNL pursued the same interactive process with domestic US companies that might be potential manufacturers of fuel cell and therefore consortium members. Tokarz would lead the effort either with Cooper, the lead technical researcher, or some other member of the team. Follow-up information would be sent with periodic telephone calls and fax or email communication. However, the domestic market never materialized, in large part owing to the inherent resistance of American corporations to change.

Qualitative Analysis

Intense and constant interaction must take place between the members of the international consortium to commercialize the technology itself. This close collaboration supersedes industrial competitive issues and conflicts. For example, in the fuel cell case presented here, several major bus and battery or control unit manufacturers are part of the international consortium. Economic competition arises between companies in the marketing and distribution of the final zinc-air technologies for vehicles much the same way as oil, gas, and vehicles are competitive. This is the interactionist perspective at the macroeconomic level and the subject of another book.

Applying the interactionism perspective on a microeconomic level, a number of specific interactions become apparent.

Interactionism Surface and Deep Representations

The first issue must clarify the surface and deep structure interactions within and between organizations forming the consortia. Using the interaction perspective outlined earlier, the following representation can be made of the basic organizational interactions.

The goal of the action researcher within an interactionism perspective is to establish business opportunities. Sets of representations provide guidelines for business actions as well as a road map to future interactive decisions. The representations must be stated, subject to further scientific verification, and applied in other cases. Ultimately, the representation could be reduced to mathematical relational symbols. Elsewhere are several sets of rules derived from the fuel cell case (Clark. 1994a,b, 1995a,b, 1996, 1997).

A second interactionist result was the clarification of representations or meaning attached to words, ideas, and concepts. Consider the example of the fuel cell case at the laboratory when the principal researcher and the program manager went to Europe to attend a conference on electric vehicles; they found that the definition of a "fuel cell" more accurately described the technology of zinc—air rather than an "advanced battery." Upon further verification with the key staff in Australia as well as at the laboratory, the conclusion was certain.

From early Dec. 1995, the zinc—air battery would be known as a fuel cell. Although there was scientific and technical evidence to support the name change for zinc—air technology, international interaction was crucial. Clearly, the technology was more of a fuel cell (that is, requiring refueling rather than recharging). Furthermore, the concept of a fuel cell itself denoted high-tech advancements and therefore had more business appeal. The decision was made. A common definition was adopted and put into use worldwide. The term "fuel cell" had been agreed upon at the deep structure level so that it could be used precisely through transformation rules to the surface structure groups and organizations in everyday language use.

A second example of creating clear and precise definitions can be seen in the creation of a new technology to parallel refueling the fuel cells. Through numerous meetings both within the laboratory and with the international parts, the concept of a ZRU with its own representations was created. Originally, the principal research described the process for the zinc to be cycled through an undefined external stationary unit that pushed or spent zinc and electrolyte into an external hopper from the fuel cell. After many hours of intensive discussion among the lead scientists, especially Cooper et al. (1995), two issues were clarified. The zinc was the issue and not zinc combined with air. In other words, the external and separate unit needed to be concerned only with zinc processing.

The other point of clarification was concerned the concept of this unit itself. Again, in checking earlier documents, the process unit was called a "refueling unit," when in fact it was a "recovery unit." The difference is significant in the context of industrial ecology. Refueling implies that a raw material (zinc) has been mined and processed so that it is ready for refueling the fuel cell. In fact, the zinc is not refueled at all. The zinc is recovered because it is reformed after being used in the fuel cell. In other words, there is little loss of zinc (unlike gasoline, for example, which is burned and needs to be refueled). The ZRU is therefore simply an external unit that takes the zinc out from the fuel cell, reconstitutes it, and pumps it back into the fuel cell after 250–300 miles of use.

Clarification of the recovery unit (rather a refueling unit) is an enormously significant deep structure issue. Aside from understanding and communication of the meaning into surface structure situations, the clear definition allows scientists to create the ZRU. In this case, the actual ZRU had not been created and designed. Now a clear technological path could be followed. The significance of the ZRU definition will lead to patentable intellectual property among other yet unknown scientific breakthroughs.

Transformations

The relationship between the deep and surface structures occurs through transformations, which are vertical interactions between the sets of rules within each structure. In the fuel cell case, there are three "nodes" or key interactive points of contact for the rules: personal or people interactions; organizational relationships; and technology operations such as the use of telephone, fax, travel, the Internet, etc. Each node is noted as a "phrase marker" because it has a definition and universal meaning.

As continual personal trust is being built through constant interaction within and between partnering organizations, personal nodes are triggered that move back and forth between deep to surface structures. For example, telephone conference calls occur (lasting hours), but then a point is reached when some of the principals within the organizations need to meet face-to-face. Earlier, one aspect of building personal trust was described as essential for establishing networks. Such trust could be built only through constant interaction among the actors.

In the fuel cell case, personal trust was firmly established within a 2-month period between the laboratory and the power utility company: once in Australia owing to a professional academic conference and once in Europe as the result of an international conference. The transformational relations moved or transformed the common definition of terms and concepts to the surface structure following the set of rules outlined previously. Other opportunities were deliberately created: the action researcher (Clark) as a participant–observer on the laboratory team traveled to Europe on two other business occasions and visited the key consortium bus manufacturer. These two meetings established an already solid interpersonal relationship through a factory tour, confidential exchange of data and information, and joint collaborative research leads to seek funding for fuel cells. Various meetings were planned in various parts of the world that would connect the key actors from each organization.

Understanding Interaction for Cognitive Economics to Be Put Into Practice

As noted, the definition of zinc–air technology as a fuel cell rather than a battery was a matter of constant interaction between the representations made on the surface structure organizations and actors who must create a common set of definitions in the deep structure. Once these terms were agreed upon, expansion into other related technologies for a ZEV became clear and far less complex to implement. The interaction rules allowed the fuel cell to be commercialized because there was a common definition of ideas, concepts, and terms. Thereafter, agreement regarding the need and application for the technology could be readily made and advanced through proposals, contracts, and seeking other consortium members.

The results of interactions between and among the actors and organizations regarding fuel cells lead directly to other cooperative technological commercialization. It is only a matter of time. There will be other technologies that the consortia will pursue as well (such as in other fuels, materials, and intelligent systems). While economists debate "market demand" and "market supply" issues, technological innovation and commercialization are very much outside these simplistic theories. Business in general, when seen as a dynamic and ever-changing collection of people (defined as a firm or company), will change as a result of the interaction of its actors. Change is the norm in business.

The process of business innovation through technological commercialization, using the transformations between surface and deep levels of representations, is interactive. It is not simply a pull or push economic approach. The use of an interactionism perspective provides for a better understanding of everyday business phenomena. Furthermore, much can be learned about conducting research and implementing business practices.

The discussion starts with how to understand the structure of what can be called the "firm." A more detailed discussion of "interactionism" is provided in the development of a "new economics" later in this volume. A few issues are worth presenting in the context of the development of the energy "market" and restructuring or creating firms in that sector. One issue that the ENRON case has clearly demonstrated is how important "knowing" people can be. That is, close friendships can be useful for political influence. In academic terms, this has been called "networks" (Håkansson, 1993, 1994).

Elsewhere (Clark and Fast, 2002, 2006), the firm is discussed and how it should be understood as the actions and knowledge of the individual actor and of the actors' collective actions over time. These actions and interactions have some consequences: i.e., the organizing of activities and knowledge. In this organizing process are moving pictures of what the firm and the environment are will be created among the actors.

The manipulation of markets by firms is rooted in the notion that the firm acts and behaves as seen earlier in the ENRON case in its (the actors') own self-interests. In this case, the firm involved and subjected its accounting firm to do its bidding. Individuals communicating in an interactive or face-to-face manner, are the basis in which the relations consist of concrete meetings between members in the firm, m order to conduct all business. To improve an understanding of the firm, it is best to start with an understanding of what constitutes the "firm": What is a firm, or rather, what is a business organization? The word "organization" is a noun, and also a myth. If you look for an organization, you will not find it. What you will find is that there are events linked together that transpire within concrete walls, and these sequences, their pathways, and their timing are forms that often are erroneously made into substance whenever anyone talks about an organization (Weick, 1976, 1979, p. 88).

When organizations are examined, especially the larger, older, famous ones, they seem solid, permanent, and orderly. This is, after all, what scholars and the public often call organizations. Images of organizations as solid, permanent, orderly entities run through many textbooks. But they tell only half the story. They obscure the other half: the chaos that looms behind the order, surfacing from time to time, such as when computer systems break down, when products are sent to the wrong destination, or when bookings are made for the wrong dates. They also obscure the immense human efforts and energies that go into keeping organizations solid and orderly.

The word "organization" is (only) a concept that is used to describe a phenomenon. It is a conceptualization of what people believe and do and what they orient toward actions. Organization is a concept in the same way as the concepts of family, a class in school, a football team, a union, etc., are all concepts. In other words, an organization is a phenomenon that people experience when and where they see more than one person involved in activities over time. Thus, organization becomes a collective arrangement in which people try to give the situation and activities meanings. In line with Blumer (and in Clark and Fast, 2002), organizations consist of the fitting together of lines of activity, the interlinking of lines of action. Actors mixing, sharing, competing, and cooperating are parts of the interactive process that defines groups and organization. That is why by definition, most organizations change and move dynamically in space and time.

By fitting together the lines (what can be called the "rule") of action and interaction as logically prior in organization, actors are discouraged from mistakenly regarding organizations as "things" or simply "solid entities" such as a building or structure. Some scholars argue that organizations are "living things" akin to the biological sciences and refer to organizations as "evolving" with "life-cycle" attributes (Graedel and Allenby, 1995).

An organization should be understood through the actors who, by their actions and knowledge, create the firm in their everyday pursuit of life. The focus and start of understanding organizations are the members or actors within the organizational life. In this, the relation between action and knowledge is the central issue of interaction.

Action specifies the behavior of people, expressed in advance by the actor whose behavior is then based on an anticipated project. "The act" expresses the result of the ongoing process or the fulfilled action. Action could be hidden or open, intervening into the outer world. All projection consists of an anticipation of future behavior with help of "fantasy" (thinking-in-future) or, as Blumer puts it, acting in relation to "the generalized other." However, it is not the ongoing action process but the fantasized act, *as it* was fulfilled, that starts all the projection.

The collective actions constitute the firm, but they are understood in relation to interpretation and intersubjectivity. The constitution is formed by the actors, but at the same time, it is a restriction for the actors through the moving picture of reality that they have built and act in relation to. This is central because the relation between the development of knowledge and the actions is the foundation for the process of interaction. All this is about how to understand everyday life processes in the actors' construction of the firm.

People understand themselves retrospectively and act accordingly, but in addition, they are thinking-in-future: What are the projects they are thinking about? In which way do they try to realize them, and how do the projects change through the process of action and interaction?

People construct their organizational reality through actions in everyday life and they build paradigms to orient themselves to their own reality. Individual actors have to relate themselves to this discussion if it is the empirical reality and not the theoretical "reality" in which one is interested. In other words, an understanding of the social construction of people's organizational life and activities is the context of their everyday business life within the firm.

The basic issue for most actors in firms is how to make the company survive and grow during any particular point in time. They must be free to move in the marketplace, but they also be secretive enough to protect its privacy. More important, firms must be concerned for others and their environment. The public good must be protected because it is in the best interests of the firm to maximize its profits for shareholders and executives alike. The protection of the public good is essential in various infrastructures and sectors (Clark and Lund, 2001). In recent months, the protection of the transportation sector must be embedded in the government. Hakansson, Hakan, and Jan Johanson noted this concern back in the early 1990s (1993). The energy crisis in California clearly showed that the energy sector is a public trust. The same could be argued for the environment, water, and even telecommunications.

CONCLUSION: THE FUTURE IS NOW

Since the early 1990s, the European Commission defined (1990) "science or technology parks," which were then and are now where research, technology, industrial focused and implemented in Europe. A science park would be established that would be located close (across the street) to universities in the

Nordic nations, Germany, and other EU nations as well as Asia, especially China. As the *Economist* noted in 1997, "Europe" was changing with the example of "Remodeling Scandinavia" (August 23, 1997: pp. 37–39).

The problem with that article and others was that the *Economist* looks at everything (business, government, academic, and families) from the neoclassical economic perspective. Other scholars in the 1990s noted that the Swedish approach to Europe was different because of both history and current events (Håkansson, 1986). Almost a decade later, the negative economic results from Sweden, following the United Kingdom and other EU nations, turned bad for the nation of Sweden and other Nordic nations (Håkansson and Snehota, 1994). Thus, the *Economist* was critical of the Nordic nations for being "socialist." They have done the same thing with Asian nations, and the *Economist* has been wrong.

This book's hypothesis is that there is enough legitimate and validated data on the new applied economics and interactive business models to formulate a theoretical perspective with science parks demonstrating the future (Clark and Decker-Ward, 1995). With that understanding, this chapter explores and formulates strategies, plans, and policies regarding how research and development can be converted into new businesses to support and assist entrepreneurship, provide programs that connect regional economic development, still demonstrate concern for the environment and climate, and create networks to enact international collaborations.

The theoretical development of science parks as a "bottom-up model" was the result of the practical experiences of research, development, and technology for over the past half century. Today, with interactionism and the qualitative economic paradigm, the creation of companies is a result of cross-disciplinary actions by people, communities, and cities. The interest in science parks today is global and is growing with national and international associations that meet annually. However, the core element in the modern development of any science park remains the firm or company results for collaboration and creation of academics, funding, and businesses. It is the concept and theory of the "firm" that has eluded scholars for decades and was certainly not the philosophy behind "classical economics." Theory began to be built about the firm and the broader area of economic development (Clark and Demirag, 2002). Yet today, a theory of the firm has still eluded economists and social scientists (Teece et al., 1996).

Science parks show some structural resemblance to universities because they both have strong and functioning internal structures, departments and faculties as independent companies (Chomsky, 1957). However, most universities have antiquated hierarchical structures, departmentalism, and systems that are overseen by governing boards in both public and private colleges and universities (Fig. 1).

Science parks provide territorial umbrella functions for diverse groups of domestic or multinational subsidiaries that have little in common except their

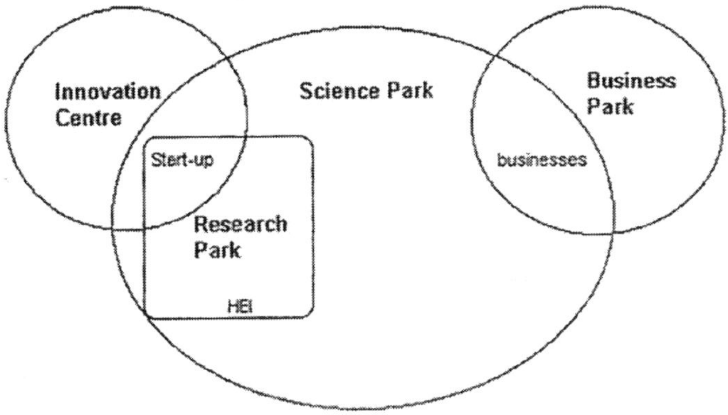

FIGURE 1 Relative positions and overlaps among science park, innovation center, business park, and research park. *Source: Tsagaris Consult.*

engagement in focus areas ranging from research to applied business activities in a common location. The flow of people, ideas, technology, capital, and industry connections between science parks and universities can be limited owing to location. In many cases, science parks have tended to locate geographically near or next to universities or cities, where land is available.

There are three key points of networks: first is the basic philosophical concepts (Chomsky, 1980). Networks must be shared resources within a triangular relationship among government, private industry, and support institutions such as education, training, and technical assistance centers. A second key point appears to be that there is an "institutional facilitator" such as DTI or network centers at many universities and in many regional communities, all of which set up the framework for the networks, provide "brokers," and continue technical assistance. A third is the collective contribution of the SMEs in terms of technology, expertise, marketing, sales, finance, or whatever is important for the creation of new firms.

A detailed analysis and description of the interactionism theory to economics has been given. Suffice it say here that the subjectivist perspective and interactionism provide a fuller picture by not making "externalities" something outside the boundaries of economic analyses, but that are instead an important part of understanding everyday business phenomena systems. The interactionism perspective yields new insights into how new companies are formed and are collaborating and competing. As these firms develop and move into these new markets, an entirely new approach to economics must be created. Business is no longer as it was. It is now "green development," global but also local for people within communities and cities.

Applying Interactionism to Regions are needed around the World: One is South America where Clark first did his anthropological field work in South America.

Central and South America

Samantha Bobo[b]
Department of Planning, Rice University, Houston, Texas

CENTRAL AMERICA

Mexico

The state of climate preservation in Mexico is a complex one. Mexico finds itself blessed with an abundance of crude oil. With the 19th largest proven reserves in the world, the country has been woefully underproducing against a rising demand for oil. This is not due to lack of resources but rather due in large part to the nationalization of the oil industry in 1938. The establishment of PEMEX as the sole player in Mexico's oil industry deprived the it of the competitive upstream investment needed to optimally develop the industry. In spite of the lag in resource development, oil retains the majority of market share in Mexico's primary energy demand. Natural gas has been steadily gaining market share since the 1990s especially as the primary feedstock for electricity generation. Low-cost and efficient, natural gas has allowed Mexico to diversify away from the expensive oil generation that dominated into the year 2000. While this diversification has allowed Mexico's energy mix to take a slightly less-carbon-emitting turn, fossil fuels continue to represent 90% of Mexico's energy mix and this does still include significant use of coal and crude oil.

In December 2013, Mexico unveiled constitutional amendments intended to execute full and comprehensive reforms to the nation's energy industry, though this did not come in the way we may expect. The main goals of the reforms, unlike many we've seen in recent history, was not necessarily expansion of renewable or clean energies. Instead, the reforms established new structures in the oil, natural gas, and electricity industries, primarily focusing on the reform and expansion of the oil and natural gas industries. Competition would be introduced into the refined product and electricity markets, and private investments would be allowed to flow into the upstream oil and gas sectors (Lajous, 2014). These threefold changes were to be executed while still maintaining the national ownership of the country's hydrocarbon resources and the nation's prohibition on concessions for development of said resources (Enriquez, 2014).

The Mexican energy reform drew upon global best practices and looked to emulate the policies put in place in the U.S and Norway. In addition, to opening up the electricity and petroleum industries to competition, secondary legislation was passed that accomplished several important pieces of the Reform. It outlined the roles of the government agencies in administering both

b. Samantha Bobo, Author, Master of Energy Economics, Worcester, Massachusetts, USA.

the Reform itself and the new energy sector that would subsequently emerge. Agencies such as the Ministry of Finance, the National Commission of Hydrocarbons (CNH), the Energy Regulatory Commission (CRE), and the Secretariat of Energy (SENER) were restructured to increase independence and transparency which in turn would help facilitate the competitive market goal of the 2013 Energy Reform. The legislation also included an overhaul of the tax code as it applied to the energy sector and put into place regulation to address both environmental and social issues stemming from the energy sector (Vietor and Sheldahl-Thomason, 2017).

While it seems the primary aim of the reforms was to advance the extractive industries such as oil and natural gas, the reforms also addressed the integration of renewable resources into Mexico's energy mix. Mexico itself is blessed with an abundance of sunshine, wind, geothermal, and hydro resources, and even prior to the reforms the Mexican government had been working to make full use of these resources. In 2008, Mexico established its first renewable energy standards, to have 35% of their electricity generated by renewables by 2024% and 50% by 2050 (Vietor and Sheldahl-Thomason, 2017). Soon after, in 2012, the Federal Government passed the General Climate Change Law. The Law outlines the steps and powers of the Federal, state, and municipal governments to take action to combat climate change and mitigate the already observed effects (Mexico, 2012).

The General Climate Change Law has seven major components:

1. Increase renewable energy use to meet the pre-established renewable energy targets by establishing a system of incentives for favor renewables
2. Economy-wide emissions reduction goals that look to reduce emissions to 30% below business-as-usual levels by 2020, and 50% below 2000 levels by 2050.
3. Establishing a public national climate change information system that requires mandatory emissions reporting and establishment of a public emissions registry of all emissions sources from energy generation to solid waste and industry.
4. Establishment of an emissions trading system that can include international transactions
5. Establish an inter-secretarial climate change commission ("CICC") composed of the heads of the various Federal government ministries to draft and execute national climate change policy
6. Establish a climate change fund to collect and channel resources to the climate change policies drafted by the CICC
7. Expanding the National Institute of Ecology's mandate to include a major focus on climate change (McCain, 2012).

While this legislation made Mexico the second country in the world to enact a comprehensive climate change law, implementation of the law has been lacking at best. While the law successfully established the climate change fund, in 2012 it was assigned only $78,000 USD for administrative operations and no other funds for mitigation projects. By 2014, even administration of the fund had lapsed. Additionally, no measures were included to ensure whatever funds were made available to combat climate change were effectively used. None of the measures to incentivize renewables, create hard and fast emissions reductions, or even establish cap-and-trade had been effectively carried out (Godoy, 2014).

The 2013 reforms included efforts to revive the policies included in the General Climate Law. It promoted and extended accelerated depreciation benefits for specific renewables as an incentive and established the Programa Nacional para el Aprovechamiento Sustentable de la Energía 2014–18 ("PRONASE"). PRONASE was made to be a roadmap for successful implementation of renewable energy and set forth industry best practices, policy guidelines, and specific checkpoints for the integration of renewables. The reforms also lead to the establishment of a carbon tax in 2014 that applied to all fossil fuels save for natural gas. This tax generated $540 million USD of revenue in its first year. Additionally, the reforms reaffirmed the clean energy target of 35% by 2024 and set forth short-term goals to achieve that level of integration.

> *In 2014, 22% of electricity came from non-fossil fuels. Energy generated from clean sources decreased by a further 3.7% in 2015, largely due to lower hydroelectric output. However, wind installation increased from 85 MW in 2006 to over 2,000 MW in 2014 … In 2015, approximately 50% of Mexico's energy was generated through combined-cycle, followed by 13% from conventional thermal, 11% from coal and oil, and 10% from hydroelectric. Around 20% (inclusive of hydroelectric) was generated from clean sources in total (Vietor and Sheldahl-Thomason, 2017).*

Current estimates show clean energy, in the form of hydroelectric, nuclear, and other renewables, making up 28.7% of Mexico's installed generating capacity (CIA, 2017a,b). In 2015, Mexico was generating 48 TWh of renewable power. A joint study done by the Mexican Ministry of Energy ("SENER") and the International Renewable Energy Agency ("IRENA") projects given Mexico's renewable resources that they could increase this 6-fold, to 280 TWh by 2030 (IRENA, 2015).

Still, the extractive hydrocarbon industries are the main driver of Mexico's energy mix and the country's economy. And these industries are having very large, very noticeable effects on both the environment in Mexico as well as the global climate. As of 2015, Mexico was ranked twelfth in the list of highest carbon-emitting countries, responsible for about 2% of the world's fossil fuel carbon emissions. And given the country's dependence on oil and gas, fossil fuel combustion and physical extraction have had observable climate effects that have directly impacted Mexico and its people.

Emissions especially are problematic considering the country's geography. Mexico City, home to about a quarter of the population, is located in a valley (CIA, 2017a,b). The surrounding mountains effectively trap air pollution generated within the city or blown in from neighboring manufacturing centers, such as Toluca and Pachuca. In 2013, a study released by the Mexico Competitiveness Institute found about 1700 of the deaths in Mexico City in any given year were likely due to air-pollution. They also found that, in 1-year, the degraded air-quality resulted in around 4200 hospitalizations, 234,000 doctor's appointments, and cost the Mexican economy $850 million USD (Grillo, 2015).

Since the early 2000s there has been viable record of climate changes in Mexico. These observations include extreme temperature fluctuations, extreme precipitation events including droughts, and super-storms. The World Meteorological Organization ("WMO") keeps detailed records of these extreme events and their occurrences. Between 2003 and 2007, there was a single extreme precipitation event per year, however, by 2010 Mexico was seeing three separate extreme events per year, some lasting multiple months at a time (Huhne and Slingo, 2011).

While 2010 was the rainiest year on record, 2011–12 turned around to be the most intense drought ever recorded. Increases in average temperature, some 0.6 C over the last 4 decades, as well as extreme rainfall and drought will have significant and lasting implications. The rise in temperature will diminish the water quality, and increased and sustained droughts will affect replenishment of reservoirs, rivers, and aquifers. Low river run-off will decrease electricity grid reliability as hydroelectric generation will be compromised. A lack of water and poor water quality directly affects agriculture, decreasing food production for use domestically and for export. Extreme rain events cause erosion and flooding that can contaminate agricultural areas, destroy homes and businesses, and allow for spread of disease (OECD, 2013). The effects are being seen directly in Mexico City.

The increase in heat and drought has increased demand for water resulting in depletion of the underground aquafers. This depletion in conjunction with the drought has caused the ground to become brittle and is causing the city to collapse in on itself at a rate 5–9 inches per year. Already 15 elementary schools have been shut down due to cave-ins and in the densely populated neighborhood of Iztapalapa, a teenage boy was swallowed up by a sinkhole that suddenly appeared in the street. Arnoldo Kramer, Mexico City's chief resilience officer states, "Climate change has become the biggest long-term threat to this city's future. And that's because it is linked to water, health, air pollution, traffic disruption from floods, housing vulnerability to landslides — which means we can't begin to address any of the city's real problems without facing the climate issue." (Kimmelman, 2017).

In September of 2017, Mexico was ravaged by two massive and deadly earthquakes, one originating from the center of the country near Mexico City, and the other originating from offshore. While these have not been to the country's oil and gas operations which exist both on and offshore, scientists all over the globe are becoming increasingly more confident in the link between oil and gas extraction and increased seismic activity. Human-caused earthquakes, also known as induced seismicity, have been increasingly linked to mining and drilling activities all over the world but not in the way that most people think. While many believe the activity of hydraulic fracturing, or "fracking" is what causes the earthquakes, in reality the wastewater disposal wells are the ones that result in the most induced seismicity. These wastewater wells exist at both fracking cites and oil wells and have been known to induce seismicity at distances of 10 miles or more from the injection point, though it should be noted that not all wastewater wells cause induced seismicity (USGS, 2017). Additionally, USGS Scientist Ross Stein believes that climate change may also be resulting in induced seismicity. Dr. Stein notes that in many areas where climate change is directly decreasing water supply, as in Mexico, people rush to dam rivers and dig reservoirs to ensure a more stable water supply into the future. Many of these reservoirs are being built on fault lines.

The water within the reservoir lubricates the faults, and filling and draining of the reservoir causes significant pressure changes and can result in cracking and instability around the fault itself, creating a higher probability for sudden shifts and therefore earthquakes. This danger is 2-fold, the earthquake itself and damage to the dam that could cause catastrophic flooding in nearby low-lying areas. While it is not suspected that the September 2017 earthquakes were induced by human action, Mexico's proximity to faults, active oil and gas sector, and already present water supply problems indicated the country is at risk of more frequent and potentially catastrophic seismic events (Hugo, 2017).

Mexico is facing significant social, environmental, and economic hardship due to the effects of climate change, and while some attempts have been made to mitigate these through passage of legislation, the Federal Government realized more was needed. In their Mid-Century Climate Change Strategy, the Government revealed its long-term vision for Mexico. The Government set forth milestones to be met at the 10-year, 20-year, and 40-year intervals to encourage sustainable development and low-emissions through efficient management of natural resources. These include environment and ecosystem conservation, public involvement in climate policy, meeting the renewable energy goal of 35% by 2024, management and reduction of emissions across all sectors of the economy such that emissions are at 50% of 2000 levels by 2056.

To accomplish this, the Government has outlined both adaptation and mitigation areas and actions to be taken. These are addressing those areas and peoples most vulnerable to climate change, reducing the vulnerability and building a resilient society, improving the resiliency of infrastructure and productive systems to climate change, encouraging ecosystem-based adaptation that increases ecological diversity, and efficient land-use planning, creating an energy-efficient energy sector that is integrated with clean technologies and promotes sustainable development, and creating an energy supply that allows for universal access to high-quality and sufficient energy at competitive prices[1] (SEMARNAT-INECC, 2016). Additionally, in the same year that the Climate Change Strategy was released, Mexico ratified the Paris Climate Accords and pledged to reduce its emissions by 25% by 2030. In addition, to their renewable energy targets, they're pledging zero deforestation by 2030, use of at least 1 million natural gas and cleaner fuel cars by the end of 2018, and creation of a functional and efficient intermodal transportation network.

At a more local level, the country's capital city has been making strides to improve sustainability. Mexico City adopted its "Green Plan" in 2007 and by mid-2009 three-quarters of the action points had begun implementation and 7% of them had been completed. The plan, supported by both the World Bank and the United Nations, succeeded in conserving 13,600 ha of sensitive environmental areas and established a reforestation program. The reforestation program itself calls for the planting of 2.5 million plants per year. Adoption of the city's climate change action plan in 2008 came with the express goal of cutting CO2 emissions by 12% by 2012 and the city expects further reductions in emissions to be achieved through their "Sustainable Light" program which plans to replace incandescent bulbs in more than 11 million Mexican homes with more energy efficient alternatives such as LEDs. Mexico City has also been the center of the country's transportation system overhaul. In 2005 the city inaugurated its first rapid transit bus system, and a second line was added to they system in 2007.

The city's climate change action plan looks to continue to upgrade and expand the public transportation system and promote alternatives to cars such as a public rental bike program and compulsory transportation of children too and from schools. To reduce water consumption, Mexico City has increased its water tariffs, with hopes of decreasing consumption by 20%. To reduce waste, the city provides incentives to companies that use biodegradable packaging, and has placed a ban on free plastic bags in shops with hefty fines for those retailers that do not comply (EIU, 2010).

In spite of these efforts, Mexico is still in need of a clear plan for how to accomplish all its climate adaptation and mitigation goals. Clear and specific programs must be created to address integral pieces of the climate puzzle such as energy efficiency and fuel standards. And these programs must include feasible implementation strategies. Especially hard will be the integration of

1. For more information and the individual action steps please refer to: SEMARNAT-INECC. 2016. Mexico's Climate Change Mid-Century Strategy. Ministry of Environment and Natural Resources (SEMARNAT) and National Institute of Ecology and Climate Change (INECC), Mexico City, Mexico. (Pages 49–55 & 85–94).

renewable technologies. In transitioning away from the fossil fuels that have been the cornerstone of the country's economy for centuries, greater focus needs to be put on the transmission grid and upgrades to it so it can handle the additional renewable generation both at a utility-scale and behind-the-meter. Mexico has long endeavored to be at the forefront of climate mitigation, and while historically its measures have not been fruitful, perhaps with the aid of the international community and its pledges to the Paris Accords it will finally emerge as a global leader and avoid escalation of the already observed effects of climate change (NRDC, 2017).

Costa Rica

Like Mexico, the Central American country of Costa Rica has also felt the long-ranging effects of global climate change. However, Costa Rica has a long history of environmental preservation and climate preservation. Successes in renewables integration, environmental regulation, and climate change mitigation have given Costa Rica the reputation as one of the most sustainable countries in the world and one who's efforts should be emulated for both the good of the planet and the quality of life for the people.

In November of 2017, Costa Rica made global news as the first country to clock 300 days where all electricity generation was from renewable sources. This is no small feat considering the need to provide significant, reliable, electricity to its population of about 5 million people. According to the National Centre for Energy Control in Costa Rica, 99.6% of their electricity production is generated from hydropower, wind, geothermal, biomass, and solar (Fig. 2).

While the country must revert to thermal power on occasion when renewable sources are not available to meet load, Costa Rica was able to carry on 200 consecutive days of renewables-only generation. Costa Rica's 300 days

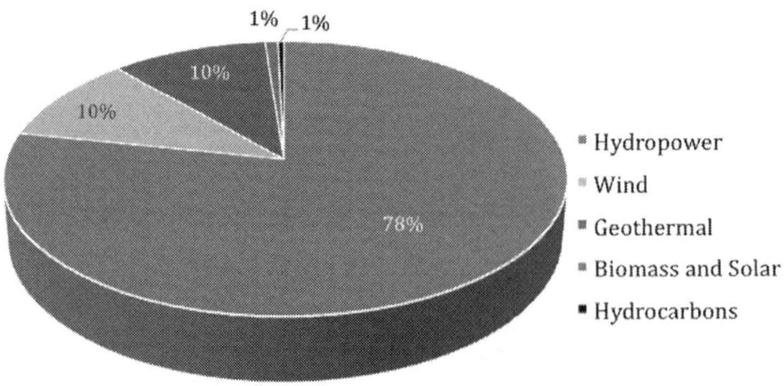

FIGURE 2 Costa Rica's energy mix as of November, 2017.

of renewable energy not only eclipsed the 250 days achieved in 2016, but also beat the previous record set in 2015 of 299 days powered by 100% renewable energy. Costa Rica, a country of lush, rich, and fragile ecosystems, has been working tirelessly to be a pioneer in climate change mitigation. Their achievements in renewable energy integration are a remarkable step forward in these efforts, and a leap forward in debunking the myth that a country must choose between social and economic development and environmental protection (Dockrill, 2017).

Costa Rica, in addition, to its commitment to renewable energy, has had a rich history of environmental conservation. Historically, it is believed that the native peoples in Costa Rica practiced significant environmental conservation techniques and that it was not until the arrival of the Spaniards in the 1500s that environmental degradation became a significant issue. The native peoples of Costa Rica limited agricultural development to 4 of the 12 "life zones," distinct local ecosystems, and only cultivated native crops. These crops, coupled with hunting and gathering allowed the native peoples to exploit the natural environment while also conserving the environment's natural ability to replenish said resources. They employed many resources management techniques such as controlled burning to replenish the fertility of the soil and seeding plants for permanent cover in areas prone to erosion (Evans, 1997).

In contrast, the arrival of the Spanish in Costa Rica saw both exploitation of native resources and introduction of European commodities such as sugar cane and livestock. The colonizers believed the native practices of agriculture less efficient than those practiced in Europe and began the slow, but continuous, process of deforestation and enslavement or displacement of the indigenous populations. Even so, the degradation was contained somewhat due to the land needs of livestock for grazing, as well as other European crops such as tobacco and cacao which were only productive in one area of Costa Rica.

Additionally, these agricultural products were being produced for local colonies and some trading in both local and regional markets. The small scale of the markets helped mitigate significant impacts from "Europeanization" of agriculture at that time. This remained the case until 1830, after Costa Rican independence, when it was found that coffee would thrive in several of Costa Rica's climactic zones. As demand increased, more and more of the previously undisturbed forests in Costa Rica were converted into coffee fields. In 1845, Costa Rica began exporting its coffee to Great Britain, and the government began providing tax breaks to local growers to incentivize coffee production. For more than 40 years following, coffee was virtually Costa Rica's only export (Evans, 1997).

Even so, Costa Rica saw conservation awareness emerging in its early years after independence in 1828. In 1833, 1846 the government passed decrees regarding forest preservation and forest preservation near cities. A decree mandating protection of watershed areas in mountains was passed in 1888, the early twentieth century saw public outcry for a national forestry code, and hunting laws passed to help conserve wildlife were passed in 1853. These legislative measures in addition, to Costa Rica's low population, less than one person per square mile in the years following independence, helped contain environmental degradation until the banana industry took off in the later 1800s and early 1900s. Unlike coffee, bananas grew in the humid areas of Costa Rica, could be harvested year-round, and were less susceptible to yield variations.

Additionally, a robust market in the United States helped bananas become the dominant agricultural product of the Costa Rican Caribbean Low-lands. Banana plantations, unlike coffee, required significant capital expenditures, specifically on labor and transportation infrastructure. The large-scale banana industry required a great deal of land area which deforested massive areas in the state. There was the clearing of areas for the plantations themselves, but also construction of railways for transport of both labor and the bananas, and cattle ranches to provide food for the growing number of plantation workers. Estimates conclude that the banana industry accounted for as much as 20% of Costa Rica's annual deforestation from 1870 through the 1980s (Evans, 1997).

Costa Rica's concerted commitment to sustainability began with biodiversity conservation when the country gained independence from Spain. Local governments were tasked with fighting for conservation and reforestation of mountains and plantations. In 1945, Costa Rica began its concerted efforts for nature conservancy by subdividing the country into different areas and different categories of environmental protection, which laid the groundwork for the country's current systems of environmental protection (Tierra Verde, 2017).

> *Currently there are 11 areas of conservation existing in Costa Rica from which 126 state-owned areas are subdivided into different categories and therefore also operated in different ways. A network of private reserves with more than 70 members are supporting nature conservation on a private initiative. For instance, alternate mechanism and management of the resources are being established through biological corridors, conservation and-drainage areas (Tierra Verde, 2017).*

Indeed, over 25% of the country is now designated as some type of protected area, including national parks, indigenous reserves, biological reserves, national forests, national monuments, and national wildlife refuges. Passed in 1969, the Forestry Law outlines the differences in the various protected areas, all of which are managed by the National System of Conservation Areas which is a division of the Ministry of Environment and Energy ("MINAE") and are divided into three different management types. Type I is "strict" protection which the objective of preserving species and reducing human intervention within the ecosystem processes. Type I areas include national parks, biological reserves, national monuments, natural reserves and wildlife refuges, as well as archaeological and historically significant sites. Type II is "partially protected" which includes forest reserves and protected zones which are open to exploitation under specific conditions. These are managed to allow varying degrees of use development including tourism and logging. Type III is indigenous reserves; "the conservation of cultures and their environments and the protection of life systems in these communities and the way natural resources are used." In spite of the ravages of colonization, coffee, and bananas, Costa Rica conservation initiatives resulted in a decrease in annual deforestation rate to 0.73% by 2000 and conservation of 1511 known animal species and 12,119 known plant species, many of which are endemic and are classified as either threatened or protected under the International Union for Conservation of Nature (Butler, 2006).

Even so, Costa Rica, with their 12 distinct climatic zones, is still susceptible to the effects of climate change. With the prevalence of high temperatures and drought seen in central America, the tropical rainforests of Costa Rica are at significant risk. Tropical rainforests in Costa Rica and elsewhere play a massive role in filtering out CO2 from the atmosphere. If climate conditions cause a decrease in these rainforests, the negative feedback loop created has implications on a global scale (Lyra et al., 2016). Additionally, sea level rise poses a significant threat to the coastal communities within Costa Rica. A rise of 3–6 feet could flood coastal communities including the port city of Puntarenas.

The seas have been rising at the rate of 0.13 of an inch since 1990, which is twice as fast as in the 1800s. While the current rate indicates that Costa Rica would not feel the effects within this century, as the Earth's temperature continues to rise, the more quickly the ice caps melt and the sooner Costa Rica and other coastal communities will begin to see these effects. Additionally, many of Costa Rica's endemic species are significantly vulnerable to climate change. Amphibians are a good indicator for the health of the environment because they are very sensitive to changes in temperature and pollution. Costa Rica is home to many endemic amphibians, specifically frogs. Several species of Costa Rican frogs have seen population reductions due to a fungal outbreak that scientists believe is attributed to global climate change. The golden toad, for example, has all but become extinct due to the outbreak, it has not been seen in Costa Rica in several years (Costa Rica Learn, 2017).

Costa Rica's national climate strategy was first put in place in 2007 with the overarching goal of carbon neutrality by 2021. Put forth by the Climate Change Department and MINAE, the goal of carbon neutrality was set to be achieved through sustainable development with low carbon emissions, climate change adaptation, enhanced economic competitiveness, and international leadership. Within this policy, the government reiterated its commitment to environmental protection and established its commitment to renewable energies. Mitigation measure within the policy are designed to "(1) reduce the sources of greenhouse gases emissions, (2) capture CO2 through forestation and (3) develop an effective carbon trading market (NAMA, 2017)," while also increasing corporate responsibility and creating products that are less carbon intensive and more environmentally friendly. According to MINAE the direct impacts of the policy are fourfold:

1. "[It] promotes a reduced impact of the individual over the environment through economic retribution.
2. Promotes shared responsibilities
3. Reduces greenhouse gas emissions into the atmosphere through financing carbon fixation forestry programs
4. Recognizes the importance of preserving the biodiversity and scenic beauty of diverse forest ecosystems (Mauri, 2009)."

In addition to the national climate change policy and carbon neutrality plans, there are additional policies that have aided Costa Rica's sustainability and climate mitigation agendas. The State Policy for the agricultural sector and rural development works to incorporate responsible environmental management into the country's second highest Nitrous Oxide and Methane emitting sector. The Environmental Services Payment System establishes Costa Rica's national carbon trading market and payment program for carbon storage, hydrological services and protection of biodiversity and landscapes. This program has been credited with reducing the deforestation rate within Costa Rica from highest in the world to net negative in the early 2000s.

Most recently, Costa Rica submitted its 'Intended Nationally Determined Contribution ("INDC")' to the 2015 Paris Climate Agreement. The INDC established Costa Rica's commitment to climate action until the year 2030 to avoid global temperatures increasing past 2°C. According to its INDC, Costa Rica has committed to "a maximum of 9,374,000 tCO2 eq net emissions by 2030, with proposed emissions per capita of 1.73 net tons by 2030, 1.19 net tons per capita by 2050 and -0.27 net tons per capita by 2100," as well as "an emissions reduction of GHG of 44%, of a Business As Usual (BAU) scenario, and a reduction of 25% of emission compared to 2012 emissions. To accomplish this goal Costa Rica would have to reduce 170,500 tons of GHG per year until the year 2030 (NAMA, 2017)." While these reductions seem lofty, given Costa Rica's long history of commitment to environmental preservation and combatting climate change it may not be out of the question for them to meet these commitments. Costa Rica has already established themselves as a leader and smashed the notion that climate change mitigation and environmental preservation can only be accomplished by developed wealthy countries. Their efforts should be used as a road map for climate preservation in developing countries across the globe.

SOUTH AMERICA

Peru

Peru, located on the Pacific coast of South America, is a country of drastic and varied landscapes. To the west there are coastal plains and to the east is the dense and lush Amazon rainforest. These two distinct ecosystems are separated by the Andes mountains, whose glaciers are of significant interest when it comes to climate change science (CIA, 2017a,b). Peru is home to about 70% of the world's tropical glaciers, small ice caps that form at high altitudes near the equator. The glaciers, however, have been disappearing due to higher than average temperatures, and the effects of the disappearing glaciers puts many citizens at risk.

In August of 2017, a dump-truck sized piece of ice broke free from the glacier on Mount Pucaranra, plunging into Lake Palcacocha and creating a wave nine-feet high that damaged the lake's emergency drainage system; an earthen dam. While this was a relatively small avalanche, scientists believe that a large one would have the same effect as "dropping a bowling ball in a bathtub," creating a 100-foot wave that would blow out the dam and send billions of gallons of ice and water roaring down the mountainside and burying the city of Huaraz with its 200,000 residents (Miroff, 2017). These incidents of calving glaciers have been occurring more recently today than in the past and climate scientists are studying the Peruvian glaciers keenly to gain insight into human adaptation to climate change (Miroff, 2017).

While Peru's high-altitude glaciers pale in comparison to the planet's polar ice caps, they are a source of fresh water, food, and hydroelectricity for millions of people. Peru's glaciers used to span across 19 mountain ranges according to the Institute for Glacier Research in Peru, one of those glaciers has already disappeared and the others are shrinking quickly causing not only an environmental issue, but an engineering challenge as well. Glacial runoff provides the country water for people, livestock and agriculture. As the runoff becomes a less reliable source of water, Peru will be forced to retool its irrigation systems, build new dams and reservoirs to effectively store water, investments in agriculture, as well as the other water-intensive industries, will need to be recalculated and flood control projects become a necessity (Miroff, 2017).

After the natural dam holding back the waters of Lake Palcacocha collapsed in 1941, the Peruvian authorities constructed an emergency spillway. As of today, the Lake has increased to 34 times its original size due to accelerated glacial melt, necessitating further flood mitigation measures and the installation of a series of pipes to siphon water out and control the water level. While installations such as these offer some assurance of flood preparedness and climate change adaptation, extreme rainfall in March of 2017 unleashed flood waters and mudslides that killed 67 people, destroyed 115,000 homes and more than 100 bridges on the coast and in the capital city of Lima. Meteorologists in Peru believe the extreme weather events, both rainfall and drought, are due to atypical warming of local sea surface temperatures by as much as 6°C from their normal, which they believe is attributed to global climate change. Government authorities, however, are wary to attribute assert increased instances of floods and more destructive storms due to global climate change (Collyns and Watts, 2017).

This is not to say that Peruvian officials are unwilling to acknowledge global climate change and the challenges it presents for the people and ecosystems within the country. Peru established it Estrategia Nacional de Cambrio Climatico ("ENCC") in 2009. The ENCC was meant to provide guidance and information on reversing the impacts of climate change at the national, sectoral, and regional development levels. Specifically, the ENCC focuses on deforestation of the Peruvian Amazon and its subsequent effects on carbon sequestration and global climate change as well as changing agriculture to protect the nearby vulnerable ecosystems (REDD, 2018).

In 2010, following the adoption of the national climate strategy, Peru began a concerted push to fight back against climate change in two areas: (1) the country's energy mix, and (2) forestry and land use. They established the country's first renewable energy production target, announcing that 33% of the electricity used within the country would come from renewable energy sources, and they would aim for the country's energy matrix to be 40% renewable by 2021. These targets were specifically to be met by non-conventional renewables such as solar and geothermal, which are in abundance in Peru, as well as hydropower. Additionally, renewable energy auctions were established to create competition and keep electricity prices reasonable for the citizens. Deforestation and poor land use management have plagued Peru for centuries, but in 2010 the Federal Government took action by actively participating in the United Nation's Reducing Emissions from Deforestation and Degradation ("REDD+") program.

The program brings together international organizations, countries, and investors to help developing countries with their forestry and land use by way of administrative efforts, teaching, and monetary contributions. "For example, Norway recently pledged to pay up to $300 million to support emission reduction projects that but deforestation in the Amazon region. Under the agreement, which runs through 2021, Norway would make payments for verified emission reductions that start in 2017 (NRDC, 2014)." A number of the REDD + pilot projects are already being implemented. The Parque Nacional Cordillera Azul has been upgraded including more rigorous protections and stakeholder outreach so it can more successfully conserve the 1.3 million hectares of forest within. In 2013 the park reported a net zero percent deforestation rate, and successful conservation over the 10-year pilot period is estimated to mitigate more than 18 million mt of CO_2 (NRDC, 2014).

Additional efforts to help climate preservation and sustainability have been made at a local level as well. From 2010 until today, the capital city of Lima has been working to implement green initiatives in all sectors including energy use, land use/buildings, transportation, waste and sanitation, air quality, and environmental governance. The projects undertaken in the city include wide-sweeping projects as large as public works projects that have installed parks and expanded green space in the city by 60% and ones as small as replacing all the lightbulbs in the traffic lights with LEDs. Lima, like the rest of Peru itself, is far from a totally sustainable, carbon neutral city, but these initiatives have helped drastically.

Waste management, which was a significant problem in the city, is now being handled by more than 200 organizations that collect solid waste, and over 400 that sort, recycle, and resell to the public and central waste collection points have been established to make it easier for citizens to take care of their waste. Additionally, at one city's numerous landfills called Huaycoloro a waste-to-energy project has been in the works. The city has also begun making a shift in its energy mix too, switching to natural gas as a primary fuel for electricity generation as well as transportation. Massive improvements have been made to water treatment and accessibility through a program called "Water for All," and the government of Lima actively involved citizens, NGOs, and other stakeholders in projects with any environmental impact (EIU, 2010).

In addition to an established national climate policy and efforts by local governments, Peru was one of the first 20 countries to pledge emissions decreases as part of the Paris Climate Accords of 2015. The Government of Peru has pledged an unconditional reduction of emissions by 20% below Business-As-Usual levels by 2030, with a 30% below Business-As-Usual levels conditional upon availability of international financing. The 20% unconditional reduction would require an emissions reduction of 77% across all sectors of the Peruvian economy (CATP, 2018).

Under the policies of the administration in power when the Paris Agreement was ratified, GHG emissions were expected to be between 149% and 169% above 1990 levels by 2030. However, the new government which assumed power after the ratification has reaffirmed their commitment to the Paris Agreement and is in the process of developing a framework law on climate change. Additionally, Peru has adopted several policies and planned a handful of projects that, in conjunction with the ENCC, are aimed to lower its GHG emissions. These policies include promotion of renewable energy with mandated targets for percentages of load met by renewable energies as well as passage of an energy efficiency law aimed at increasing efficiency in both the public and private sectors of the Peruvian economy.

Other policies look to improve energy accessibility in rural areas as well as establish a market for biofuels as a way to diversify Peru's fuel mix. As an extremely high emitter of GHG, the Federal Government has worked to improve the transportation sector within Peru by working to establish a public transportation network in the capital city of Lima as well as facilitate the conversion to natural-gas-powered vehicles. The adoption of a Sustainable Construction Code in 2015 as well as energy efficiency labels being fixed to household appliances have aided in new, more sustainable construction (CATP, 2018).

Another area of policy the Peruvian Government has looked to address with less success is forestry. Peru is home to 750,000 square kilometers of Amazon rainforest, the second largest area of Amazon located in one country, and the health of the Amazon has serious implications in global climate change due to its ability for carbon sequestration. However, the Amazon has been deforested and degraded. In 2012, the Amazon had already shrunk by 20% and Peru has made a sizeable contribution to the destruction. The destruction of the rainforest in Peru is largely due to the expansion of the palm oil business and other agriculture, however, there is a fair amount of illegal logging and mining that occur there as well.

The Government has tried to put in place several policies to help combat the deforestation, including a law and executable decree for compensation for services to ecosystems and policies to increase sustainable forest management and the establishment of protected areas. Unfortunately, under the current policies, emissions from deforestation are still projected to increase by about 80% from 2012 levels by 2030, and any emissions reductions due to these forestry policies is unknown (CATP, 2018).

While Peru seems to be heading in the right direction with their climate policies, they do have some serious work to do in order to meet their pledged goals. Climate Action Tracker is an organization that ranks countries based on their climate policies, projects, and agendas with regard to the Paris Agreement as well as projects outward to 2030 the effect of policies on GHG emissions. Climate Action Tracker believes that Peru's current policies are insufficient to meet both their overall climate strategy goals as well as their commitment to the Paris Agreement by 2030.

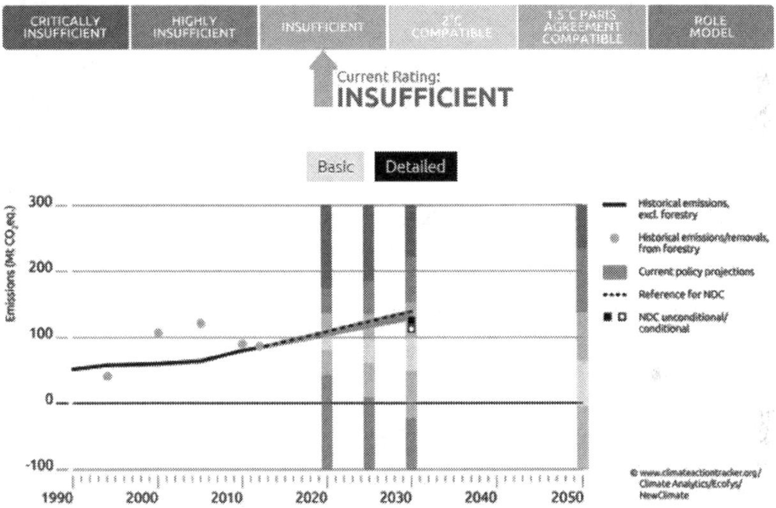

Land use, land use change and forestry represent half of Peru's emissions and these are set to double by 2030 under the current policies and "Peru will need to implement additional policies to reach its proposed targets. The "Insufficient" rating means that Peru's climate commitment is at the least stringent end of what would be a fair share of global effort, and is not consistent with the Paris Agreement's 1.5°C limit, unless other countries make much deeper reductions and comparably greater effort (CATP, 2018)." As it stands right now, Peru is a prime example of a country that is living the common struggle of a developing country hard-pressed to find balance between economic and social development and climate and environmental preservation. The effects of global climate change are clear and present within Peru and this seems to have motivated national and local governments as well as the citizens whose livelihoods are directly affected by the changes. Peru is making great strides in the right direction in an attempt to do its part in combating climate change, only time will tell if their efforts will be fruitful in making Peru a sustainable country.

Brazil

The largest country of South America is not isolated from the effects of climate change that have so affected its neighbors. Similarly, to the issues seen in Costa Rica and Peru, Brazil's bustling agricultural sector has been hit hardest by the effects of climate change as far back as the early 2000s. In 2004, Hurricane Katrina seriously damaged agricultural production in the rural south of the country, while in 2005 a drought in the Amazon region left the country with the lowest rainfall totals seen in half a century while in the northern plains that same year temperatures soared to record highs, maxing out near 113 degrees Fahrenheit. These extreme conditions had significant impacts on the Brazilian economy. Soy production decreased by more than 10% in 2005 due to the hot and dry stretches. Soy, a valuable export, is also a valuable domestic commodity used for livestock feed and bio-fuels. Other crops have suffered as well. Coffee production has also begun to decline due to drought and high temperatures. Brazil is the world's leading producer of coffee, and in its "coffee belt," the commodity has been grown and exported since the 1950s. Production, however, is slowing. Production of Robusta, a coffee strain used to produce espresso and instant coffee, was down 30% in 2016 due to the "coffee belt" experiencing extreme drought, the most recent of which has lasted over 3 years (Garcia-Novarro, 2016).

Following the storms and droughts of the early 2000s and the closing of the UN Climate Change Conference in Copenhagen in 2008, President Luiz Inacio Lula da Silva signed Brazil's National Climate Change Strategy into law (Robinson, 2010). The goal of the Plano Nacional Sobre Mudanca Do Clima ("PNMC") was to create a plan for addressing the country's current and futures GHG emissions, improve the country's ability to adapt to climate change, and address potential climate change mitigation strategies (Gov. of Brazil, 2008).

The strategy gives an explanation of 32 emissions-reducing programmes that were already in place in Brazil, as well as those anticipated to begin in the near future. Some initiatives, such as expansion of hydroelectric generation and the National Ethanol Program were quite successful. However, there were some significant issued inherent in the PNMC and its emissions-reducing programs. Many of the initiatives were only recommended rather than mandatory, had no specific checkpoints for progress, and lacked general implementation measures or even specific authorities in charge of implementation (Gov. of Brazil, 2008) (Robinson, 2010).

The same law that establishes the PNMC, however, does help to fill in some of the gaps. Brazilian Law 12.187 also includes voluntary emissions reductions of between 36.1% and 38.9% by 2020, a lofty goal considering that Brazil at the time was the worlds fifth largest GHG emitter. The law also provides estimates for the emissions reductions by sector for this goal to be met, and allows for further reduction targets to be specified in the future. Additionally, the law requires climate mitigation actions to be quantifiable and verifiable such that upon review by international officials it may be determined whether or not emissions reductions have taken place (Robinson, 2010).

The law, however, was accompanied by an interesting veto by President Lula. The President rejected the section of the policy that pushed for "gradual abandonment" of the use of fossil fuels. This veto was due in part to the fact that Brazil's energy mix was in transition towards a more fossil-fuel-based supply and that vast offshore oil reserves had been found in the deep water off Brazil's coast. While other legislation was passed allocating oil revenues to funds for the alleviation of poverty and environmental sustainability, this tension between emissions reductions and the available fuel supply remains a challenge for climate policy within Brazil today (Robinson, 2010).

Brazil is still a developing country, and must work to balance economic and social development with the care and conservation of both the local environment and the global climate. About 4% of the Brazilian population is below the extreme poverty line, and 3.7% more are below the poverty line (CIA, 2018). While this is a small percentage of the population, alleviating such poverty without increased GHG emissions or environmental degradation still proves challenging. Weakening enforcement coupled with an economic recession that has forced many in low-income rural areas to resort to illegal deforestation as a source of income has made it near impossible for the country to maintain its climate-conscious land use, land-use change, and forestry ("LULUCF") policies. While the country has pledged zero deforestation by 2030, the recent reversals of the progress made following the implementation of the national climate strategy are reflective of a lower ambition to address climate change (CATP, 2018).

Additional changes need to be made to the country's energy supply sector if Brazil is to come anywhere close to meeting its Paris pledge. Historically, Brazil's generation mix has been characterized by a high share of renewables, especially hydropower. However, rapid economic development and expansion of the transportation sector have resulted in significantly increased demand for both fossil fuels as fuel but also as additional electricity generation. Current government targets wish to have 45% of Brazil's electricity generated by renewable energies by 2030, however, recent policy developments seem to be at odds with the plan to decarbonize the power sector.

he market share of electricity load met by renewables has declined from 50% in 1990 to about 39% by 2014, due in part to both increased energy demand and the decreasing reliability of hydropower due to the extreme droughts the country has been facing (CATP, 2018). However, the current government energy infrastructure plan looks to diversify the energy mix to lower dependency on hydropower by increasing investments in coal, oil, and natural gas generation. The plan looks to increase fossil fuel investments to 70.6% of the total energy investments in 2024 (CATP, 2018). Additionally, the proposed bill that looks to privatize the power sector would also allow result in subsidies for the modernization of the existing coal-fire generation as well as the construction of new coal generation facilities. The government went further by cancelling the only reserve energy auction for solar and wind in 2016, sending negative signals to solar and wind developers looking to invest in the country. All of these developments favoring fossil fuels as the energy of choice for Brazil limits the country's ability to achieve their emissions reduction commitments as well as long-term decarbonization of the Brazilian economy as a whole (CATP, 2018).

At a more local level, the city of Sao Paulo has made significant strides to reduce its climate impact and increase its sustainable development. By 2012, the city generated all of its electricity using hydropower; generating zero CO2 emissions from the electricity sector. With adoption of a city-wide climate policy in 2009, Sao Paulo committed to reduce emissions of all GHGs by 30% of 2005 levels by 2012 and established a climate change committee to monitor progress towards that goal. The city was able to make significant strides to meet their reduction commitment using waste-to-energy conversion plants as well as closing the cities two main landfills, and constructing thermoelectric plants to burn the methane generated by the waste on-site. Burning the methane supplies 350,000 MWh of electricity annually while cutting CO2 emissions by 11 million tonnes through 2012 (EIU, 2010). Additional green initiatives put in place under the city's climate change policy include robust land-use and building construction policies.

Local law requires all new buildings to include solar-powered hot-water heating while other energy-efficiency standards were put in place in 2009 setting standards for new buildings as well as outlining the process of retrofitting older buildings to be more energy efficient. The city itself has 55 square meters of green spaces per person, which is a significant accomplishment for a city of around 12 million people. Adoption of Agenda 2012 in 2009 set forth 223 short and medium-term goals for the city, including construction of 50 new neighborhood parks and planting 800,000 trees. With regard to transportation, city ordinance prohibits a large percentage of cars from entering the city during peak times to help cut down on traffic congestion and therefore emissions. Another of the city's initiatives has helped to cut down on illegal dumping of waste in the city's streets, and the city provides centralized collection points where citizens may dispose of the items, such as furniture and yard waste, that do not fit in their residential recycling bins (EIU, 2010).

In 2016, Sao Paulo was ranked 79th on the Arcadis Sustainable Cities Index, the first highest ranked South American city, followed immediately by Buenos Aires. While the city struggles with expansion and the subsequent congestion, Sao Paulo's Vision 2030 looks to drastically overhaul the city right down to the infrastructure. The vision "includes housing programs, geographically dispersed businesses, improved mobility, an environmental agenda, and initiatives that generate cultural and economic vibrancy." However, the city faces the significant issues to the sustainable development of the city through their 2030 vision of attracting investment and adoption of new technology.

Brazil ratified the Paris Agreement on September 21, 2016 pledging a reduction in carbon emissions to 1.3 gigatonnes by 2025 and 1.2 gigatonnes by 2030; 37% and 43% below 2005 emissions levels respectively. However, in order to meet these targets, Brazil will have to implement additional policies, as projections under the current regulatory regime have emissions continuing to rise in most sectors until at least 2030. Weakening climate change policies have already begun to undermine the strides made by implementation of the national climate change strategy (CATP, 2018). Forestry emissions mitigation, which had declined rapidly following the strategy's implementation, has begun to trend back upwards. Deforestation has increased over the last decade and the government states a lack of resources to address the illegal logging trade that plagues the entire national territory. Spending cuts have undermined the enforcement capacity of forestry authorities and in 2017 the Federal Government cut the Environment Ministry's budget by over 50%. It would seem that more progress in the right direction is being made on the local front and that if other communities within Brazil follow the lead set by Sao Paulo, the country may take steps towards climate preservation, however, the setbacks of the current national administration will prove a significant challenge for the country as it is pulled in two opposing directions with regard to global climate change (CATP, 2018).

Argentina

The most notable impacts of climate change in Argentina have been in the southern-most areas near Patagonia where glaciers and ice are most prevalent. In Patagonia, the average temperature has risen 1° over the past 50 years, twice as much as the average increase for the rest of the country (Koop, 2014). Retreat of the glaciers has been ongoing and some, like the Upsala Glacier, have been shrinking since the 1800s. The Upsala Glacier's retreat has sped up in the last century due to both rising temperatures of both air and seawater. Over the period of 1997–2003 the front of the glacier is purported to have retreated some 13.4 square kilometers (Clarin, 2004). The drastic temperature changes have also wreaked havoc on animal species in Argentina as well.

The Magellanic Penguin is one species that has been hit particularly hard by climate change. The threat is twofold, drenching rainstorms drown chicks who do not yet have waterproof feathers, while extreme heat will kill the chicks as they cannot yet use the water to cool off (McGraw Hill, 2014). Of the 40% of chicks, on average, that die each year, 7% of that number die due to climate change related exposure. "Climate variability in the form of increased rainfall and temperature extremes, however, has increased in the last 50 years and kills many chicks in some years ... in 2 years it was the most common cause, accounting for half [of] the dead chicks, and 43% in another (McGraw Hill, 2014)."

Additionally, "the number of storms at the breeding site when the chicks are less than 25 days old, increased between 1983 and 2010 (McGraw Hill, 2014)," and changes in fish behavior resulting from warmer water temperatures has caused starvation and changes in the penguins breeding schedules (McGraw Hill, 2014). Further effects of global climate change are seen in the increased variability of water and temperatures. Iguazu Falls, one of the new natural wonders of the world, dries up more frequently due to periods of extreme high temperatures and drought (Alvarez, 2013). All the while extreme weather events have become more frequent and intense across the country and the number of days with heat waves and tropical temperatures increased, especially in the provinces to the north and east of the country causing significant local water shortages and harming crop yields. Argentina's environmental authority expects that these effects of climate change will only get worse, and the temperature increases will be most drastic and effects most devastating in the Northeast of the country (Koop, 2014).

Climate change related legislation has been difficult to enact in Argentina due to the nature of the country's economy and economic development. Argentina was struck by a massive recession that lasted from 1998 until 2002, during which nearly 60% of the population sunk below the poverty line. As such, the last decade has been marked by the federal government's focus on economic recovery rather than environmental protection and climate change mitigation. Experts feel that climate change is not an important issue for the government who believe that it is the developed nations that should be taking action. The government believed they are addressing climate change as best they can, and that their policy strategies are positive legislation in the right direction (Koop, 2014).

When it comes to implementation of environmental protection and climate change legislation, each Province maintains jurisdiction over environmental issues and natural resources while the Federal Government is responsible for creating and carrying out a framework of environmental protection. It is the Ministry of Environment and Sustainable Development is in charge of development and implementation of climate-change and decarbonization initiatives (FIEL and OECD, 2017). Like so many other countries in South America, Argentina is struggling to balance sustainable social and economic growth with climate preservation. The country ratified the Kyoto Protocol in 2001, however, its status as a developing country meant that it had no emissions commitments under the accord (Townshend, 2013).

At a local level, Buenos Aires is ranked 80th on the ARCADIS sustainable cities index of 2016, immediately following Sao Paulo, Brazil, and being the second highest ranked South American city in the index. This ranking is for good reason. In 2009, Buenos Aires established a climate change action plan looking to reduce CO_2 emissions to 30% of 2008 levels by 2030. This action plan was accompanied by a series of initiatives focused on energy efficiency, including replacing all street lights with LEDs, establishing efficiency standards for household appliances, and launching a public-awareness campaign to promote energy conservation. Establishment of the "Energy Efficiency Programme in Public Buildings" facilitated energy assessments of municipal buildings with the goal of reducing building energy consumption by 10% in 2010, and to 20% of 2007 levels by the end of 2012 (EIU, 2010).

The energy assessments of these building results in energy efficiency plans tailored to each individual building so as to maximize energy savings. There is also a plan to revitalize the city's green spaces connecting them all with green space bike lanes. To increase energy efficiency within the city's transportation sector and increase mobility, Buenos Aires unveiled their first electric hybrid bus in 2010 with hopes to replace the entire flee with hybrid vehicles. Additionally, the separating of public and private transportation into their own lanes within the city streets has cut travel times by up to two-thirds and reduced both congestion and emissions from idling vehicles. The incorporation of the city-wide master water management plan has also aided in the city's sustainability by installing water usage meters across the city and setting water-usage reduction targets (EIU, 2010).

While Buenos Aires is making some significant strides towards lowering emissions and combating climate change, there are still many sectors of the city that need increased efforts to become sustainable. Air quality is still a major issue in Buenos Aires, due primarily to vehicle traffic. Air quality monitoring is sub-par and the city lacks a comprehensive air-quality policy. In 2010 the city had lower than average levels of nitrogen-dioxide, however, particulate matter and Sulphur-dioxide levels were unmonitored and therefore unknown (EIU, 2010). Additionally, while the city's Environmental Protection Agency performs regular environmental monitoring and has undertaken many activities in the interest of environmental preservation, the city lacks the legal capacity to implement its own environmental legislation. Additionally, the drafted legislation shows a lack of public and NGO involvement in the process and the environmental reviews conducted of the city do not address areas of fundamental sustainability shortfalls. For example, the review done in 2010 failed to address sanitation, waste, and land-use, three areas in need of sustainable development within the city (EIU, 2010).

Instead of a set national climate policy, Argentina has a series of independent legislation that address aspects of the economy related to climate change. The National Program for Rational and Efficient Energy Use looked to reduce energy demand by 2400MW by 2015 resulting in a reduction in CO_2 emissions by 28 million tonnes between 2006 and 2015. The law also requires the blending of biofuels into all liquid fuel types used for transportation and states a target renewable energy percentage for electricity production of 10% by 2017.

Further independent legislation related to climate change includes prohibitions the importation and commercialization of incandescent lightbulbs for residential use in order to cut energy demand for lighting and creation of the National Argentine Carbon Fund to incentivize projects that will directly mitigate CO_2 emissions (Townshend, 2013). Lack of a cohesive climate strategy has made it difficult for Argentina to make large strides towards sustainable development and climate preservation. With the pressing need for economic growth to help alleviate the poverty caused by the recession, Argentina's climate policies or at least implementation and enforcement of said policies seems to have taken a back seat.

In 2016, Argentina, much like other Central and South American countries, ratified the Paris Climate Accords. In a revised commitment, Argentina put forth two absolute emissions reduction targets for 2030. The first unconditional target is a limiting of emissions to 22% above 2010 levels in all sectors except land use, land-use change, and forestry ("LULUCF"). The conditional target, including LULUCF, is a reduction of emissions by 7% below 2010 levels.

> *The new government of President Mauricio Macri has introduced new policies, including a renewable energy law and a biofuels law. These policies are projected to contribute to emissions reductions of around 16–17 MtCO2e in 2030. The new renewable energy law, published at the end of 2015, aims to increase the share of renewables (including hydro smaller than 50 MW) in total power generation to 20% by 2025. Further, the new biofuels law (Decree 543/16) adopted in March 2016 requires a minimum 12% of bioethanol blend in transport fuels from 2016 (CATP, 2018).*

Additionally, in spite of deforestation activity and the high levels of methane emissions due to the agricultural sectors, LULUCF emissions are expected to decrease going forward due in part to the government's improved accounting methodology which has had a significant impact in decreasing LULUCF sectoral emissions. In spite of the pledged emissions reductions and a comprehensive list of mitigation measures to be undertaken in order to achieve the targets, emissions from all sectors of the Argentine economy are expected to increase. Overall emissions are expected to be 50% greater than 2010 levels by 2030 under the current policy regime and energy related emissions are expected to increase by about 60% between 2010 and 2030 while agricultural emissions are expected to increase by 30% in the same period (CATP, 2018). Additional policy measures and resources for enforcement and implementation will be needed for Argentina to meet its Paris pledges.

REFERENCES

Alvarez, A., 2013. Argentina-Brazil: Iguazu Falls Thundering Again after Scary Encounter with Climate Change. Retrieved from: http://www.onewater.org/stories/story/argentina_brazil.

Bailey, E.M., Wolfram, C., June 18, 2012. A whole different kind of innovation. The Journal Report: Innovations in Energy, Wall Street Journal. www.wsj.com.

Berger, B. (Ed.), 1991. The Culture of Entrepreneurship. Institute for Contemporary Study Press, San Francisco, CA. - "Introduction," B Berger, pp. 1-12. - "The Culture of Modern Entrepreneurship," B Berger, pp. 13-32.

Blau, P., April 1960. Structural effects. American Sociological Review 25 (2).

Blumer, H., 1969. Symbolic Interaction – Perspective and Method. Prentice-Hall, Englewood Cliffs, NJ, 1986.

Brown, R.H., September 1978. Bureaucracy as praxis: towards a political phenomenology of formal organizations. Administrative Science Quaterly.

Buchholz, T.G., 1989. New Ideas from Dead Economists. Penguin Press, New York.

Butler, R., 2006. Costa Rica. Mongabay. Retrieved from: https://rainforests.mongabay.com/deforestation/archive/Costa_Rica.htm.

Climate Action Tracker Partners (CATP), 2018a. Climate Action Tracker: Peru. Retrieved from: http://climateactiontracker.org/countries/peru.html.

CATP, 2018b. Climate Action Tracker: Brazil. Retrieved from: http://climateactiontracker.org/countries/brazil.html.

Chomsky, N., 1957. Syntactic Structures. Mouton & Co., The Hague.

Chomsky, N., 1975. Reflections on Language. Pantheon Books, NY.

Chomsky, N., 1980. Rules and Representations. Columbia Un. Press, NY.

Chomsky, N., 1988. Language and Problems of Knowledge. MIT Press, Boston, MA.

Chomsky, N., May 8, 2012. Plutonomy and the Precariat: On the History of the US Economy in Decline. TomDispatch.com, pp. 1–5.

CIA, 2017a. Mexico. CIA World Factbook. Retrieved from: https://www.cia.gov/library/publications/the-world-factbook/geos/print_mx.html.

CIA, 2017b. Peru. Retrieved from: https://www.cia.gov/library/publications/the-world-factbook/geos/pe.html.

CIA, 2018. Brazil. Retrieved from: https://www.cia.gov/library/publications/the-world-factbook/geos/br.html.

Clarin, 2004. A Patagonia Glacier Was Reduced 13.4 Km2 in the Last 7 Years. Retrieved from: https://www.clarin.com/sociedad/glaciar-patagonico-redujo-134-km2-ultimos-anos_0_BkxxCGRJRte.html.

Clark, W.W., Bradshaw, T., October 2004. Agile Energy Systems: Global Lessons from the California Energy Crisis. Elsevier Press, Oxford, UK.

Clark Jr., W.W., Decker-Ward, J., July, 1995. Encyclopeadia Britannica: case of successful technology transfer from national science foundation: science curriculum in the public schools. In: Technology Transfer Society. Annual Conference and Proceedings, Washington, DC.

Clark Jr., W.W., Paolucci, E., July 1997. An International Model for technology commercialization: fuel cells into vehicle process and design manufacturing. Journal of Technology Transfer.

Clark II, W.W., Demirag, I., 2002. Enron: the failure of corporate governance. Journal of Corporate Citizenship 8, 105–122. Winter.

Clark, W.W., Fast, M., 2008. Qualitative Economics: Toward a Science of Economics. Coxmoor Publishing Company, Oxford.

Clark Jr., W.W., Lund, H., November, 2001. Civic markets. International Journal of Global Energy Issues X (X).

Clark II, Woodrow W., Michael, Fast, 2002. Innovation and Capitalisation. International Journal of Technology Management 24 (4), 391–418. Inderscience, UK.

Clark II, Woodrow W., Michael, Fast, July 2005. Qualitative Economics: toward a science of economics. Presented at the Western Economics Association Conference, San Francisco, CA and published in 2006.

Clark II, W.W., Sorensen, O.J., December 2001. Entrepreneurship: theoretical considerations. Inderscience, London, UK International Journal of Technology Management.

Clark Jr., W.W., Bline, D., Demirag, I., 1998. Financial markets, corporate governance and management of research and development: reflections on US managers' erceptions. Chapter in Book. In: Demirag, I. (Ed.), Comparative Capital Systems. Pinter, London.

Clark Jr., W.W., Fall 1994a. Defense conversion: lessons from the American and European cases. Journal of Business and Industrial Marketing.

Clark Jr., W.W., 1994b. Networks and International SMEs. Report in preparation on the Danish Technological Institute "Network Program.".

Clark Jr., W.W., April 1995a. Sharing the American and European economic experiences in technology conversion. The Journal of Technology Transfer. US Department of Energy. Washington, DC.

Clark Jr., W.W., July, 1995b. International technology transfer: lessons learned. In: Technology Transfer Society. Annual Conference and Proceedings, Washington, DC.

Clark Jr., W.W., July 1996. A Technology Commercialization Model: the case of fuel cells for zero emission vehicles. The Journal of Technology Transfer. Washington, DC.

Clark Jr., W.W., 1997. Publicly-funded Environmentally Sound Technologies: The Case of the US. Report for United Nations (One of Ten Industrialized Country Studies). New York, NY and Geneva.

Clark II, W.W., 2003a. Science parks (1): the theory. Inderscience, London, UK International Journal of Technology Transfer and Commercialization 2 (2), 179–206.

Clark II, W.W., 2003b. Science parks (2): the practice. Inderscience, London, UK International Journal of Technology Transfer and Commercialization 2 (2), 179–206.

Clark II, W.W., 2009. Sustainable Communities. Springer Press.

Clark II, W.W., June 2010. Sustainable Communities Design Handbook. Elsevier Press.

Clark II, W.W. (Ed.), December 2017a. Sustainable Communities Cities Design Handbook, second ed. Elsevier Press.

Clark II, W.W. (Ed.), 2017b. Agile Energy Systems: Global On-site Distributed and Central Grid Systems, second ed. Elsevier Press.

Collyns, D., Watts, J., 2017. Peru floods Kill 67 and Spark Criticism of Country's Climate Change Preparedness. Retrieved from: https://www.theguardian.com/world/2017/mar/17/peru-floods-ocean-climate-change.

Cooper, J.F., Clark II, W.W., 1996. Zinc/Air Fuel Cell: an alternative to clean fuels in fleet electric vehicle applications. International Journal of Environmentally Conscious Design and Manufacturing 5 (3–4), 49–54.

Cooper, J.F., Fleming, D., Hargrove, D., Koopman, R., Peterman, K., 1995. A Refuel able Zinc/Air Battery for Fleet Electric Vehicle Propulsion. SAE Paper 951948.

Costa Rica Learn, 2017. How Is Global Warming Affecting Costa Rica? Retrieved from: https://costaricalearn.com/how-is-global-warming-affecting-costa-rica/.

Davis, G.G., January 31, 2002. Press Release: Letter to FERC Chairman Wood.

Dockrill, P., 2017. Costa Rica's Electricity Has Run Entirely on Renewables for 300 Days in 2017. Science Alert. Retrieved from: https://www.sciencealert.com/costa-rica-s-electricity-run-entirely-renewables-300-days-2017-power-green.

Economist, August 23, 1997a. The Puzzling Failure of Economics, p. 11.

Economist, August 23, 1997b. Europe: Remodeling Scandinavia, pp. 37–39.

Economist, September 20, 1997c. Editorial: "The Visible Hand", p. 17.

Economist, December 6, 2001. The Amazing Disintegrating Firm.

Economist, February–March, 2002a. RE: Enron.

Economist, March 4, 2002b. Social Science Comes of Age, pp. 42–43.

Economist, January 19, 2002c. The Real Scandal.

Economist Intelligence Unit (EIU), 2010. Latin American Green City Index. Siemens, Lima.

Enriquez, D., 2014. Mexico's Energy Reform: A Game Changer in the Nation's History (An Upstream Perspective). Wilson Center, Mexico Institute. Retrieved from: https://www.wilsoncenter.org/sites/default/files/Enriquez_Mex_Energy.pdf.

Evans, S., 1997. The Green Republic: A Conservation History of Costa Rica, 1838–1996: Part I. The University of Kansas, KU Scholar Works.

Fast, M., 1992. The internationalization subjectivity and intersubjectivity: a development of a hermeneutic understanding of companies internationalization as an alternative to mainstream theory. In: Institute of Development and Planning, International Business Studies, Aalborg University.

Fast, M., September 1993. Internationalizationas a social construction. In: Proceedings from the 9th IMP-conference, Bath UK.

FIEL, OECD, 2017. Argentina. Retrieved from: https://www.oecd.org/environment/cc/g20-climate/collapsecontents/Fiel-argentina-growth-and-low-carbon-transition.pdf.

Firth, A., 1990. Lingua franca" negotions: towards an interactional approach. World Englishes 9 (3).

Fortune, January 16, 2002. The Letter to Ken Lay.

Freeman, C., March 2–7, 1994. Lecture Series: "Interfirm Cooperation and Innovation; Technical Change and Unemployment: The Links Between Macro-economic Policy and Innovation Policy; Schumpeterianism: Between History and Economic Theory". University of Aalborg, Denmark.

Garcia-Navarro, L., 2016. Coffee and climate change. In: Brazil, a Disaster Is Brewing. All Things Considered; NPR. Retrieved from: https://www.npr.org/sections/thesalt/2016/10/12/497578413/coffee-and-climate-change-in-brazil-a-disaster-is-brewing.

Godoy, E., 2014. Mexico's Climate Change Law – More than Just Empty Words? Retrieved from: http://www.ipsnews.net/2014/04/mexicos-climate-change-law-just-empty-words/.

Government of Brazil, 2008. Plano Nacional Sobre Mudanca Do Clima. Retrieved from: http://www.mma.gov.br/estruturas/smcq_climaticas/_arquivos/plano_nacional_mudanca_clima.pdf.

Graedel, T.E., Allenby, B.R., 1995. Industrial Ecology. Prentice Hall, Englewood Cliffs, NJ.

Grillo, I., 2015. Global Warming Is Making Mexico City Smog Worse. GlobalPost. Public Radio International. Retrieved from: https://www.pri.org/stories/2015-06-01/global-warming-making-mexico-city-smog-worse.

Gullestrup, H., August 1994. Evaluating Social Consequences of Social Change in the Third World Countries. Development Research Series, Working Paper No. 30. Department ofDevelopment & Planning, Aalborg University, Aalborg, DK.

Hakansson, H., Johanson, J., 1993. The network as a governance structure: interfirm cooperation beyond markets and hierarchies. In: Grabher, G. (Ed.), The Embedded Firm: On the Socio-economics of Industrial Networks. Routledge, London, UK, pp. 35–51.

Håkansson, H., Snehota, I., March 1994. Developing Relationships in Business Networks. University of Uppsala, Sweden.

Håkansson, H., 1986. The Swedish approach to Europe - introduction - international marketing strategies. In: Turnbull, P.W., Valla, J.-P. (Eds.), Strategies for International Industrial Marketing. Kent.

Håkansson, H., April 1994. Lecture: "Network theory". Aalborg University Center, Denmark.

Hofstede, G., 1980. Culture's Consequences: International Differences in Work-related Values. Sage Publications, Beverly Hills, CA.

Hofstede, G., 1991. Cultures and Organizations: Software of the Mind. Intercultural Cooperation and Its Importance for Survival. McGraw Hill, Europe, Berkshire, UK.

Hugo, K., 2017. The Weird Way that Climate Change Could Make Earthquakes Worse. Retrieved from: http://www.newsweek.com/weird-way-climate-change-could-make-earthquakes-worse-669922.

Huhne, C., Slingo, J., 2011. Climate: Observations, Projections and Impacts-Mexico. Retrieved from: https://www.metoffice.gov.uk/binaries/content/assets/mohippo/pdf/c/6/mexico.pdf.

IRENA, 2015. Remap 2030: Mexico. Retrieved from: http://www.irena.org/DocumentDownloads/Publications/IRENA_REmap_Mexico_summary_2015.pdf.

Isherwood, William, Ray Smith, J., Aceves, Salvador, Gene, Berry, Clark, Woodrow W., 2000. II with Ronald Johnson, Deben Das, Douglas Goering and Richard Seifert, "Economic Impact on Remote Village Energy Systems of Advanced Technologies". University of Calif., Lawrence Livermore National Laboratory. UCRL-ID-129289: January 1998. Published in Energy Policy, Fall.

Jones, C., 1996. Careers in project networks: the case of the film industry. In: Arthur, M., Rousseau, D. (Eds.), The boundaryless career. Oxford University Press, New York, NY, pp. 58–75.

Jones, C., Borgatti, S.P., Walsh, K., 1998. Career Competencies in a cultural industry: sources of individual advantage and industry influence. Organization Science, special issue.

Kimmelman, M., 2017. Mexico City, Parched and Sinking, Faces a Water Crisis. The New York Times. Retrieved from: https://www.nytimes.com/interactive/2017/02/17/world/americas/mexico-city-sinking.html.

Koop, F., 2014. Patagonia worst-hit by climate shifts. Buenos Aires Herald. Retrieved from: http://www.buenosairesherald.com/article/176083/patagonia-bears-the-brunt-of-climate-change.

Kuttner, R., January 1–14, 2001. The enron economy. The American Prospect 13 (1).

Kuttner, R., March 25, 2002. The road to enron. The American Prospect 13 (6).

Lajous, A., 2014. Mexican Energy Reform. Center on Global Energy Policy. Columbia University (Retrieved from).

Lyra, A., Imbach, P., Rodriguez, D., Chan Chou, S., Georgiou, S., Garofolo, L., 2016. Projections of climate change impacts on central America tropical rainforest. Climate Change 141, 93–105.

Mauri, C., 2009. The national climate change strategy and the carbon neutrality challenge. Ministry of Environment, Energy, and Telecommunications.

McCain, C., 2012. Mexico's Historic Climate Law: An Analysis. Environmental Defense Fund. Retrieved from: http://blogs.edf.org/climatetalks/2012/07/16/mexico-climate-law-analysis/.

McNeill, D., Freiberger, P., 1993. Fuzzy Logic: The Discovery of a Revolutionary Computer Technology– and How it Is Changing Our World. Simon & Schuster, New York.

Mead, G.H., 1934. Mind, Self, & Society – from the Standpoint of a Social Behaviorist, vol. 1962. The University of Chicago Press, Chicago.

McGraw Hill, 2002. PowerWeb. "Entrepreneurship". http://register.dushkin.com.

Mexico, Government of, 2012. General Law on Climate Change. Retrieved from: https://www.iea.org/media/workshops/2015/15thghgtradingworkshop/GeneralClimateChangeLaw_Englishversion.pdf.

Miroff, N., 2017. A Flood of Problems: Peru's Glaciers Have Made it a Laboratory for Adapting Too Climate Change. Washington Post. Retrieved from: http://www.washingtonpost.com/sf/world/2017/08/07/perus-glaciers-have-made-it-a-laboratory-for-adapting-to-climate-change-its-not-going-well/.

Nardi, B., 1993. A Small Matter of Programming: Perspectives on End User Computing. MIT Press, Cambridge, Massachusetts.

Nardi, B. (Ed.), 1996a. Context and Consciousness: Activity Theory and Human-computer Interaction. MIT Press, Cambridge, Massachusetts.

Nardi, B., 1996b. Studying context: a comparison of activity theory, situated action models and distributed cognition. In: Context and Consciousness: Activity Theory and Human Computer Interaction. MIT Press, Cambridge.

Nationally Appropriate Mitigation Actions, 2017. Climate Change: Costa Rica's Climate Change Strategy. Retrieved from: http://www.namacafe.org/en/costa-ricas-climate-change-strategy.

Nelson, R.R., 1990. "Capitalism as an Engine of Progress". Research Policy. Elsevier Science Publishers B.V., North Holland.

National Resource Defense Council (NRDC), 2014. Acting Now on Climate: How Is Peru Addressing Climate Change and Moving toward a Low-carbon Future?.

NRDC, 2017. The Road from Paris: Mexico's Progress toward its Climate Pledge. Retrieved from: https://www.nrdc.org/sites/default/files/paris-climate-agreement-progress-2017-mexico-ib.pdf.

OECD Secretariat, 2013. Mexico: Climate Change Impacts on Water Systems. Retrieved from: www.oecd.org/env/resources/waterandclimatechange.htm.

Perkins, L., 1996. John, senior physicist, Lawrence Livermore national laboratory. "What Is Physics and Why Is it a "Science"?", Lecture Presented at Graduate Physics Seminar. University of California, Berkeley.

Pfeffer, J., Salancik, G.R., 1978. The external control of organizations - resource dependence perspective. Harper & Row, New York.

Porter, M.E., 1980. Competitive Strategy: Techniques for Analyzing Industries and Competitors. Free Press, New York.

Porter, M.E., 1990. Competitive Advantage. Free Press, New York.

REDD, 2018. National Climate Change Strategy (Peru). Retrieved from: https://theredddesk.org/countries/plans/national-climate-change-strategy-peru.

Reinert, E., 1994. Symptoms and causes of poverty: underdevelopment in a schumpeterian system. Forum for Development Studies (1–2), 73–109.

Reinert, E., 1995. Competitiveness and its predecessors – a 500 year cross-national perspective. Structure Change and Economic Dynamics 6, 23–42.

Reinert, E., 1996. Economics: 'the dismal science' or 'the never-ending frontier of Knowledge?: on technology, energy, and economic welfare. Norwegian Oil Review (7), 18–31.

Reinert, E., 1997. The Role of the State in Economic Growth. Working Paper #1997.5. University of Oslo, Centre for Development and the Environment, pp. 1–58.

Reinert, E., 1998. Raw materials in the history of economic policy. In: Cook, G. (Ed.), The Economics and Politics of International Trade. Rutledge, London & NY, pp. 275–300.

Robinson, K., 2010. Brazil's Global Warming Agenda. World Resources Institute. Retrieved from: http://www.wri.org/blog/2010/03/brazils-global-warming-agenda.

Schumpeter, J., 1934. The Theory of Economic Development. Harvard Un. Press, Cambridge, MA.

SEMARNAT-INECC, 2016. Mexico's Climate Change Mid-Century Strategy. Ministry of Environment and Natural Resources (SEMARNAT) and National Institute of Ecology and Climate Change (INECC), Mexico City, Mexico.

Smith, A., 1937. The Wealth of Nations. Modern Library, New York, NY.

Sørensen, O.J., Nedergaard, A., 1993. Management Decision Making in an International Context: The Case for Intuition and Action Learning. University of Aalborg, International Business Economics, Denmark. No. #2 Reprint Series.

Staff, Governor Gray Davis, March 21, 2002. Talking Points. Enron, Sacramento, CA.

Suchman, L., 1987. Plans and Situated Actions: the problem of human/machine communication. Cambridge University Press, London, UK.

Suchman, G.I., Evered, R.D., December 1978. An Assessment of the Scientific Merits of Action Research. Administrative Science Quarterly 23.

Suchman, L., 1994. Working relations of technology production and use. In: Computer Supported Cooperative Work, vol. 2. Kluwer Academic Publ, Netherlands.

Teece, D.J., October-November, 1996. Graduate Seminar Lectures on Economic Theory. Haas School of Business. University of California, Berkeley, USA.

Tierra Verde, 2017. Conservation Efforts in Costa Rica-History, Projects, & Organizations. Retrieved from: http://adventure-costarica.com/travel/sustainable_tourism_costa_rica/conservation_efforts_history_projects_organizations.htm.

Townshend, T., 2013. Argentina's Climate Change Laws. Retrieved from. http://www.climatechangenews.com/2013/01/29/argentinas-climate-change-laws-in-focus/.

USGS, 2017. Induced Earthquakes: Myths and Misconceptions. Retrieved from: https://earthquake.usgs.gov/research/induced/myths.php.

Vietor, R.H.K., Sheldahl-Thomason, H., 2017. Mexico's Energy Reform. Harvard Business School (Retrieved from:).

Weick, K.E., mars 1976. Educational organizations as loosely coupled systems. Administrative Science Quaterly 21.

Weick, K.E., 1979. Cognitive processes in organizations. In: Research in Organizational Behavior, vol. 1.

FURTHER READING

California Energy Commission, February 2002. Report on Summer 2001. Sacramento, CA.

Clark, W.W., 1977. On formalism. Section Five, Anthropology and Education.

Clark Interview at Hanergy Corp: All-solar Cars, July 2, 2016. http://v.qq.com/x/page/o0313vci9m6.html.

Clark II, W.W., Cooke, G., November 2011. Global Energy Innovation. Preager Press.

Clark II, W.W., Cooke, G., 2016. The Green Industrial Revolution. Elsevier Press, 2014 and shorter earlier version in Mandarin.

Clark II, W.W., Paulocci, E., 2001. Commercial development of environmental technologies for the automotive industry: toward a new model of technology innovation. International Journal of Environmental Technology and Management 1 (no. 4), 363–383.

Clark II, W.W., Paulocci, E., Cooper, J., 2002. Commercial Development of energy – environmentally sound technologies for the auto-industry: the case of fuel cells. Special Issue Journal of Cleaner Production.

Grabher, G. (Ed.), 1993. The Embedded Firm: On the Socioeconomics of Industrial Networks. Routledge, London, UK.

Gullestrup, H., 1992. Kultur, Kulturanalyse og Kulturetik - eller hvad adskiller og forener os ? Akademisk Forlag, København.

Gullestrup, H., 1997. Culture, Cultural Analysis and Cultural Ethics" (Extracted Chapters 2 and 3). Aalborg University Press, Aalborg, DK, pp. 1–54.

Hanergy Energy Company, July 2, 2016. Beijing, China. All Solar Car Event. http://www.hanergy.com/showCar/carshow.html.

Hill, M.G., 2002. PowerWeb, "Entrepreneurship". http://register.dushkin.com.

IEA., 2016. Mexico Energy Outlook.

Lundvall, B.-A., 1993. Explaining interfirm cooperation and innovation: limits of the transaction-cost approach. In: Grabher, G. (Ed.), The Embedded Firm: On the Socioeconomics of Industrial Networks. Routledge, London, UK, pp. 52–64.

Rand Report, Bernstein, M., January 2001. "California Energy Market Report", Winter 2000–2001. Unplublished paper presented to Staff in California State Legislature, Sacramento, CA.. Rand Corporation, Santa Monica, CA.

Reinert, E., Daastol, A.M., 1997b. Exploring the Genesis of Economic Innovations: the religious gestalt-switch and the duty to invent as preconditions for economic growth. The European Journal of Law and Economics 4 (3/4), 1–58.

Sacramento, B., April 2001. Gorging Profits. from California Energy Commission, Sacramento, CA.

Schumpeter, J., 1942. Capitalism, Socialism and Democracy. Harper & Brothers, New York, NY.

State of California, Jan. 31, 2002. Governor Davis Calls for Federal Investigation of Enron's Role on California Energy Crisis. Press Release.

Teece, D.J., Bercovitz, J.E.L., De Figueiredo, J.M., 1994. Firm Capabilities and Managerial Decision-Making: A Theory of Innovation Biases. Unpublished. University of California, Berkeley, USA. Haas School of Business.

University of Pennsylvania, 2007. Global Warming and its Impact on Brazilian Agriculture. Public Policy. Retrieved from: http://knowledge.wharton.upenn.edu/article/global-warming-and-its-impact-on-brazilian-agriculture/.

Weber, M., 1964. Social and Economic Organization. The Free Press, New York.

Chapter 7

Finance, Economics, and Sustainability

Woodrow W. Clark, II [1], Rich Gibson[2], James Barth[3], Danilo Bonato[4]

[1]*Research Professor in Economics, Pepperdine University Graziadio School of Business (PGSB), West Los Angeles, CA Campus, Los Angeles California;* [2]*Sustainable Partners, Temple, Arizona;* [3]*Professor of Economics, Auburn University, Auburn, Alabama;* [4]*Managing Director ReMedia, Milan, Italy*

Chapter Outline

Climate Preservation In Urban Communities Case Studies. https://doi.org/10.1016/B978-0-12-815920-0.00007-1

INTRODUCTION

It is clear that the experts, regulators, economists, environmental groups, and elected and utility officials who started to change the energy gird system in the early 1990s got it all wrong. Their deregulation of the California power system, based on the "economic" philosophy of then US President Reagan who was influenced by UK Prime Minister Thatcher, did not make energy more economical and efficient. The "market" economics did not exist then. Nor has western economics that followed the Adam Smith classical economics over decades ever been a reality. Classical economics has the "balance" between "supply" and "demand" then, historically and now today into the second decade in the 21st century ever been a reality. It has only been a theory.

There is growing evidence that the failed policy then, which became an "energy crisis" in the late 1990s for California, was orchestrated through collusion of a small group for private economic gain. Nonetheless, in the end, the chaos of the energy system in California and then other states and nations certainly provided ample quick gains for the largely out-of-state power companies who gamed the state and local energy grid systems for huge private financial gains. While these marketers impacted the energy system at least economically in the short run, their impact was felt in the long run, even today in 2017. And many would argue now that energy crisis contributed to the Recall of Governor Davis in October 2003, which was not his creation. He inherited it from two prior Republican governors over two terms of 8 years each that lasted over a decade and half.

During the decades in which energy deregulation was considered, the state's investor-owned utilities were not the major proponents of the new policy. Basically, the deregulation plan legally forced them to dismantle the state's three largest central grid utilities with large parts of their generation business sold to competitors. Only hydroelectric and nuclear plants would remain under their ownership. Almost all the remaining power generation would be sold to out-of-state energy market companies.

Furthermore, their control over procurement of power and long-term planning was taken over by the power exchange (PX). The PX went bankrupt within only 3 years of its operation. Another "market mechanism" was the new California Independent System Operator (CA ISO), which is responsible for scheduling, operating, and monitoring the energy transmission system. The CA ISO still exists today. Back then the energy utilities did benefit greatly from the lucrative sale of energy by gas-fired plants. And they also recovered costs associated with their expensive nuclear plants and other stranded assets.

A new key concept was used in the late 1990s, known then as "clean energy," which is "natural gas" as it is cleaner than oil but still a fossil fuel that emits greenhouse gases and pollutes the atmosphere. However, the utilities were reluctant participants from the beginning rather than champions of the changes initiated by the regulatory agencies and the economic market experts for renewable energy systems as well as financing them. The result is that

California in 2017 is 56.4% dependent on natural gas and limited to under 1% power from both oil and coal resources with only 3.1% from nuclear power which will be 0% by 2020. Hydropower is about 11%–27% renewable energy today. This pattern can be seen around the United States and other nations today. Oil and coal have declined as energy sources, but natural gas, another fossil fuel source, has grown dramatically. As CA ISO reports at the end of 2016:

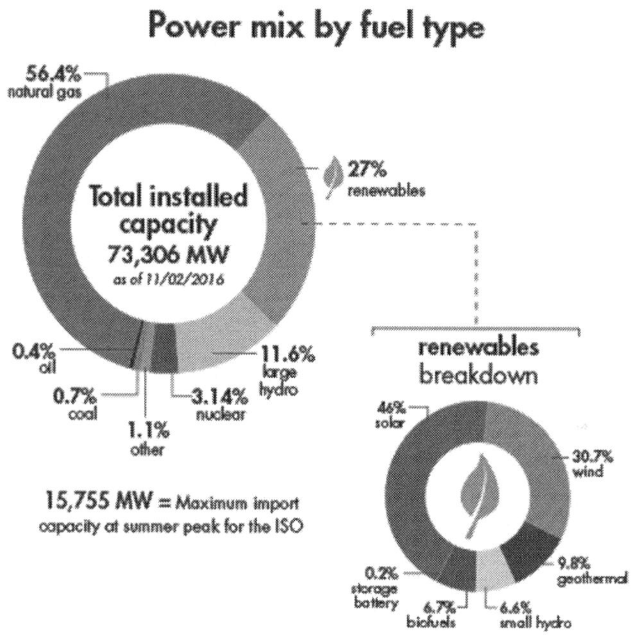

In fact, two major regions of California remained outside the deregulated energy sector: the vast Los Angeles region in Southern California (Los Angeles Water and Power District) and the area surrounding the State Capitol in Sacramento (Sacramento Municipal Utility District). During the energy crisis, starting in San Diego in the spring of 2000 through 2002, these two districts did not suffer from shortages in large part due to long-term guaranteed contracts.

RISKS AND UNCERTAINTIES

The neoclassical or conventional model suggests that the benefits of competition are in who assumes the risks in the power business, such as estimating price and demand, choosing technological options, planning effective management, and obtaining credit. Under regulation, these risks are monitored by the regulators and passed on to the consumers. Under competition, these risks are not monitored except by the owners of the system—" they will pay for mistakes or profit from good decisions and management".

In a system with rich and robust competition, it is reasonable to assume that the private firms will have to absorb the losses from bad decisions and that the firm will benefit when it makes good decisions. Moreover, in order to have improved efficiency in the system, we have to assume that over the long run the firm's gains from good decisions will be passed on to customers rather than used to escalate prices. However, in a less than perfectly competitive system (which has to be the starting assumption in electricity), efficiency gains will tend to be retained. Furthermore, in a less than perfectly competitive market, it is reasonable that the electric utilities will pass on the cost of their mistakes (or at least some of the costs) to consumers. If all the competing utilities make the same mistakes, there will be no price mechanism for firms to absorb the costs of being inefficient. In the United States alone by 2016, electric generating capacity by fuel type has changed dramatically as seen below (Bloomberg New Energy Finance, February 2017).

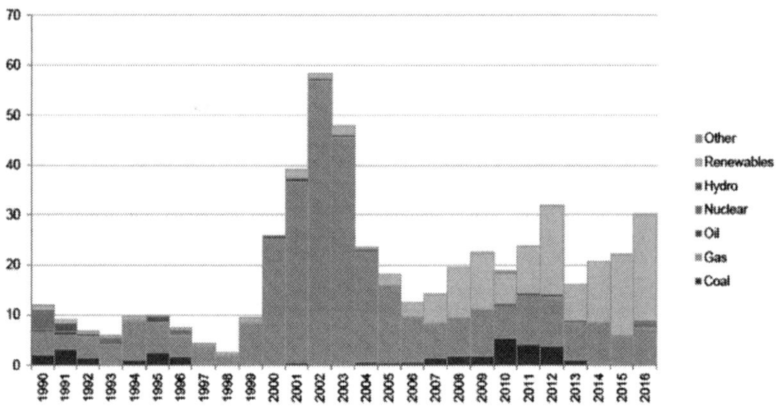

USA Energy Generation Capacity. *US Energy Overview (US Department of Energy, December 2016).*

The point of these data is that "market forces" and conventional economics have had little to do with US energy generation. The unquestioned assumption that competition will shift the risk from consumers to the firms supplying power is very attractive to political leaders and to the consumers, as long as it works. Much of the theoretical premises have been modeled in what has been for effective management, operations, and obtaining credit.

The problem with game theory (aside from it being just a theory and now using computer models) is that some companies were always winners while others were losers. In short, one or two companies won the competition and hence controlled the market or playing field. The fact that the utilities supported deregulation indicates that they were not particularly worried that they were going to lose a lot of money by reducing rates. How wrong they were: Pacific Gas and Electric Company (PG&E) went into bankruptcy, while Southern California Edison barely survived at the turn of the 21st century.

In addition, competition adds a number of layers of companies into the power industry. Each company has its management and overhead costs. The issue for increased efficiency and lower prices is to assure that the multiple firms do not add more costs to the system than an integrated utility under one management. Yet even more destructive to the electricity sector in California from deregulation was the elimination of research and development from within the energy utilities.

Instead of innovation coming from the companies or vendors, the burden fell as a compromise onto the CEC, Public Interest Energy Research program. This program, while useful and strategic in some ways, is nonetheless hindered by bureaucracy and more significantly with an assumption that it will conduct R&D unless legislatively directed. The consequences over the last 5 years in California have been a lack of statewide energy planning, no consideration of renewable energy generation, advancement of new technologies, or consideration of hydrogen. According to both CEC commissioners and staff, "there had been no legislative mandate or authorization." Only due to the state energy crisis and the results of both legislation and executive orders did the CEC change its mission and role in the energy sector.

MARKET SUPPLY AND DEMAND FROM NEOCLASSICAL ECONOMICS

The limitation to a fair and efficient competitive market in network industries and regulated utilities is the potential for market power, the economist's term for abuse of monopoly or oligopoly power to raise rates to obtain unreasonable profits. It is clear that Enron and other generators and electricity brokers manipulated the system and used their ability to withhold power from the grid, to create congestion on critical transmission lines, and to otherwise bypass competition in order to raise prices on the system.

The presence of these market manipulations is relatively unquestioned, but there remains no clear resolution to the question of how much of the manipulation was due to criminal collusion and how much was an unfortunate confluence of circumstances.

Prior to passing the legislative plan to restructure the California electrical system in 1996, economists were concerned about creating a system that would not have the problems of market power that emerged. Joskow and Schmalensee noted this in 1983, recognizing that the amount of competition in generation would vary greatly from place to place and that without reducing the advantage of existing utilities that exercise of market power might not be avoided (2002:199–202).

The solution in California and other deregulating industries has been to force the existing utilities to sell part or all of their generating capacity to other entities so that ownership is decentralized and a competitive market is fashioned. There remains a substantial debate in the literature about what measure of dispersion is adequate. The range extends from no firm having over 10% to no firm having more than 40% of the total ownership in a competitive market, and the means to achieving this goal is forced sale of generation assets.

In California, existing utilities were given incentives to sell all their fossil fuel plants (mostly natural gas) and retain their nuclear and hydro resources. Despite the quick divestment of power plants by the utilities, California's system did not really become competitive. The bulk of the power plants were owned by four companies, and they were linked with brokers who marketed their power for them. The small number of competing firms made it easy for them to manipulate the system even without explicit collusion when supply was limited and huge profit could be made by driving prices even higher by strategic withdrawals of supply at critical times (Fig. 1).

CENTRAL GRID POWER SYSTEMS AT TURN OF 21ST CENTURY

The real issue is whether neoclassical competition can solve the problems with the electric system or whether competition will aggravate or create new problems. The issue often painted by the advocates of competition who came to California is that competition is good and regulation is bad, and that policies that adopt competition and reject regulation are beneficial (Fig. 2).

As the California economy developed, electricity was taken for granted. However, over the last two decades (1990s to 2010s) with the transition from central grid to energy deregulation that assurance no longer holds since companies could do whatever they wanted to do with energy such as ENRON did at the turn of the 21st Century (Clark and Demirag, 2002). Agile energy systems, which are both central power and on-site power systems (Clark and Bradshaw, 2004), document this change in California. The simplistic perspective on the relationship between energy and the economy is that ample power and low prices are good for the economy. Hence, anything a locality can do to assure that it has cheap power, the better. Yet, when agile energy systems as seen globally (Clark, 2017a,b), economic development is not so simple and straightforward. Some of the key changes and challenges

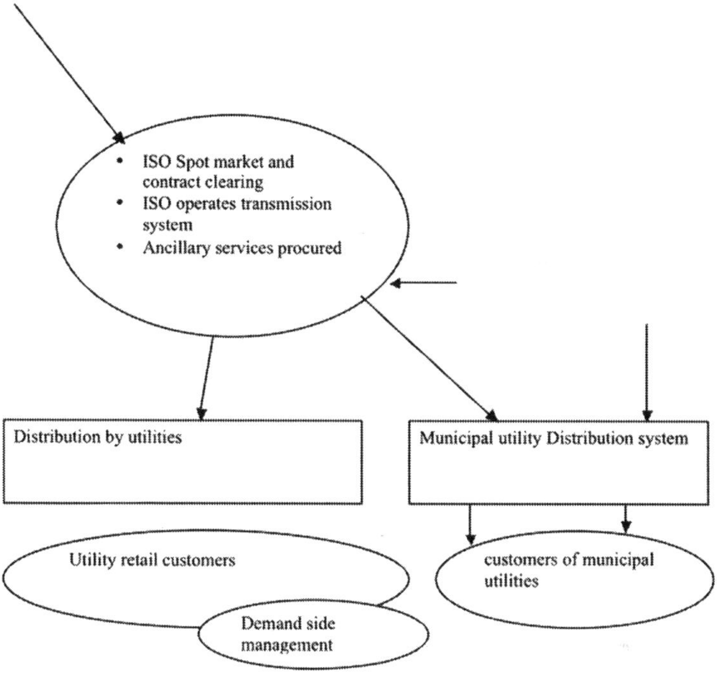

FIGURE 1 2003 competitive generation retail market coordinated through ISO after energy crisis.

that result from such simple neoclassical economic theories are what has created a new economic paradigm.

GREEN DEVELOPMENT AS CIVIC CAPITALISM: KNOWLEDGE, MANAGEMENT, AND ECONOMIC PLANS FOR AGILE ENERGY SYSTEMS

It is clear that the experts, regulators, economists, environmental groups, and elected and utility officials who impacted the energy system in the early 1990s got it all wrong. The deregulation of the California power system did not make it more economical and efficient. What went wrong? The argument can be made that the lessons from the failed California energy deregulation experiment are that modern power systems have become so complex that neither experts nor the public fully understand the way the systems operate or their consequences. Hence, there is a need to explore the gulf between what was known and what needed to be known to have avoided the crisis in California. More importantly, what needs to be learned from the California crisis appeals to other states and nations. Understanding what happened in California and its

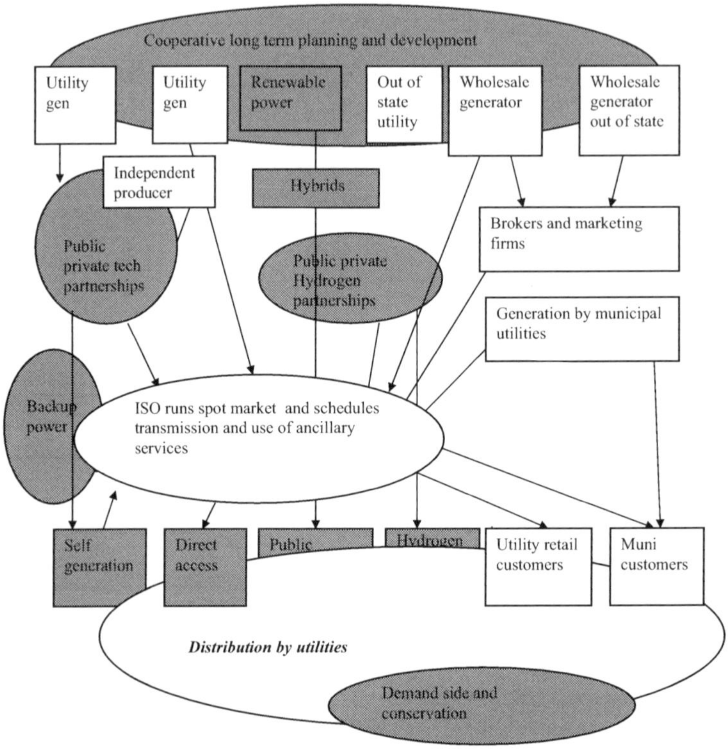

FIGURE 2 Cooperative long-term planning and development. Note: The shaded parts of this diagram are the agile functions that operate for the public good. These agile functions in this diagram represent system components that are already available or achievable within 1 year. Over the next 10 years, the shaded areas are expected to grow substantially in proportion to the remainder of the system. Note, only representative arrows are shown.

actions to correct energy crisis can apply to other states and nations today for them to move forward. Clearly, the need for an "agile system" that reaches beyond centralized power grids to predict, control, and diversify energy supply must be implemented.

Solutions to Complex Problem: Government Never Has Been an "Invisible Hand"

Leaving energy, water, environment, or waste, among other infrastructure sectors, to the "market" or "competitive forces" of supply and demand was wrong in the first instance. The predictable results were monopolies of supply. Instead, all governments must adhere to a higher standard for the public good. The basic issue in other words is a "philosophical" and hence political one. Governments cannot allow monopolies of vital infrastructure sectors like energy to be solely in the hands of the companies and their executives. None of this has anything to do with the "market." It has more to do with the companies financing these companies. Today, this pattern is seen from financial and other investors with "cap and trade," which does the same thing for business people to profit from gross revenue.

History has shown that there is a need to consider the public good in certain infrastructure sectors. Energy has emerged as critical by the end of the 20th century. The association of profits and funds from oil and gas follows what was the basic energy supplier for almost 100 years: coal. The money derived from the energy sector is used to fund and supply other areas of concern today such as ISIS and into the future for land development, corporate expansions, and control of international financial resources. Equally important, there are business opportunities in these sectors, as when electricity and transportation sectors are integrated and hence can share the need, leverage resources, and lower prices (Lund and Clark, 2002). Energy and environment sectors such as water and waste, needed to look out for the public good (Clark and Chung, 1997) when first looked at by the end of the 20th Century and into the 21st Century when California experience a massive energy crisis that had nothing to do with supply and demand. Other states soon had the same energy problems too.

ENERGY INDEPENDENCE FOR COMPLEX MODERN COMMUNITIES

To reduce the energy dependence on petroleum and natural gas, California need to increase its emission standards, support the development of fuel cell propelled vehicles, and invest in a hydrogen-fueled transportation infrastructure. California did all of this and more. Laws were passed whereby states needed to reduce carbon from cars through hybrid (e.g. electric and regenerative braking) cars that reduced emissions and the need for fossil fuels. Then all electric cars were given tax breaks and even special driving routes on the state highways. State commissions and boards such as The California Air Resources Commission were empowered with rules to enforce carbon reduction.

Energy production and consumption are inextricably linked with the state's natural resources. Choices about energy use of land, water, and air impact transportation, waste, telecommunication, and the sustainability of all ecosystems that affect one sector from the use of energy which affect the sustainability of the economy, the environment, and the quality of life.

For example, energy needs can be met in a variety of ways to mitigate the negative impact on our natural resources, including the development and use of natural gas and renewable resources. Some communities are looking to distributed generation (DG) facilities because of remote location, the need for emergency backup generation, or simply the desire for a specific type of onsite energy generation, such as solar, biomass, or wind or the commercialization of new technologies such as microturbines, fuel cells, and hydrogen.

On the transportation side, communities make decisions about roads, rail, transit, bike, and pedestrian pathways which all reflect local, regional, and state funding and approval decisions about California's energy needs. While these local and regional decisions meet the concerns of regional citizens' needs, they must also be transparent in an overall transportation system, reflecting current needs while building toward the "hydrogen economy" of the future.

ECONOMICS OF AGILE ENERGY SYSTEMS

The premise of this chapter is that agile energy systems are good, if not the best, for economic development at local, regional, and nation-state levels of society. The process of encouraging the expansion of renewable, dispersed, and locally appropriate energy resources is both a source of jobs and a creator of business opportunities at the local level. The process whereby state officials convene, cochair, and facilitate problem-solving at the local level yields important energy savings and generation possibilities, as well as combining significant synergies that create jobs and economic well-being.

Local economies are vulnerable to crises and benefit from opportunities originating outside their jurisdictions. However, not all communities are equally vulnerable. Communities can control the extent to which external events alter local preferences by building a diversified economic base and being attentive to flexible responses to changes as they occur.

Since the Gold Rush, California communities have seen and experienced the opportunities derived from new emerging technologies. Communities across the State have historically benefited from rapidly growing technology commercialization into entrepreneurial businesses that offered new jobs, career ladders, and tax revenue. More recently, California has had its share of crises.

The energy crisis deeply threatened confidence in the state. The "dot-com" crash further eroded that self-confidence. New technologies, many of them dependent on electronics supplied by the state's energy infrastructure, now led to elimination of thousands of companies and tens of thousands of jobs. The falling stock market after the dot.bombs and then with the 9—11 attack on the World Trade Center in New York led to declining public tax revenues.

The California government, like 43 other states, was faced with huge unreimbursed electricity costs and spending that greatly exceeded revenues, leading to an unprecedented fiscal crisis with a \$35 + million budget deficit. These problems continued to spiral the economy into a weaker position as private business earnings fell, investment slowed, and jobs continued to be lost. What was once an energy problem became a more general social and economic problem which, in the end (fall 2003), helped to bring on the Recall of Governor Davis.

Nonetheless, local communities should look to the energy infrastructure systems as an opportunity for broader local economic revitalization beyond just getting enough power so that blackouts are avoided. DG, especially with renewable and hybrid technologies, along with extensive conservation create local economic benefits regardless of whether they are done by investor-owned utilities or some other mix of public and private entities.

An added benefit is that local communities can address long-standing problems of environmental justice associated with old and existing power production. Issues of environmental justice have been concentrated in finding legal solutions for the poor and ethnic minority communities.

Agile energy systems allowed the California state government to view economic revitalization of local power systems as a place for civic markets to work. Public—private partnerships were the only way to reverse the economic downturns as well as provide solutions for the energy crisis. The State provided high-level officials for civic market interaction with the local community activists, private companies, and government leaders. The very presence and engagement of the State officials with private sector companies and local government together symbolized how to work out problems and find solutions collaboratively.

AGILE ENERGY INFRASTRUCTURE AT THE LOCAL LEVEL—COMMUNITIES AND REGIONS

Clearly as the Congressional Budget Office noted in September 2001, California's response to the energy crisis was to formulate a new role for government in the power sector. While deregulation did not work, nor did regulation work such that

> *the state's response to the crisis and its efforts to secure adequate electricity supplies and control volatile wholesale process are leaving California with a new market structure. (2001: 27)*

We will sketch a different model of technological innovation (Clark and Paulocci, 1997). Examples exist in other high-technology industries. See, for example, insulin; the biotech industry; the computer mainframe industry; and the pharmaceutical industry where there are important aspects in technology development and commercial transfer, concerning coordination of innovative labor among firms.

The main point is that communities must consider innovations requiring that firms "assemble" information and knowledge coming from different scientific fields with different and sometimes conflicting methodologies, theories, test procedures, and the like, that are part of technical knowledge base. Furthermore, the innovation process is creative by definition, requiring constant change by its very nature. Often, sources of knowledge are located inside different organizations, which constitute a "locus of innovation" where significant and high-priority information can be stored and processed.

More often than not, innovation and creativity reside in the head of the team and individuals of researchers. This "knowledge capital" is key to the success of any research as Narin (97a and b) demonstrates in his research on intellectual property. Nevertheless, as Clark showed in a study of energy and environmentally sound technologies in the United States, private sector research funds often were not available. The primary source for research funding for these two areas was the American government, especially the US Departments of Energy and Defense along in recent history with the US Environmental Protection Agency. Support came from the National Science Foundation and the National Health Centers, but proportionally less than other high-technology fields.

According to this perspective, it is impossible to treat innovation as made up of homogeneous parts, with well-defined borders, where economic and technological aspects can be separated. The key element is the mutual interaction among these parts that change patterns of technological evolution: depending on how interfirm interactions are made, starting/ending point of a technological trajectory may remain unchanged, but there will be different ways in reaching the "full" development of an innovation.

At the same time, the legal aspects, concerning patenting and property rights attribution, cannot be considered as sufficient for completely determining the movement of an innovation as it enters into the market. Elsewhere (Cooper and Clark, 1996; Clark and Paolucci, 1997), these facts are analyzed by explicitly considering the role played by organizations involved in "zinc–air fuel cell" (ZAFC) development and commercialization. However, this role changes with different stages of innovation. This model is nonlinear rather than chain-link linear one because it explicitly considers all sources of knowledge and of scientific research at the same time.

It is not just a matter of tacit knowledge but also of related forms of division of innovative labor. If knowledge is really idiosyncratic and difficult to be transferred, it cannot be acquired simply by "waiting" for spillovers to happen, and during the development process it is better to permanently involve organizations that own it. At the same time, it is impossible for a single organization to develop all the technological innovations required to "transform" a ZAFC into a commercial product.

Another aspect concerns the structure of the innovation's life cycle. Existing models consider a life cycle where incremental and/or radical innovation improves product performances (or trade-offs among variables defining a technological trajectory). In the case of most advanced new technologies, the main issues are (1) the market structure before and after "discontinuities," (2) shakeouts caused by introduction of innovations, (3) emerging of a dominant design, (4) period of new entrant's success, and/or (5) incumbent displacement. Consider this plan for on-site green, smart, and healthy distributed community that is also connected to the central power grid.

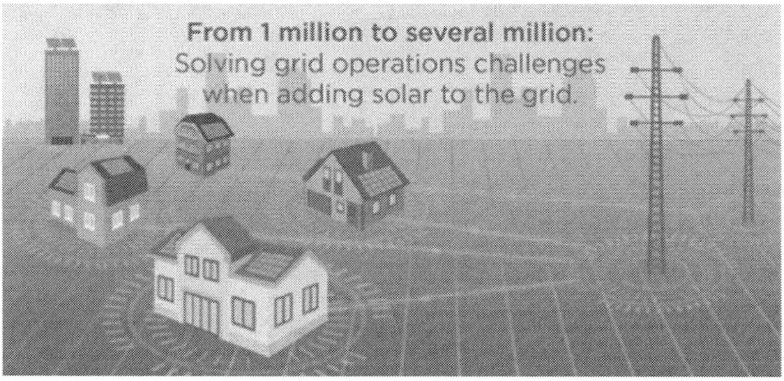

Conversely, it is far better to concentrate, at least in early development stages for technology development, refinement and costs that are not usually in the product definition. In order to introduce technologies that are "pulled" by environmental regulations, innovation life cycles must be considered as made of different steps, each focused on different products and market segments.

New markets, where technology is introduced, can be different from the traditional ones for which the technology was designed. The introduction to "target markets" may take place in later stages, when there is larger diffusion of the new technology. Hence "learning by using" in new industries and application fields can be created and well established.

The Next Economics

It is no wonder that the public policy for the electricity industry has problems given the complexity of the electrical power system and the number of issues that affect it. The power industry has a nearly unique relation to the public since it is so capital intensive and its product cannot be conveniently stored. In the process of deregulating the electricity industry in California, the state envisioned changing the foundation of a publicly regulated industry that had been operating quite well for many years. The industry faced a number of significant new challenges due to technological change and the availability of many independent power producers both in the state and out of state.

The massive changes introduced into the California system were catastrophic because they were not simple and they led to unanticipated and unprecedented conclusions, based entirely on false assumptions about the market. While some analysts recognized the danger of the scale of changes proposed in California, and while in retrospect most everyone should have been more cautious, no one person or group is in a position to point fingers at others.

The system complexity is not a cause of the failure in California because everyone knew that the system was complex. What caused the failure was that the solution drove the problem. The myth of deregulation and competition was the solution to a very complex problem that was not fully discussed, analyzed, and considered independent of the proposed solution. A simple solution required a complex implementation, and the complex policy that resulted worsened the problem rather than fixed it.

Complex problems can be addressed by effective policy, but the more complex the problem is, the more attention needs to be given to finding agile solutions where the benefits outweigh the costs. J. A. Casazza (2001, p. 42) proposes that the costs and benefits be assessed.

> *The entire restructuring process failed to investigate the costs and benefits resulting from the policies being adopted. Unlike our environmental procedures, an impact statement was not required from those proposing major changes in how electricity was to be produced, distributed, bought, sold, and priced.*

The California crisis suggests that the failure to understand the complex economic and technical characters of the electrical system has potentially catastrophic consequences. The fact that the California policy failed was not an accident but was largely predetermined. When public policy is made over vital societal infrastructure sectors through political compromise, the end result will be failed mechanism for implementation. This is especially true when the political basis for decision-making rests with economic theory rather than business realities. This is why circular economics is critical.

PUBLIC OVERSIGHT

The Role of Government Must Be Transparent: Reregulate and Oversee

The option to "reregulate" the energy infrastructure is not feasible for two reasons. First, the replacement of the regulatory system would be different because now the utilities no longer own most of the market forces that control the energy generations. Thus, the utilities have little control over their largest cost factor, which is the fuel or primary source of power. Similarly, the long-term energy generation contracts entered into by the state have fixed the prices of power at a relatively high level that leaves little room for regulatory flexibility to achieve public goals.

Secondly, the drive to reregulate the power system neglects the fact that political conflicts, tensions, and vested economic interests in the regulated system caused the pressures for infrastructure deregulation in the first place. What this means is that the same pressures and stakeholders who demanded change during the mid-1990s and resulted in the energy crisis remain powerful forces today. These same market forces (for-profit and not-for-profit) would lobby against and certainly destabilize a return to an integrated and regulated system.

For example, many energy consumers will want either very low electricity prices or opting out of the regulated system entirely. Often, consumer needs are better met by self-generation or contracting with private generators without any utility hookups except as standby power. In addition, the municipal utilities or power suppliers still can undercut regulated utility prices because they can avoid some high costs of a regulated utility system such as operations, maintenance, and transmission lines. The regulated utilities remain unable to capture the increasing efficiency of small-scale operations and distributed power generation systems. Furthermore, they are limited because they need to reduce prices in order to retain large power users. In short, reregulation will return the state to the same tensions that it started with 10 years earlier.

The "proper deregulation" model is no more feasible than reregulation. Energy infrastructures are not suitable to the competition market forces, although some parts could be competitive, especially at the local and regional levels. Such a negative view of the deregulation option is based on two considerations.

First, the failed deregulation system was fundamentally flawed in ways that more deregulation would not fix. Consider the lessons worldwide where power systems universally tend toward monopoly. Even if more retail competition were present, little market force would counter the possibilities of major anticompetitive consolidation taking place. In fact, one business school "exit strategy" that is universally taught is to start, grow, and then sell or merge the company.

Additionally, full deregulation is both problematic and theoretical because market systems have little incentive to pursue policies for the civic good. The corporate goals are to get a good return on the investments. Yet, the most problematic parts of any energy system are peak power and pursuing demand reduction (conservation and load shifting) strategies. The public good is not a market force concern or consideration.

When power generation plants are used but a few hours on only the highest peak demand days, there is little incentive for any firm to pay the high costs involved with having this capacity available. Peak power is already high priced, but looking after proper capacity reserves for use only on the hottest summer afternoons is not attractive to purely market forces—reliable peak power has been one of the most remarkable successes of the regulated system.

Similarly, the public interest is well served by conservation and demand reduction programs. However, market forces reward competitive firms for selling the maximum amount of power, not for selling less. This has been a persistent problem that California addressed with a large publicly run conservation program paid for by a charge to ratepayers on all power sold. Such programs, however, leave little room for efficient pursuit of these goals by private firms looking to maximize local opportunities.

The second reason why "more deregulation" is not feasible is that we have no models of deregulated private competitive markets that have really succeeded anywhere in the world. While some would argue that United Kingdom is a success, recent critiques and events there suggest otherwise.

Growing evidence has shown that the British attempt to deregulate or "privatize" or "liberalize" their energy infrastructure (and perhaps other public sectors) is a failed tragedy. Earlier, the issues surrounding the nuclear power plant bankruptcy in the United Kingdom were discussed. While the United Kingdom has had some positive side effects in the energy sector, such as experimentation with renewable energy generation systems such as wind and advanced technologies like fuel cells, even these innovations were well under way before "privatization" took place.

The questions surrounding the British energy sector are far deeper and more profound. In 2003, other European countries (especially Norway, France, and Italy) began to experience energy problems. Norway had a rainfall shortage in the winter of 2003, which left its hydroelectric system at its lowest levels in decades. The result has been for the Norwegian government to accelerate its pursuit of new energy generation. The economic issue for the Norwegian is the fact that they are one of the world's largest exporters of oil and gas.

Hence, much of their energy generation could be supplied by increased generation from these fuel sources. However, the country has a limited supply of these fossil fuels (most estimates are at 20–25 years maximum) and the resulting emission pollution problem violates both European and Kyoto Accord Standards to which the Norwegians are historically staunch supporters. Accelerated efforts to become a "hydrogen economy" with more and more emphasis on renewable energy generation began in late 2003.

Denmark has moved into agile energy systems as a PhD student (Lixuan Hong) showed in her thesis and as many of the cities and towns in Denmark have done.

Dr. Hong has gone on to work in China with some of these and other related ideas that she learned about at Aalborg University in Northern Jutland, Denmark.

Italy suffered some of the worst energy blackouts in the summer of 2003. While France recorded over 12,000 deaths associated with its energy shortages over the summer of 2003, little action has been taken to date to basic problems. Italy, however, had realized that dependence on 25% of its energy coming from nuclear powered plants in France is both risky and not good public policy.

The country had long ago (late 1980s) engaged in changing from a government regulated monopoly energy infrastructure to something else. Like California, Italy is in transition with many market changes. What appears to be a parallel set of public policies to California is the move toward agile energy markets with the core being a hydrogen economy. The regions of both Lombardy and Piedmont have been leading the Italian transformation into the civic market structures. Italy now has a national government commission for green energy and focuses on "circular economics."

Nonetheless, it has been the British government, for example, that has placed strong pressure on other European Union member states and new applicant nations to be "competitive" in energy and other sectors. Yet at home, the privatized energy markets have failed. Moreover, within the European Union a growing skepticism has arisen while nations watch California experience their own energy problems and monitor the entry of a dozen new members to the European Union in 2004. What is clear for the practice of deregulation and privatization, and not the theory, is that market forces should not replace publicly regulated energy suppliers and services. The civic market must be established and provide the framework for regional energy public policy and practical market mechanisms.

Two additional points need to be made. First is the basic issue of more privatization or deregulation being advanced, primarily by economists and some policymakers as the final solution to energy issues, needs, and future supply. Yet, there is no evidence that the market will even consider the public needs for the future, let alone reduce prices through competition. And why should they? Energy, among other public sector resources, does not lend itself well to competition and "market forces." The time has come to move away from such empty and purist arguments.

Consider, secondly, some interesting energy models of "privatization" which already exist in Europe and the United States that might be far more useful and practical. In reality, this is not one model. Yet civic markets operating as public–private partnerships have been successful through documentation and evaluation matrices for years (Clark and Chung, 1997; Clark and Jensen, 1994, 1997, 2002). For example, some European countries prefer to maintain partial government ownership in the energy companies, including shares, board membership, and regulatory oversight. This model appears to be working well in the Nordic countries as well as in Germany and France. Italy is experimenting with it now.

CIRCULAR ECONOMIC MARKETS: COLLABORATIONS AND PARTNERSHIPS IN "AGILE ENERGY SYSTEMS"

The transition from the traditional vertically integrated utility to the emerging "agile power system" is a struggle between two different perspectives on how to increase the public good—on the one hand "privatization" and the other hand circular economics. Deregulation was proposed under the premise of privatization, which is the neoclassical economist view that competitive markets reduce prices and allow maximum choice, while providing optimum conditions for technological innovation. The belief has been a cornerstone of Prime Minister Margaret Thatcher public sector reforms during the 1980s as well as the philosophy behind relaxing regulations in many other industries from banking to airlines. In all these cases, proponents look at the declining prices and innovation but also have to acknowledge gaps in service and occasional scandals such as the collapse of the American Savings and Loan institutions. The failure of the California deregulation can be added to the list of failures.

The philosophical premise that the public good rests at the core and is the theoretical basis for any analysis of market options is key to any discussion of the energy sector. The public good is a concept linked to community-wide, shared benefits by people that transcends special individual or group interests and hence benefits everyone collectively. Put another way, public good is a shared benefit that in aggregate exceeds those that might otherwise be retained as exploitive benefits for only a few individuals.

We think that the model for resolving the tension between those who want to see the energy sector reregulated and those who wish for more deregulation lies in the pursuit of civic markets that encourage programs for the public good. The public good in the energy infrastructure sector for power systems is represented by six electricity benefits:

1 cheap,
2 reliable,
3 predictable in terms of future availability and costs,
4 safe with few negative social costs (externalities) or health impacts on local communities,
5 environmentally friendly in protecting air quality and natural resources, and
6 oriented to the future needs of society as a whole such as convenient transportation, pure water, waste removal, and renewable energy generation.

These characteristics are the cornerstone of most of the restructuring proposals that have recently been published such as the Bay Area Forum and the California Policy Institute. These characteristics are generally well represented through consumer pressure and public advocacy for government oversight of the energy sector. Regulatory systems generally have had an easy time assuring these goals because the goals are largely identical to the marketing goals of the utilities which are being regulated. The key issue for regulators is to keep the price reasonable, while preventing the utility from making excessive profits.

On the other hand, some aspects of the public good are not so easily represented in the market. Individual consumer interests, for example, tend not to be closely aligned with utility interests. To be specific, the public good is enhanced by environmental and equity issues that do not have traditional cost–benefit market values for utilities. Environmental issues are important because of the scale of the electric power industry and the fact that power generation creates a number of types of pollution, including smog, particulates, and other airborne wastes that contribute to global warming, ash that needs disposal, and nuclear waste that needs long-term isolation, heat pollution, and visual pollution from plants and power lines.

These same environmental problems impact the public good in another way: health. A growing amount of literature in "environmental justice" documents the health costs for communities surrounding power plants. San Francisco (SF) has been active in meeting the concern through its Bayview Hunters' Point community action group which seeks to close two local power plants and replace them with renewable energy generation. The concern then for public good translates into high medical costs for society. Community-based action groups (Citizens for Better Communities, among others) have provided technological expertise and contact networks to provide solutions to local community clean energy needs.

Some of these energy–environmental problems can be addressed by regulatory prohibitions that reduce or eliminate some of the offending behaviors.

However, the record, both of utility regulations and environmental laws, shows that much progress can be made by requiring that technologies be developed and installed that reduce pollution, even if the market would not voluntarily select them. Japan, as noted later in the book, has many agile energy systems. The focus is also on "green energy" from solar as seen below in Japan where a solar farm is in a community for its power.

In addition, a range of environmental problems need to be addressed by experimental technologies and considerable investment in alternatives that may cost more at the early stages, until the economies of scale make them more price competitive. The requirement for including advanced technologies that are not immediately competitive in the energy sector mix until they reach enough volume that they become cost-effective is one of the key objectives of progressive regulatory agendas. Clark and Chung (1999) prove this argument with an international study of environmentally sound technologies. Moreover, California has pursued this strategy during the energy crisis through several civic market partnerships, collaboratives, agency, and commission programs (CSFCC, 2002) discussed earlier about advanced technologies.

Other technological issues also benefit the public good in ways that need public encouragement and regulation to assure that the technological advances are provided. The three most important of these are renewable energy generation (grid, DG, and on-site), peaking power, and conservation. The public sector through government programs such as regulation, standards and codes, as well as public procurement policies can assure that these benefits are met.

California has demonstrated this approach since 2001 numerous times. For example, in order to conserve energy, the California Consumer Agency (CCA) launched a public awareness campaigns, called "Flex Your Power" (CCA,

April 2001) that has proven to be so successful that the public and private sectors reduced energy consumption by an average of 12% over the critical summer months of 2001.

The same Agency has spearheaded a "Sustainable Public Building" program (CCA, December 2001) whereby State, and now local, government buildings are installing solar PV systems, net metering, and conservation devices. Procurement policies aimed at purchasing and installing such system have had a direct impact at encouraging the marketplace. All these programs have helped drive down the consumer price as the market becomes cost competitive.

The key now is the circular economic model noted earlier in this book. As the picture of solar systems in Japan show, the generation of power from one source impacts others. And to be both financially responsible and transparent, it is best to have the other sources working together in a community. ReMedia Corporation (that recycles electronic and other systems such as TV, computers, washing machines, etc.) in Milan, Italy, has done throughout Europe is now leading the European Union into enacting a circular economic model.

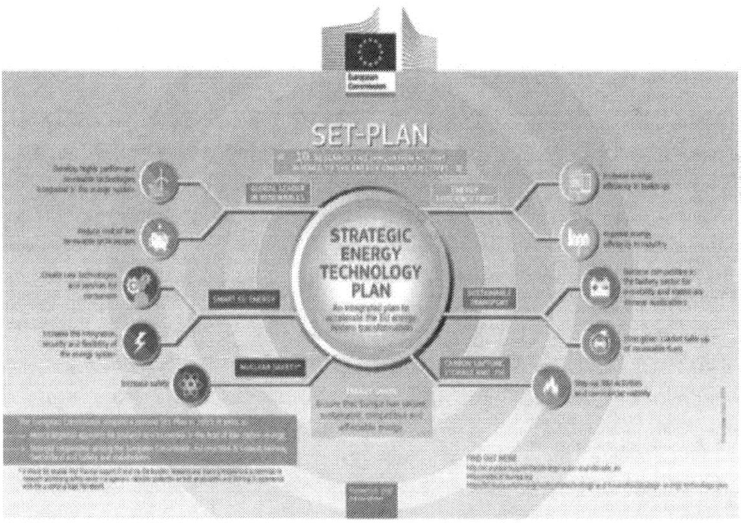

The equity issue also remains one for nonmarket regulation. The problem in power is that it needs to be available to a range of people who cannot afford it but who need it for their safety. Similarly, some groups of people have very strong needs but the cost of service is greater than they can afford at least initially. For example, rural populations were last to get power because of the high cost of providing lines and distribution networks for power over much longer distances on average than in urban areas, and rural populations tended to be poor. This led to public rural electrification programs in most countries. Both of these issues contribute to public well-being, safety, and security.

The results of the local community energy plan will be shutting down polluting power plants as well as putting online clean energy generating facilities. More importantly, the local community will be able to train workers, employ them at this and related facilities, as well as create and grow new businesses in the energy and environmental sectors. The community is calling this their "94124 Plan" after their local postal zip code and is strongly supported by the local SF community college staff and students. With some leveraged support from the State, local government and the private sector, the Plan appears on its way to success for a "sustainable community."

COMMUNITY CONTROL OVER RESOURCES FOR THE FUTURE: ENERGY, WATER, LAND, TRANSPORTATION, AND OTHERS

SF has continued to lead the way to clean energy with several voter initiatives passed in 2002. Among the most significant was the Vote Solar which yielded the city $100 million for solar systems for its buildings. A number of programs support the effort with major new construction ramping up daily. The city has had plans to create its own energy power company which would be the holding company for local energy generation from communities, public, and commercial buildings.

Where local control can be most effective is in the promotion and development of small or on-site energy generators. Permitting of power plants under 50 MW in size is subject to local review under the California Environmental Quality Act. Plants over 50 MW are required in California to be permitted by the California Energy Commission in Sacramento.

The sitting of these smaller power plants requires that a local agency assumes the lead in determining what the environmental impacts of a proposed power plant are and providing for their mitigation. Hence, it is possible for the city to either own smaller power plants through the Hetch Hetchy dam (about 70 miles east of the SF Bay Area) water and power system or enter into contracts for power. Very small generators, particularly renewable energy generators, are within the domain of oversight of the city.

Promote Opportunities for Economic Development

The choices SF makes about its energy future will have a bearing on the local economy. Not only is it important that homes and businesses have reliable and affordable energy but also that some of these businesses become part of the local energy supply infrastructure. This goal would consider to what extent different energy technologies are able to keep dollars in the local economy by supporting local community or city manufacturing, production, distribution, and installation services. Local employers would need local workers trained at the community colleges and available for work and advanced education.

Some advanced technologies are available outside the local economy such as combustion and combined-cycle turbines, microturbines, fuel cells, and large wind turbines. However, with the exception of wind turbines since they can be installed, operated, and maintained, the fuel needed to run these other technology systems involves revenues leaving the city. Solar technologies, peak load management systems, and energy-efficient products are all available locally in the SF Bay Area and can reduce spending on fuel.

Communities within SF saw new opportunities for the local economic development and growth in an increased demand for solar products and energy efficiency, especially with the passage of local, citywide energy propositions. An expanded and growing market in SF would support new business enterprises, encourage training and education programs, and create jobs while reducing air pollution.

The basic strategy for the civic markets process in SF was to create and implement a local "agile energy system" that consisted of clean energy generation, lower energy demand through conservation and efficiency, and providing a new economic basis for a sustainable community with job training, advanced education, and business creation.

ECONOMIC DEVELOPMENT STRATEGIES FOR AGILE SYSTEMS

From a national and global position, the economic development implications of an agile power system are huge. California's electricity sector is a multibillion dollar business, with the total number unknown when considering all the communities, cities, and counties. The national electricity sector is trillion+ dollar business, so that changes in regions or nationally are significant. The cities, states, regions, and countries that lead in the changes being discussed.

Programs and the people who run them can make an agile power system work in ways that contribute to the economy rather than draw from it. Bradshaw and Blakely (1999) argued that third wave economic development has three key components that set it apart from other forms of economic development, and these are applicable to the challenges of implementing a viable future for agile energy systems. Examples of on-site distributed energy systems are growing, which include companies and government buildings as these photos demonstrate for rooftop solar panels.

The first is leadership which in this application refers to the ability of individuals and organizations in public–private partnerships to set priorities, establish a vision, communicate an agenda, and mobilize resources to make it happen. The notion of leadership is not one of position, but one of ability to establish and implement a vision. Leadership expands the network of people involved in a project and helps to set priorities.

The second component of an effective economic development strategy for the future is information. In a classical economy, people are assumed to have adequate or perfect information in order to act rationally, but we all know that most decisions are made with inadequate information; some of the actors involved in a decision are easily exploited because they lack key pieces of information such as the regulators who were manipulated by private companies during the blackouts.

In addition, information is often better collected by the community and made available to the widest number of possible firms and investors. For example, knowledge of available wind sites had not been mapped in California prior to the early efforts of the CEC and their charting of wind areas became available to private and public entities that could develop the state's wind fields. The public role in making this information available was essential.

The third key component is brokering. In complex society someone with a leadership vision and adequate information is needed to put critical interdependent parts together. The broker role is essential because it recognizes that connections between necessary parts are not formed in the most efficient way by accident, but they need to be facilitated. Brokers are essential to create the networked industrial fabric of the new economy. Their role is to solve problems, form coalitions, identify markets, and expand the network of people in an industry to gain strategic advantages (Fig. 3).

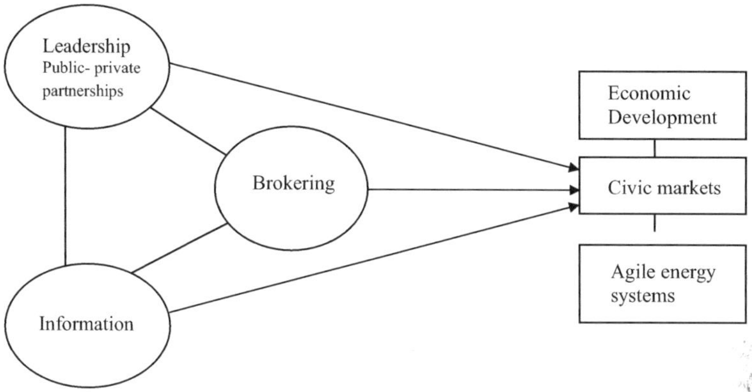

FIGURE 3 Economic development for agile energy systems.

The three components of an effective third wave economic development program link directly to the creation of a civic market. This is a partnership in this case with the public and private firms that can jointly solve economic development objectives while creating the framework for a successful energy system.

The Hunters Point, SF, example clearly illustrates these interconnected factors in creating a civic market for renewable energy systems that will also create jobs and employment in the local community. The leadership came from community members, nonprofit organizations, colleges, and government in identifying environmental justice issues and striving to find a collaborative way to resolve problems and gain benefits for a wide variety of organizations and firms.

The information role was played by the state agencies who shared understandings and facts. The broker role was played by senior members of Governor Davis' administration and their associates who kept the process of target, identified state resources that were available, and helped the group to see the value in what they were doing.

This process at the neighborhood level can be extended to the state or regional level where the legislature and governor set goals and strategies such as for the renewable portfolio standard, based on widely shared information, and the key players are brought together to make it happen. One should note that the documents of this process are often not made public at the early stages of the process of forming leadership, collecting and disseminating information, and brokering. The process of partnership building is iterative and good leadership assures that many pieces are working together.

Local Leadership for Green Economic Development

The leadership that led to the stationary fuel cell collaborative is a good example here. California is pushing ahead with research and development in fuel cells in a way that will have enormous economic development implications. While the future is not yet determined, it appears that fuel cells will be an essential part of the agile energy system. This will lead to the emergence of a huge industry that will take its place alongside oil, automobile, computer, and other significant industries, and the strength of the industry will be concentrated in a relatively few regions.

In the same way that semiconductor industry concentrated in Silicon Valley and Boston, as well as a few other places around the world, one way of looking at the fuel cell development is to anticipate where it will be located. What we know about the location of industrial clusters now is that they concentrate around the universities and research centers where the initial discoveries are being made, that the top people in the early industry have strong ties to the centers, that money flows to only a few areas, etc.

The establishment of the Fuel Cell Collaborative is important to California's effort to be open centers for this industry in the future. Therefore California needs to gather and form a set of global leaders together in a region (city, county or area) of the state with top researchers and advocates of fuel technologies is no accident. Such a strategy positions the region to compete for prominence in the emerging green car industry.

Clearly, California is not the only state aspiring to be the leader in fuel cells. Ballard is located in Canada, and a number of other areas have fuel cell firms and leadership. First, we can see that the vision for this emerging industry did not come just out of the private industrial sector. While fuel cells have been known longer than the internal combustion engine, advanced research has just begun on how new materials and applications work together in a way that will not simply provide affordable power but will achieve other social goals as well.

Second, leadership is needed to secure and expand investment. The federal investment in fuel cells has been important largely because of space applications. But more recent investment has been in finding clean and inexpensive sources of electricity. The University of California has contributed to the project. Some of the private sources of investment that are available to expand fuel cell commercialization are large interested firms, small businesses, and venture capital.

The Fuel Cell Collaborative provided leadership in another way of being a visible center for discussion. The scientists and entrepreneurs at the collaborative are now known around the world as being a place where important things happen in fuel cells, and visitors from every part of the globe working in fuel cells pay a visit to the Collaborative.

Another critical function for economic development is the role of leadership to remove real and perceived barriers in the way of development, often regulatory barriers. The classic example of this was the leadership role that the federal government played in promoting nuclear power where legislation was passed to limit liability of firms to disasters associated with their plants.

Another example is the failed leadership of the regulatory agencies, especially Federal Energy Regulatory Commission to take a lead in stopping the worst part of the energy crisis when rates soared. Finding equitable solutions to barriers as they emerge is an important leadership role as the photo of solar panels on a company shipping building shows.

The leadership function is essential to economic development of emerging industries because perception is everything. The Governor's statements about renewable energy standards, advocacy of the state signing on to greenhouse gas reduction standards, and continuing publicity about the state leadership in finding a new energy direction have all contributed to the perception that the state is on the right track for a potential investor and entrepreneurial leadership. Such collaborative activities help accomplish these five areas to succeed:

1. define opportunities, white paper, and road maps or tariff mechanisms,
2. develop consensus on vision and direction,
3. focus on goals and objectives,
4. stimulate investment, and
5. set priorities and implement programs that can be evaluated.

Information for Expanding Industries

The collection of information that is needed for economic development in the agile energy system is coming available. This information is collected by the public for private business or public–private partnerships that can expand energy development. Information about what can be done is the first step to assembling the resources to do it.

1. Technical assistance standards

2. University research
3. Professional and industrial associations are essential. Conferences and demonstration projects
4. Information centers bulletin boards, etc.
5. Resource and site assessments

Brokers Create Networks for Business Success

The third component of the economic development effort to increase economic development efforts involving agile energy systems is to serve as a broker in making the network viable and effective. For example, the establishment of public–private working groups involving key parties. In the competitive model, there were few roles for brokers, but in the agile energy system there is so much cooperation and interdependence that brokers are essential. The broker acts as a catalyst in making links, solving-problems, and identifying resources that are needed.

Economic development brokers often do not think of themselves as economic developers, as they may play this function as planners, regulators, or just administrators. However, they are typically marginally involved in the project being developed, at least over the long run. The broker function, however, is essential to making the connections that allow a renewable producer to identify and become a hybrid company, to finding appropriate markets, and to adapting a technology such as biomass to the needs and resources of a particular area.

The brokering role has had great success already. The facilitation of the state in setting standards for intermittent production is a state example. In addition, at the local level the key has been to help communities solve siting problems and to engage the community in setting standards and programs for conservation.

Another brokering function is helping to create markets for innovative programs and firms that build the local economy. One of the most important findings of economic development is that without a reliable and well-functioning market for whatever is being produced, the innovative companies have a hard time getting started.

The changing power system bridges community desires for reliable inexpensive energy that improves rather than degrades the environment and the community desire for economic stability and growth. This chapter has shown both that economic interests can be resolved at the local level and that the pursuit of advanced energy systems is a classic economic development challenge. The tools of economic development are parallel to the tools used to fashion good energy policy, and the good news is that this is something economic developers understand.

PACIFIC GAS AND ELECTRIC COMPANY CENTRAL ENERGY SYSTEMS

No organization has had a greater impact on the design of SF's electric system than has the PG&E. PG&E was formed in 1905 to harness the hydroelectric potential of rivers in the Sierra Nevada like the Feather, the Mokelumne, and the Pit. The state granted monopoly status to PG&E as a supplier of electricity in Northern California by the 1930s. By the 1950s, the basic architecture of the high-voltage transmission system bringing power into the Bay Area was formed. SF, being at the tip of a peninsula, gets most of its electricity over power lines swooping across the Central Valley and then winding their way around and across the south bay.

PG&E, like other investor-owned utilities in the United States, thrived for decades under a system of state regulation. A stable legal and economic structure allowed electric utilities to earn a predictable return on their investments in exchange for providing universal service with regulated rates. Under this regulatory structure, PG&E was able to build major fossil fuel power plants in Pittsburgh and Moss Landing in the 1950s and 1960s to serve the Bay Area. In SF, PG&E built smaller plants at Hunters Point in 1958 and Potrero in 1965 to provide reliable service at the end of its transmission lines.

In the late 1960's PG&E, like many other utilities, including Southern California Edison, made a commitment to develop nuclear power. Both California utilities faced formidable challenges during construction that caused significant delays and cost escalation. At the beginning of the 1980s, slippage in the construction schedules at Diablo Canyon and San Onofre triggered concerns about possible shortages of power in California. Because of these concerns, the California Public Utilities Commission (CPUC) aggressively promoted nonutility development and ownership of power plants.

Taking advantage of a 1978 federal law, the Public Utility Regulatory Policy Act (PURPA), California led the nation in encouraging the development of geothermal, biomass, wind, and solar thermal generating capacity. By the early 1990s, about 11% of California's electricity came from nonhydro renewable energy sources. In addition to incubating the development of renewable energy technologies in California, PURPA also stimulated the mainstreaming of cogeneration power plants that very efficiently use fuel to produce both heat and power.

At the same time, the CPUC ordered PG&E to institute demand-side management (DSM) programs, which encouraged the implementation of energy-efficiency measures as a means of meeting new demand for power. PG&E's DSM efforts enabled the utility to cost-effectively defer investments in new power plants and brought nationwide recognition of the role of electric utilities in managing energy-efficiency programs.

In 1994, the CPUC announced that it wanted to end the monopoly power of the state's investor-owned utilities, eliminate government supervision of electric resource planning, and give consumers a choice of electricity suppliers. By 1996, the California legislature unanimously passed legislation designed to accomplish these goals. Under orders from the legislation, PG&E sold off most of its fossil fuel and geothermal power plants to private companies not regulated by the state. Its transmission lines were placed under the control of a new nonprofit organization called the Independent System Operator (ISO) that was given responsibility for assuring that all power plants would have nondiscriminatory access to the grid and thus to customers. A separate organization called the PX ran short-term spot markets that set the price for power that PG&E and the other electric utilities bought.

PG&E started in the first decade after the turn of the 21st century a program for on-site power distributed power. One of the first companies to do this was Google.

As can be seen here, all the parking areas and most of the buildings are covered with solar panels. Also on the bottom left are the cofounders of Google recharging their electric car. So the solar-powered buildings provide power to the buildings as well as to electric vehicles. Additionally, Google has two fuel cell storage plants for power (which is not seen above in the overview of the Google company offices) generation and storage in the absence of sunlight.

To ensure that utility managed energy-efficiency programs would not fall by the wayside, a Public Goods Charge (PGC) was added to customers' electric bills to fund the continuation of energy-efficiency and renewable programs. Since 1997, PG&E and the other California investor-owned utilities have been collecting the PGC and using these funds to administer energy-efficiency programs under the direction of the CPUC. Below image shows (and several Appendices get into the actual examples) how communities and cities can be "agile" as well as "green, smart, and healthy."

Small-scale DG as illustrated above is designed to power single buildings or connected structures. To reduce the amount of load on the system, the facilities are usually energy-efficient as well. In general, renewable energy sources are the cleanest form of electric generation and have a number of suitable applications in the city. Solar panels are the most readily available technology for reducing city electricity demand in the near term. Small-scale wind turbines that can be installed on roofs or on property near a building are now on the market, but standard wind technology is more appropriate outside the city. While the electricity generated from outside would not contribute to SF load reduction, it would raise the percentage of renewable power feeding the transmission grid.

Solar, water, wind, geothermal, and biomass projects, where they can be sited, have proven successful in the SF. But more appropriate urban technologies, such as fuel cells, will take several years before becoming commercially available and affordable. Other technologies now in the development stage, such as tidal and marine current energy, would be well suited to SF and could support a city's longer-term goal of zero fossil fuel generation.

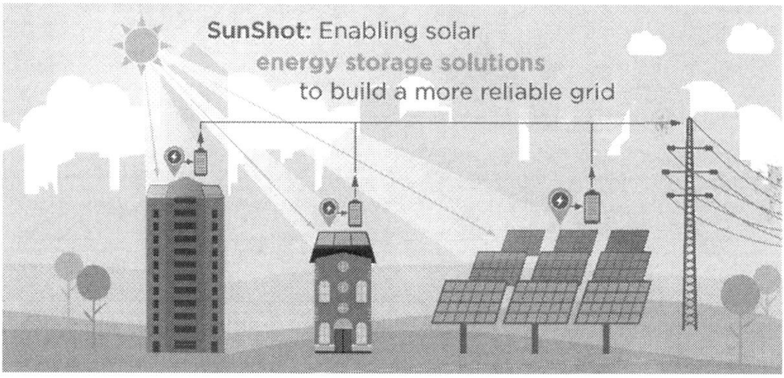

Cogeneration, or combined heat and power systems, can be an efficient source of DG for facilities that have heating/hot water needs, such as hospitals and hotels. Some of these systems do use fossil fuels, however.

By definition, DG reduces the need for central power plants, but not all DG are cleaner. Emergency generators at hospitals and in many office buildings burn diesel fuel, which emit higher number of pollutants per kW than new natural gas power plants. Despite the ability of these generators to meet certain reliability situations, their use is patently discouraged in this plan except in genuine emergencies. Like energy efficiency, renewable energy and cogeneration projects are being funded by state programs. There are both programs that provide financial incentives for installation and others that help finance projects.

Energy Efficiency and Conservation

Energy-efficiency measures create long-lived reduction in electricity use because it is built into the equipment, rather than being dependent on human behavior. According to the San Francisco Public Utilities Commission (SFPUC) reducing demand can be achieved by:

- installing equipment that is more efficient whenever it runs, e.g., the newest high efficiency fluorescent lighting uses 40% less energy than older versions.
- turning off or turning down equipment during the peak hours, e.g., turning off computer monitors when not in use.
- installing systems to use energy off-peak to avoid use during peak periods.

Demand reduction also:

- saves the city, businesses, and residents money on their monthly utility bills, which support the economic development goal to make more dollars available for local investment and job creation.
- avoids or defers building and operating generation systems, even renewable ones, which reduces the environmental impacts of construction.
- reduces the stress on the transmission and distribution system improving reliability.

Energy efficiency can involve new appliance standards and energy codes or incentive and educational programs. Typical programs are rebates for the purchase of energy-efficient products, energy auditing services, and incentives for upgrading the operational efficiency of buildings, processing plants, and other facilities. There are also programs that encourage the purchase of new high-performance homes and others that provide design and engineering support for qualifying commercial buildings and multifamily complexes.

PG&E has education and support service centers in SF. One concept tried in other communities has been local core-satellite outreach education—service centers where local residents are trained in energy conservation. They can service and assist home owners and renters alike in installing new energy-efficient lights, meters, and surge protectors. Moreover, this "energy team" can be trained, and education for new jobs in the local power plant, energy companies, and businesses can be created for the operation, maintenance and service of the systems.

CASE: SAN FRANCISCO FOR SMART GREEN HEALTHY CITY INITIATIVES

In November 2001, the electorate of SF voted on energy-related ballot initiatives that would provide financing for renewable energy and energy-efficiency projects. Proposition B authorized the issuance of $100 million for the acquisition, construction, and installation of projects for city agencies, departments, and enterprises. Proposition H simply authorized the SF or city board of supervisors to issue revenue bonds for renewable energy and energy-efficiency projects for private sector or city projects without requiring a vote of the electorate.

The SFPUC has already identified specific solar photovoltaic and wind projects that would qualify for financing under Proposition B. Both SFPUC and SFE will be working with the private sector and appropriate city agencies to facilitate the installation of renewable and energy-efficient technologies in commercial and residential properties.

In January 2002, the Mayor and Board of Supervisors joined other cities in setting a goal of reducing SF's greenhouse gas emissions to 20% below 1990 levels by 2012. Actions to achieve this ambitious goal will include increasing the amount of renewable energy in the SF portfolio and ramping up energy-efficiency efforts as well as supporting transportation and solid waste programs. A plan needs to be written such as this one that is also healthy for everyone. As Norman Cousins put in his dozen or more books, "laughter is good medicine." Body and mind together are needed in healthy communities, such is the point in this program.

Big Heart Intelligence in healthy workplaces and sustainable communities. *Source: Alliances for Discovery by Julian Gresser and William Moulton (May 2017).*

This *Electricity Resource Plan* reaffirms the will of the voters and city leaders with a prescribed path for reducing our dependence on electricity generated from fossil fuels. Its implementation would lend substance to commitments made to communities, especially neighborhoods in the southeast. There are challenges, but there are also viable solutions. These are discussed in the following chapters.

In SF more than 60% of electricity is used for commercial purposes, while residential use accounts for 27%. From 1994 to 2000, consumption of electricity grew by 9%. Growth in consumption in both office buildings and residential buildings was slightly greater than in remaining uses. The fastest rate of growth was in the communication sector, which increased by 29% between 1998 and 2000. Most of that growth occurred appears to have been associated with the expansion of the Internet. Overall electricity use decreased by about 2.5% in 2001.

Based on comments received during the public hearing process, eight goals were selected to guide the plan. As noted above, the goals expanded on the specific scope of the city ordinance due to new public concerns over developments directly related to SF's energy future. The broader goals tend to lend even stronger support for a plan that seeks solutions to the health and environmental impacts in southeast SF as called for in the ordinance. The goals are as follows:

1. Develop renewable power
2. Maximize energy efficiency
3. Assure reliable power
4. Support affordable electric bills
5. Reduce air pollution and prevent other environmental impacts
6. Support environmental justice
7. Increase local control over energy resources
8. Promote opportunities for economic development

RENEWABLE (GREEN) ENERGY SYSTEMS

Renewable energy options currently available to the city include solar, wind, biomass, and geothermal. Emerging renewable technologies such as wave, tidal current, and ocean thermal energy may be available in the near future. Each of these resources has unique opportunities, advantages, and sometimes disadvantages.

Solar

The city's first large solar power development will be at the Moscone Center. With approximately 90,000 square feet of perfectly flat unshaded roof, this football field–sized showpiece will significantly reduce Moscone's purchase of power and provide a solar showplace for visitors from all over the world.

The SFPUC has installed radiometers at 11 sites on city buildings and schools to collect data about the availability of sunlight. The variability in solar incidence is based on microclimate and geography, and when cross-referenced with availability of appropriate space limits the application of solar technologies in some areas of the city. To develop a well thought out strategy of implementation, city needs to understand the resource and develop it where it is most cost-effective.

If sufficient participation by commercial and residential customers is obtained, at least 50 MW of solar could be installed in SF. Price of systems is a major consideration in achieving this magnitude of installation. A sustained program to develop solar in SF can help, if done in an orderly manner, reduce the overall cost of solar technologies.

Wind

Wind has been used for centuries to create mechanical power for uses such as water pumping and the milling of grain. In recent years, wind turbines have been developed that produce electricity. The technology is now well developed and can be used to generate significant amounts of relatively low-cost power. Modern wind turbines have increased in size and output to megawatt-scale machines in recent years. SF could obtain significant amounts of wind power in areas such as the Altamont Pass, where wind speeds are high enough, and where other conditions like proximity to transmission can be met.

The SFPUC is currently looking at several sites including those adjacent to its own Bay Area reservoirs. The estimate potential for wind development in the greater Bay Area for SF's use could exceed 150 MW. Electricity from these projects would require transport using PG&E's transmission lines. There may be additional opportunities for developing small-scale wind projects in the city itself.

Biomass

The combustion or gasification of wood, agricultural waste, and other forms of biomass offers options for SF. The SFPUC is currently reviewing several potential biomass projects. Last year, the SFPUC installed a small reciprocating engine to use biogas recovered from the Oceanside Water Treatment Control Plant. This year, a 2-MW biogas plant will be operating at the Southeast Water Treatment Control Plant. Both of these plants use methane gas produced by the sewage that would otherwise be flared-off.

Fuel Cells

Fuel cells are a developing technology that is expensive and not yet readily available. Fuel cells do not burn fuel, they chemically convert it, much like a battery chemically stores electricity. Fuel cells need hydrogen and produce water and heat, though current models use the hydrogen contained in natural gas or gasoline because those fuels are readily available. As this technology becomes mass produced, the cost will reduce, and it will be a popular option because there are essentially no air emissions.

Marine Energy

Surrounded on three sides by water, the technical potential of marine energy technologies is enormous. The energy of the ocean is stored partially as kinetic energy from the motion of waves and currents and partly as thermal energy from the sun. Although most marine energy is very diffuse, in special situations it could be cost-effectively captured for practical use.

Among the marine energy technologies that are currently being investigated are those that convert the energy in waves, tidal/marine currents, ocean thermal gradients, salinity gradients, and marine biomass fuels. Tidal and marine current is one of the most serious of the marine energy resources to be studied. The technologies are in development with demonstration projects now being installed in Europe, Canada, and the United States. The city will initiate partnerships with appropriate agencies to develop demonstration projects.

Geothermal

Geothermal energy has been used commercially for over 70 years both for electricity generation as well as direct use. A significant area of geothermal development has been the Geysers region in Northern California. More recently, large-scale geothermal development is taking plant around the Salton Sea in Southern California. Several geothermal developers have approached the city with competitive offers for power purchase agreements. A challenge for SF in utilizing geothermal energy is arranging for cost-effective transmission into the city.

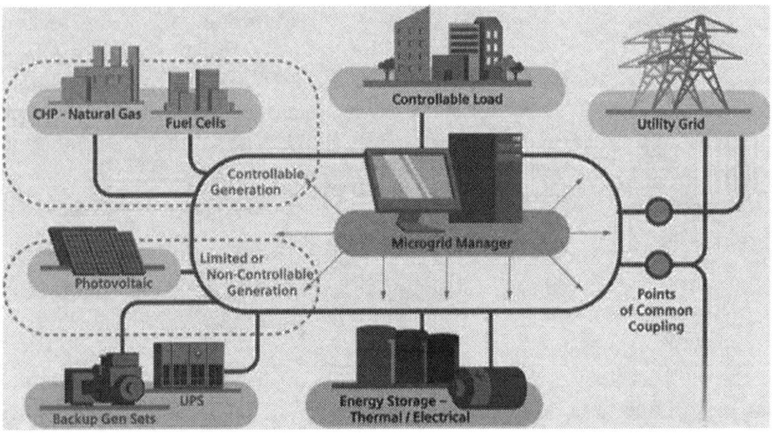

The key is to combine technologies so that the communities can be green, smart, and healthy. There are many companies now that are providing the technologies. And many such as Siemens knows (see diagram above from Siemens) that show the application of the technologies.

CONCLUSION: GREEN ECONOMICS IN THE GREEN FUTURE

Economics should be, as a science, concerned with formulating theories of ideas and reality that produce descriptions of how to understand phenomenon and create experiences, hypotheses generation, and data which need to be proven or disproven through testing and further analyses. Today economics is focused on numbers and statistics. The "field "lacks theory(s), other than those rooted in the neoclassic paradigm, and it rarely generates hypotheses that are documented, tested, analyzed, and explained. Moreover, economics today infrequently describes and discusses the definitions, meaning, and forecast of the numbers. The financing of technologies, systems and other solutions to climate change means that there must be clear definitions of words, numbers, formulas, and digital processes. Finance (hence economics) needs to be used to reverse climate problems in order to save the earth and protect it.

Hence, economics needs to be a science. Climate change has been proven through science. Now is the time that economics and its related fields need to be a science as well. But to become a science, economics must be able to handle the essence of finance, economics, and human actions with group interactions. Academics and scholars need a scientific perspective that can document, hypothesize, understand, and analyze human dynamics. And that is where qualitative economics (QE) becomes a science. The economic actions of people, groups, and their networks are organizations about people interacting in everyday of life, trying to construct the future, and making sense of the present. The science of economics is about the monetary, financial, and numbers that transpire between and within people, their homes, organizations, and local government.

Economics on the macrolevel is one kind of interaction between people and groups, countries, and regions. Yet, economics on the microlevel is different form of interaction. Often, microeconomics can be seen in business and interpersonal transactions. Economics takes place among people: macro and micro. While we acknowledge and recognize the value of quantitative approaches in such macro- and microinteractions, there is a significant missing element: qualitative interactions of people that turns into hypotheses, testing, and analyze which then makes the entire field of economics scientific. Within the philosophical interactions between people and with groups at any level is the core basis for qualitative economics.

As noted, QE is based on the formation of questions that result in definitions and meanings. Thus, the definition of the zinc–air technology as a fuel cell rather than a battery was a matter of constant interactionism between the surface structure organizations and individuals, which directly impact the lexical meaning of the technology itself in the deep structure. Once these terms were agreed to, the expansion into other related technologies to creating a zero emission vehicle becomes clear and far less complex to implement. The interaction rules allowed the ZAFC to be commercialized since there was common definition, and hence agreement, on the need and application for the technology. Furthermore, a detailed generic proposal (with different modifications for various consortia members) was written with exact tasks, costs, and timelines outlined. The key is to have a plan and with everything in it defined. That is what interactionism is and QE is. Case in point below is a home.

Clearly the distinctions between competence and performance are significant and are *not* just numbers. The basic paradigm remains the same but is now globally well established. Critical to this discussion is the basic notion instead that meaning cannot be separate from structure. However, the discussion below is not aimed to review or enter into the current linguistic debate over these issues. Instead, the business cases of advanced technologies demonstrate that meaning and structure must be combined within a linguistic theoretical framework. An argument is presented that uses linguistic formalism for describing and understanding the actions of actors in everyday business life. Linguistic formalism allows scholars and observes to see relationships in the context of micro- and then macrotheories. Indeed for that reason alone, transformational linguistic grammar theory is a very powerful theoretical paradigm.

Linguistic theory from Chomsky provides the framework for combining meaning and structure in human interactionism. In short, linguistic theory allows the study of business economics to be scientific. Above all, the importance of performance is paramount and acknowledged as a key element in everyday business life. Everyday life interactions are the basis of what Blumer argues makes humans human — human beings reflect through the "generalized other" on their actions and those of other humans. In shorts, actors think; they do not just react to stimuli or events. Thus, Blumer established the basis of understanding how and why actors interact with one another. His approach today would be called nonlinear and parallel processing of actions rather than a simplistic linear or causal sequence of activities and events.

The ability of the economics field to theoretically predict economic events has been limited. Quantitative models and simulations have not been very successful. Even the interests in applied areas of economics like "game theory" fall far short of providing predictability. The interactionist approach provides a qualitative perspective on the economics of business. As such and through its use of theories from symbolic interactionism in sociology and generative grammar in linguistics, the paradigm states predications that, like science, come true. Note the case below and also other cases in the Appendices.

QE is a significant part of economics in order to make it a science that lends itself well to the analysis of hypotheses, economic issues, and financial problem solutions. The "green revolution" is a current example. Hence, the reason that China has now used "green development" as a critical policy for the entire nation. There are other examples. The key is to understand the meaning and definitions of numbers, words, terms, and sentences. From there, quantitative economics can be used. The statistics and data from gathering of historical and current numbers (once defined and verified) then make sense. They also can prove or disprove an economic theory or financial plan with its set of hypotheses. In short, economics becomes a science.

A Case for Financing: Smart, Green, and Healthy Communities

Rich Gibson
Sustainable Partners, Temple, Arizona

Necessity has often been the mother of invention. According to the World Economic Forum, the United States places tenth globally for its overall infrastructure. Furthermore, research from the American Society of Civil Engineers revealed that the United States faces a funding shortfall of more than $2 trillion dollars to make any meaningful inroads into repairing the country's infrastructure problems. These inescapable problems are what prompted the creation of Sustainability Partners, and infrastructure reform company headquartered in Tempe, Arizona.

Sustainability Partners' CEO Tom Cain, one of the earliest CleanTech venture capitalists, has pioneered a new way of delivering effective and impactful energy, water, and fuel-efficient systems to municipalities, universities, schools, and hospitals by accelerating traditional project and procurement processes—which often takes years—and utilizing an innovative, value-added approach called Sustainability-as-a-Service, or trademarked as SaaS.

SP removes the traditional barriers and risks, enabling customers to implement the latest energy, water, and fuel-efficient technologies and realize immediate savings while retaining critical capital for more mission-specific needs. After all, our cities, schools, and hospitals—as well as many corporations—have more than a trillion dollar problem. The inside infrastructure has aged to the point of being obsolete, wasting massive amounts of energy and costing billions of dollars from inefficiencies.

Cain conceived of a model very similar to how utility companies charge clients. You do not pay for the equipment providing your Internet cable or electricity. You pay for usage fees on a monthly basis. Now this model can be applied to the internal infrastructure of any company or municipality. Sustainability Partners implements all the new high-tech energy saving equipment at no cost. This not only saves organizations' money but also spares them the risk of equipment failure.

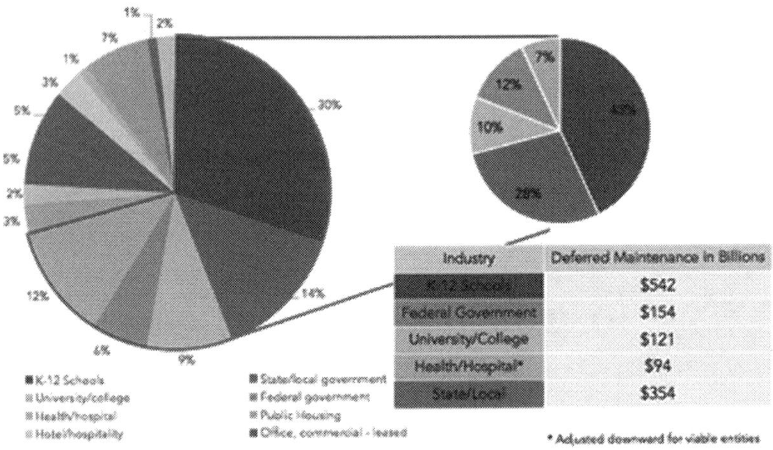

Industry	Deferred Maintenance in Billions
K-12 Schools	$542
Federal Government	$154
University/College	$121
Health/Hospital*	$94
State/Local	$354

* Adjusted downward for viable entities

For example, if your organization is a university or hospital with decaying equipment, or a municipality with crumbling infrastructure, executives may realize the need to repair and fix these looming problems.

What do these executives or administrators do? Budgets are finite. To replace these aging systems of HVAC, chillers, water pumps, and sputtering engines often requires hundreds of millions of dollars. How do you explain giving the lion share of precious funds to infrastructure systems few will see but everyone will need?

For example, nearly 75% of US school system facility infrastructures are in need of immediate repair or replacements with a total cost of more than $270 billion, or nearly $5000 per student *and* the majority of civic buildings and prisons are 25- to 50-years old with the original decaying facility.

Executives are also wired to prioritize their budgets to assure fiscal survival. They prefer to allot these precious funds for staffing—luring the most brilliant professor or most coveted researcher in their field. They want to revamp computers or office equipment or purchase cutting edge MRI machines or cancer fighting technologies. They need the option to invest in research and development to keep their organizations current and relevant.

Cain observed the critical need for a platform economy, as infrastructure modernization is continuously deferred to everyone's detriment. This problem is shoved aside while upgrades on aging devices are indefinitely delayed, and deferred maintenance costs continue to rise and accelerate every year. And when something breaks, this band-aid approach backfires causing the organization to hemorrhage money. These municipalities, universities, schools, and hospitals—as well as corporations—are forced into emergency spending just to keep everything running. This does not solve the problem. The need to eventually upgrade the decaying system is currently at the point no one can ignore as evidenced by national news headlines ranging from Penn Station's motors breaking down in New York to the Minneapolis Bridge collapse.

The American Society of Civil Engineers' Report card for America's infrastructure in 2017 was a D +. That same failing grade (the report is modeled after a school grading system) was also given to schools, wastewater, roads, and energy waste. In addition, there is the problem of procurement. Even if an organization finds these emergency funds, the process between identifying the need and getting the new equipment often takes years due to the internal back and forth jockeying and lack of decision-making which causes gridlock and fatigue.

In a recent interview, Gary Cohn, the former chief operating officer of Goldman Sachs, and director of the National Economic Council, told CNBC that currently the " infrastructure approval process takes seven to 10 years to build relatively simple roads. We need to streamline that to less than 2 years."

However, although the government wants to raise $1.5 trillion to address infrastructure needs, the reality is that there is no agreement on how to raise those needed funds. A company like Sustainability Partners is already getting the job done with their model. Launched in 2016 by engineers and infrastructure experts with 30 years of experience, the company has already seen results with its initial projects (Gibson, 2017).

Since Sustainability Partners pays the total cost of replacing the old equipment from raising funds in the equity market, these problems are eliminated. Furthermore, the costs are now off the balance sheet paving the way for organizations to use needed funds to further other goals. The savings have already been proven in some of their early implementations.

HOW DO WE PAY FOR IT?

Financially Self-Sufficient	Monitor & Verify Our Upgrades	Direct Meter Billing
We have our own capital (just like any other utility company).	We meter all of the upgrades by installing our own usage meters on each major device.	We are repaid by invoicing you directly from the new product meters.

	Cost/KwH	KwH Usage	Billing	Total	
Old Lighting	$0.09	71,964	$6,476.76	$6,476.76	
New Lighting Utility	$0.09	30,000	$2,700.00		58% Energy Savings
SP SaaS	$0.03	30,000	$900.00	$3,600.00	
			Total Savings	$2,876.76	44% Utility Cost

You keep the benefits from reductions in utility bills, maintenance costs, disruption to operations and improved systems!

As an added bonus, clean energy, which is environmentally responsible, can be implemented as a result of this SP business model, which has far-reaching potential to shift these global problems into a viable and positive solution.

REFERENCES

Bradshaw, T., Blakely, E.J., August 1999. "What Are 'Third-Wave' State Economic Development Efforts? From Incentives to Industrial Policy. Economic Development Quarterly 13 (3), 229–244.

California Stationary Fuel Cell Collaborative (CSFCC), 2002.

Casazza, M.L., 2001. San Luis Drainage Feature Re-evaluation. Environmental Impact Statement.

Clark II, W.W., December 2017a. Sustainable Communities Cities Design Handbook, second ed. Elsevier Press.

Clark II, W.W., 2017b. Agile Energy Systems: Global On-site Distributed and Central Grid Power, second ed. Elsevier Press.

Clark II, W.W., Bradshaw, T., 2004. Agile Energy Systems: Global Lessons from the California Energy Crisis. Elsevier Press.

Clark II, W.W., Lead Author/Director with RaeKwon Chung, "Transfer of Environmentally Sound Technologies from Developed to Developing Countries". Public Funds for technology project. Framework Convention for Climate Change, UN: Study of economics of "green" energy in six countries. Nov 1999.

Clark II, W.W., Chung, RaeKwong, 1997. Transfer of Publicly Funded R&D Programs in the field of Climate Change for Environmentally Sounds Technologies (ESTs): From Developed to Developing Countries – A summary of six country studies. UN Framework Convention for Climate Change, Report, Bohen, Germany.

Clark II, W.W., Demirag, I., Bline, D., 1998. Financial Markets, Corporate Governance and Management of Research and Development: Reflections on US Managers' Perspectives. In: I. Demirag, Corporate Governance, Accountability and Pressures to Perform: An International Study. Oxford Un Press, Oxford, UK.

Clark II, W.W., Demirag, I., Winter, 2002. Enron: the failure of corporate governance. Journal of Corporate Citizenship 8, 105–122.

Clark II, W.W., Jensen, J.D., 2002. *Capitalization of Environmental Technologies in Companies: economic schemes in a business perspective,* International Journal of Energy Technology and Policy, 1. Inderscience, London, UK nos 1/2.

Clark II, W.W., Jensen, J.D., 1997. Economic Models: The Role of Government in Business Development for the Reconversion of the American Economy.

Clark, W.W., Jensen, J.D., 1994. The IO Fund: International Alliances for Developing SMEs. Paper on the Danish Government Fund for business development in transitional economies.

Clark II, W.W., Paolucci, E., 1997. Environmental Regulation and Product Development: issues for a new model of innovation. Journal of International Product Development Management.

Cooper, J.F., Clark II, W.W., 1996. Zinc/Air Fuel Cell: an alternative to clean fuels in fleet electric vehicle applications. International Journal of Environmentally Conscious Design and Manufacturing 5 (3-4), 49–54.

Gibson, R., 2017. Sustainable Partners" Presentations. www.sustainabilitypartners.com.

Lund, H., Clark II, W.W., Nov. 2002. *Management of Fluctuations in Wind Power and CHP: Comparing Two Possible Danish Strategies*. Energy Policy, Elsevier Press 27 (5), 471–483.

FURTHER READING

Aubrecht, G., 1994. Energy. Englewood Cliffs, NJ. In: Supplies of Energy and Related Materials Form the Basis of the World Economy, second ed. xiii.

Banks, F., October 03, 2004. Economic Theory and Update on Electricity Deregulation Failure in Sweden. Unpublished paper presented at Arne Ryde Memorial Conference on Nordic Electric Markets. Lund University, Sweden.

Blumer, H., 1969. Symbolic Interaction – Perspective and Method. Prentice-Hall, Englewood Cliffs, NJ, p. 1986.

Bollman, N., 2002. Regionalism: Public Policy Perspective for the State of California. Public Policy Institute, San Francisco, CA.

Brookings Report, December 8, 2016. Growth, Carbon, and Trump: State Progress and Drift on Economic Sgrowth and Emissions 'decoupling.

California Air Resources Board (CARB), 2001. Secondary Benefits of Zero Emission Vehicle Rules, Sacramento, CA, Sept 01.

California Sustainable Building Task Force, 2002. Building Better Buildings: A Blueprint for Sustainable State Facilities (Blueprint), Governor in Executive Order D-16–00, Feb 02. http://www.ciwmb.ca.gov/GreenBuilding/TaskForce.

CalPirg. Renewables Work, June 2002. Job Growth from Renewable Energy Development in California. Sacramento, CA.

California Consumer and Power Authority (CCPA), 2001.

Chomsky, N., 1968. Language and Mind. University of California Press, Berkeley.

Chomsky, N., 1975. Reflections on Language. Pantheon Books, NY.

Chomsky, N., 1980. Rules and Representations. Columbia Un. Press, NY.

Chomsky, N., 1988. Language and Problems of Knowledge. MIT Press, Boston, MA.

Circourel, A.V., 1964. Cognitive Sociology – Language and Meaning in Social Interaction. Free Press, New York, 1974/1984.

Clark II, W.W., Bradshaw, T., 2017. Agile Energy Systems: Global Lessons, 2n Ed. Elsevier Press.

Clark II, W.W., Cooke, G., November 2011. Global Energy Innovations. Praeger Press.

Clark, W.W., Fast, M., 2004. Qualitative Economics. Coxmoor Press.

Clark, W.W., Li, X., 2003. Social Capitalism.

Clark II, W.W., Sowell, A., November 2002. Standard Practices Manual: life cycle analysis for project/program finance. International Journal of Revenue Management. London, UK: Inderscience Press.

Clark, W.W., Kune, A., Feinberg, T., Kaplan, A., October 2003. Sustainable Environmentally Sound Technologies (EST); the Role of Public Policy. California Labor and Workforce Development Agency unpublished.

Clark II, W.W., Isherwood, W., January 2010. Inner Mongolia Autonomous (IMAR) Region Report, Asian Development Bank, 2008 and Parts Reprinted and Expanded on in Utility Policy Journal. Special Issue. environmental and energy sustainable Development, China.

Clark II, W.W., 2002a. Greening technology. International Journal of Environmental Innovation Management. Inderscience, London, UK.

Clark II, W.W., 2002b. Entrepreneurship in the commercialization of environmentally sound technologies: the American experience in developing nations. In: Kuada, J. (Ed.), Culture and Technological Transformation in the South: Transfer or Local Development. Samfundslitteratur Press, Copenhagen, Denmark.

Clark, W.W., January 2004. The California Hydrogen Economy. Data and Forum Summary released on CD.

Clark II, W.W., 2009. Sustainable Communities. Springer Press, NY, 2nd Ed Elsevier Press 2017.

Clark II, W.W., June 2010. Sustainable Communities Design Handbook. Elsevier Press, NY.

Coase, R.H., November 1937. The nature of the firm. Economics 4, 386–405.

Coase, R.H., 1988. The Firm, the Market, and the Law, Chicago. University of Chicago Press.

Coase, R.H., Winter 1993. The nature of the firm: influence. In: Williamson, O.E., Sidney, G. (Eds.), The Nature of the Firm: Origins, Evolution, and Development. Oxford Un Press.

Davis, G., January 31, 2002. Press Release: Letter to FERC Chairman Wood.

Deming, W.E., 1994. The New Economics: For Industry, Government, Education, second ed. MIT Press.

Environmental Goals Policy Report, October 2003. Governor's Office of Planning and Research, Available through the Local Government Commission. http://www.lgc.org. Sacramento, CA.

Fitzgerald, G., Bolinger, M., Wiser, R., September 2003. Green Buildings: The Expanding Role of State Clean Energy Funds. Lawrence Berkeley National Laboratory, US Department of Energy, pp. 1–11.

Gresser, J., Moulton, W., May 2017. "Alliances for Discovery Diagram" Big Heart Intelligence in Healthy Workplaces and Sustainable Communities.

Hawkins, P., Lovins, A., Hunter Lovins, L., 1999. Natural Capitalism: Creating the Next Industrial Revolution. Little, Brown and Company, Boston, NYC.

Isherwood, W., Ray Smith, J., Aceves, S., Berry, G., Clark II, W.W., Johnson, R., Das, D., Goering, D., Seifert, R., January 2000. UCRL-ID-129289: Published in Energy Policy. Remote Power Systems with Advanced Storage Technologies for Alaskan Village, vol. 24. University of Calif., Lawrence Livermore National Laboratory, pp. 1005–1020.

Jawetz, P., et al., 2003. Renewable Energy Technology Diffusion. Final Report. San Francisco, CA, pp. 1–32.

Kenidi, R., June 2003. Personal Communications.

Perkins, J., 1997. The Scientific Paradigm.

Porter, M.E., 1980. Competitive Strategy: Techniques for Analyzing Industries and Competitors. Free Press, New York.

Porter, M.E., 1990. The Competitive Advantage of Nations. Free Press, NY, NY.

Saxenian, A., 1994. Regional Advantage: Culture and Competition in Silicon Valley and Route 128. Harvard Un. Press, Cambridge, MA.

Schultz, D., 2001. Standard Practices Manual (SPM) State of California. Governor's Office of Planning and Research, Sacramento, CA.

Schwarznegger, G., January 18, 2004. State of the State Inauguration Speech.

Solar Opportunity Assessment Report, 2003. Solar Catalyst Group and Clean Edge Inc., Berkeley, CA, pp. 1–67.

Southern California Edison, 2003a. Economic and Development. Rosemead, CA. www.sce.com/ebd.

Southern California Edison, 2003b. Financial Resources Manual: A Guide to Capial Access and Financial Information. Economic and Development, Rosemead, CA. www.sce.com/ebd.

Southern California Edison, 2003c. Selecting the Right Technology for Your Business: A Small Business Guide. Economic and Development, Rosemead, CA. www.sce.com/ebd.

Teece, D.J., Spring 1998. Capturing value from Knowledge Assets: the new economy, markets for know-how, and the intangible assets. California Management Review 40 (3), 55–79.

Vesper, K.H., 1990. New Venture Strategies, Rev. Ed. Prentice Hall, Englewood Cliffs, NY.

Williamson, O.E., Sidney, G. (Eds.), Winter 1993. The Nature of the Firm: Origins, Evolution, and Development. Oxford Un Press.

Williamson, O.E., 1996. The Mechanisms of Governance. Oxford Un Press.

Wiser, R., Bollinger, M., 2003. Lawrence Berkeley National Laboratory, Berkeley, CA.

Ziman, R., December 2003. Acceptance Speech for "Sustainable Business Award", California League of Conservation Voters. Tenth Annual Meeting, Los Angeles, CA.

Chapter 8

National Solutions at Local Levels: Case of Japan

Andrew DeWit
Professor, School of Economic Policy Studies, Rikkyo University, Tokyo, Japan

Chapter Outline

Consider What Japan Has Done: Japan's Smart Cities and Communities

Andrew DeWit[a]
Professor School of Economic Policy Studies, Rikkyo University, Tokyo, Japan

The Ernst & Young (EY) authors point out that, in policy terms, Japan's smart cities were given a significant boost by the 2012 Japan Revival Strategy as well as the 2013 Abe administration's Japan Reconstruction Strategy. There

a. Andrew DeWit PhD

Climate Preservation In Urban Communities Case Studies. https://doi.org/10.1016/B978-0-12-815920-0.00008-3

certainly was ample finance: the Abe government more than tripled its Information and Communications Technology (ICT) spending from FY 2013's roughly JPY 300 billion to over JPY 1 trillion in fiscal year (FY) 2014.[1] The Abe cabinet also set up a government "chief information officer" (CIO) agency within the cabinet secretariat in June of 2013, specifically to coordinate ICT projects and avoid overlap and incoherence due to silos in the bureaucracy and business worlds.[2] The CIO agency has since maintained a relatively steady posting of information up to October of 2017. But the CIO's site has yet to itemize and publish a listing of smart cities and their budgets. Nor does it offer a guide to policy integration on smart and compact city projects.[3]

In the absence of an official list of Japanese taxpayer-financed smart cities, the EY study had to search for and add up all the central government initiatives it could find. The EY authors determined that Japan's national government agencies alone were undertaking over 160 projects. The EY results would appear to confirm a Japan Research Institute survey of April 2013 that determined Japan's government-sponsored smart city projects had increased from 22 prior to March 11, 2011, and reached well over 100 as of April 2013.[4] Moreover, the EY count of 160 projects did not include those being led by the prefectural as well as the smaller local government level as well as private sector—led initiatives. Many of these latter have some kind of direct or indirect subsidy (e.g., tax breaks), so they arguably should be included.

A lot of projects thus were not included in the EY assessment. Among Japan's many local government—led smart city projects were the "Smart Hikari Town Kumamoto Project" being undertaken in the newest of Japan's designated cities.[5] The prefecture of Saitama, just north of Tokyo, was also building "eco-towns" centered on smart grids and distributed generation (primarily renewable but also fuel cells) and conservation through home energy and business energy management systems as well as smart meters. Indeed, a January 2014 survey by the Kanto-area METI (the regional bureau)

1. On this spending increase, see (in Japanese) Okui, N., June 2, 2014. Details: The Big Picture on the 2014 ICT-Related Budget. In: *Nikkei BP ITPro*. http://itpro.nikkeibp.co.jp/article/COLUMN/20140527/559602/.
2. On this promised role of the cabinet in coordinating the ICT projects, see (in Japanese) p. 6 "This Time It's Different: Cabinet Leadership" in LDP Dietmember and ICT Strategy Special Commission Chief Hashimoto Gaku's June 14, 2013 presentation "The Abe Regime's ICT Strategy". http://ga9.cocolog-nifty.com/130614Interop.pdf.
3. The CIO's portal is here. http://cio.go.jp/policy.
4. Sato, K., (in Japanese), April 30, 2013. Issues and Projects Toward Realizing the Smart City. Japan Research Institute, Paper, no. 2013-02. https://www.jri.co.jp/MediaLibrary/file/report/researchfocus/pdf/6743.pdf.
5. Kumamoto City became Japan's 20th designated city on April 1 of 2013. The smart city project, a joint effort of Kumamoto City and Kumamoto Prefecture, is described (in Japanese) available at: http://www.hikarikumamoto.jp/index.html.

determined that among the roughly 500 local governments in the entire region, about 10% (44) were undertaking smart community initiatives.[6]

In addition to the subnational government-led projects, Japan featured a significant number of private sector–led smart cities, towns, and communities. These latter include 16 "smart towns" under development by the conglomerate Sekisui House.[7] Another significant development was Panasonic's Fujisawa SST "smart city" initiative.[8] Hitachi also has, not surprisingly, a smart city development underway in Hitachi City in addition to multiple other projects in Japan and overseas.[9] Toshiba, in turn, lists several initiatives, such as the greenfield Ibaraki City Smart Community Project in Osaka Prefecture among its ventures.[10] Mitsui Fudosan also had several projects underway, particularly within the metropolitan region of Tokyo.[11] If all of these public and private sector projects are added up, it is clear that in 2014, Japan had well over 200 smart communities. It is also clear that the projects were nationwide, even though there was significant clustering in Tokyo and the broader Kanto region.

There also appeared to be plenty of scope for growth. A contemporary survey of Japan's smart communities was undertaken by EcoNet Tokyo 62 (an association of the 62 governments in the Tokyo Metropolitan Region) between June 19 and July 4 of 2014.[12] The survey sampled all 62 of the EcoNet Tokyo 62 governments, and all of them responded. Among the survey results was evidence of a dramatic increase in regional awareness of smart communities. Moreover, whereas only two of the area governments were in the midst of deploying a smart community in 2012, the figure had risen to 10 by 2014. All

6. On the survey and its results, see (in Japanese) p. 6 of Kanto METI Natural Resources and Energy Division, "Towards the Establishment of a Kanto-Area Smart Community Collaborative Association (Tentative Name)," February, 2014. http://www.kanto.meti.go.jp/seisaku/smacom/data/20140219shiryo_3.pdf.
7. The Sekisui House "smart community" promotional literature includes a convenient map of their projects. http://www.sekisuihouse.co.jp/bunjou/smarttown/.
8. The project's promotional website was opened on April 26, 2014. http://fujisawasst.com/JP/.
9. Hitachi provides an overview, in Japanese, of several of the projects as "case studies" in a September 2015 special edition of its corporate magazine *Haitakku*. http://www.hitachi.co.jp/products/it/portal/info/magazine/hitac/document/2015/09/1509.pdf.
10. Toshiba's Ibaraki City project appears to have been in planning since 2008, according to (in Japanese) "Redevelopment of North Region's Demolished Factory Area Accelerates, with Universities, Hospitals, Residences and Other Elements Planned Between Hitachi and Panasonic," Construction News, March 18, 2014. http://constnews.com/?p=1552. Toshiba lists its domestic and global smart city projects at (in Japanese) "Toshiba Smart Community Centre". http://www.toshiba.co.jp/vs/casestudies/company/scc.htm.
11. Mitsui Fudosan Group's smart city strategy began in 2012, and is detailed (in English) here. http://www.mitsuifudosan.co.jp/english/corporate/csr/2013/special/smartcity/index.html?id=global.
12. The survey results (in Japanese), presented to the EcoNet Tokyo 62 research commission on October 8, 2014, are available at: http://all62.jp/saisei/meeting_h26/meeting_h26_03/meeting_h26_03_05.pdf.

told, in 2014, 22 of EcoNet Tokyo's 62 member governments were either initiating projects or preparing to, versus a total of 14 in 2012.

The results of the EcoNet Tokyo survey also showed a consistent focus on energy throughout, even among governments that were simply thinking about smart communities. In 2012, there were 156 replies (with multiple choices allowed) on the desired goals of the smart community. Of these, 40 sought increased residential energy efficiency, 31 increased office building energy efficiency, 17 community economic stimulation, 8 enhanced tourism, 11 increased industrial development, 8 area energy independence, and 36 opted for greater resilience of schools, hospitals, and other facilities. In other words, fully 79 of the 156 replies focused on energy, either through efficiency or generation.

By 2014, the total number of replies in the EcoNet Tokyo survey had risen to 176. Of these, 38 sought increased residential energy efficiency, 29 increased office building energy efficiency, 15 desired community economic stimulation, 11 enhanced tourism, 11 increased industrial development, 29 area energy independence, and 36 opted for greater resilience of schools, hospitals, and other facilities. That is, 86 of the 176 replies focused on energy, either in the areas of efficiency or generation. Resilience remained a steady concern. Moreover, the desire for localized energy independence (*eria ni okeru enerugii no jiritsuka*) leapt from 8 in 2012 to 29 in 2014, showing by far the greatest increase among all categories.

THE MASUDA REPORT: INTEGRATING DEMOGRAPHIC CHALLENGES

Simultaneous with the expansion of smart city policies and projects, the Japan Policy Council published a study on May 8, 2014 titled (translating directly) "Stop Declining Birth Rates: The Local Revitalization Strategy."[13] The report sought to galvanize debate and policy making on the intersection between aging and the economy. The report is often referred to as the "Masuda Report" after its chairman, Masuda Hiroya. The Masuda Report attracted widespread public attention with its stark warning that 896 local governments—roughly half the total—risked "extinction" by 2040 through further declines in their populations of young women. The report argued that the best response to regional decline would be a strategy of building "regional cities attractive to young people" by forging a "new structure of agglomeration" and a "choose and focus" strategy of investment. The report emphasizes the need to make these regional cities into the nodes of networks that function to "dam" the flow of younger people into the big cities. The Masuda Report was presented to Japan's Council on Economic and Fiscal Policy, the Industrial Competitiveness Council, and other ranking policy-making organs. Its warnings and

13. See (in Japanese) "Stop Declining Birth Rates: The Local Revitalization Strategy," Japan Policy Council, May 8, 2014. http://www.policycouncil.jp/pdf/prop03/prop03.pdf.

recommendations, and subsequent interventions by the same Japan Policy Council, became a core part of the Abe government's revitalization strategy.

The context of the Japan Policy Council report merits examination. The Masuda Report reflected a broad-based consensus among reformist politicians and policy makers in the national government and bureaucratic district of Kasumigaseki. Theirs is an ongoing technocratic effort to overcome sectionalism and systemic inertia, and thus deserves scrutiny based on the details that have emerged in Japan's quality press and specialist media. Masuda Hiroya, the report's lead author, is a former ranking bureaucrat from the Ministry of Land, Infrastructure and Transport (MLIT), who, from the mid-1990s, went on to become a notable governor of Iwate Prefecture and later a Minister of Internal Affairs and Communications (MIC). When Masuda was first elected governor of Iwate in 1995, he was 43 and the youngest-ever prefectural governor in Japanese history. Masuda subsequently served three full 4-year terms until 2007, building for himself a strong reputation as a fiscal reformer. During the reformist years of the former PM Koizumi's cabinets (2001–2006), Masuda made deep cuts to prefectural spending on public works as well as other areas of administration. Masuda also went so far as to apologize in the prefectural assembly, on March 2, 2005, for past spending on public works. Masuda also highlighted what he regarded as the core factors responsible for the prefecture's fiscal difficulties. Masuda emphasized the deleterious role of repetitive rounds of public works—the post-1980s bubble economy fiscal stimulus—as well as the special fiscal measures introduced in the Local Allocation Tax (Japan's primary mechanism for interregional fiscal redistribution) in order to compensate for inadequate revenues in its base structure.[14]

Masuda then served as MIC (and special minister for decentralization) between August 27, 2007 and September 26, 2008. During that time, Masuda turned his attention to localizing sustainable development, stressing demography. For example, Masuda established a measure that focused local policymaking resources to allow midsize cities more means to build attractive alternatives for young people and thus limit their incentives to relocate to large urban centers. This initiative became the "autonomous settlement region" (*teijuu jiritsuen kousou*), which the MIC inaugurated in 2008. The measure is anchored on "core" regional cities of at least 40,000 residents, building on transport, ICT, and other networks to link them with surrounding towns and villages and rationalize the region's distribution of health, education, and other services. In February 27, 2015, there were 85 of these regions, and that number had grown to 119 by October 5 of 2018. The program is financed with special incentive measures in the special Local Allocation Tax (LAT) (ordinarily used

14. Horiguchi, H., (in Japanese) "Masua Hiroya, Iwate Prefecture Governor: How I Put the Prefectural Primary Balance into the Black," *Foresight*, April 2007: http://www.fsight.jp/3341.

for emergencies). From FY 2014, these incentives were increased to JPY 85 million for the core city and JPY 15 million for each area community.[15]

Masuda also presided over the implementation of a "provisional measure" in the LAT. This measure sought to relieve intergovernmental fiscal inequalities through pooling revenues from a portion of the "local business tax" and the "corporate residence tax." These revenues were then distributed preferentially to fiscally weak local governments. Importantly, this measure was adopted after it proved impossible to secure the Ministry of Finance's agreement to move business taxes to the national level in return for a higher share of the consumption tax for local governments.[16]

Masuda's background and stance on issues related to local revitalization are of great significance. For one thing, his career's work indicates that he straddles the middle ground between bottom-up decentralization to encourage local initiative and top-down technocratic national planning. Masuda's background is reflected in the character of ongoing local revitalization policies, with their emphasis on incentives plus equity.

Moreover, Masuda also had ample influence in policy-making circles. Masuda not only had strong personal connections within the Abe government; his eponymous 2014 report saw the Abe government turn its attention to "local revitalization," something that was not prominent on its agenda prior to the Masuda Report. Abenomics thereafter was rebranded as "local Abenomics."

Masuda indeed came to be depicted as the "shadow minister for local revitalization" and an agent for driving change among bureaucrats and politicians. Japan's respected business daily, the *Nikkei* newspaper, emphasized these points in a June 9, 2015 article.[17] The article describes how Masuda was the core element of a broad-based initiative within the higher councils of government, bringing together ranking officials steeped in decades of public works–centered nation-building but now seeking means to construct a decentralized and sustainable political economy. Masuda's contacts in the Abe Shinzo cabinet include the chief cabinet secretary Suga Yoshihide, the most influential actor in the Japanese government aside from the prime minister. Also, Suga was himself MIC (and special minister for decentralization) between September 26, 2006 and August 27, 2007.

The political context suggested that Japan's local revitalization strategy would gain more headway were it linked to resilience and energy. Integration

15. Details on the programme and its financing are available (in Japanese) at the Ministry of Land, Infrastructure and Transport's regularly updated internet website on "autonomous settlement regions". http://www.soumu.go.jp/main_sosiki/kenkyu/teizyu/index.html.
16. This reform remains deeply controversial. See, for example, Yamamoto, I., (in Japanese), July 9, 2016. Masuda Hiroya and the TMG's loss of JPY 1 trillion in tax revenues. Yahoo News Japan. https://news.yahoo.co.jp/byline/yamamotoichiro/20160709-00059809/.
17. See (in Japanese), June 9, 2015. "Shadow Minister for Local Revitalization" Masuda Hiroya's Strategy for Incentivizing the Bureaucratic-Political Spheres. *Nihon Keizai Shimbun*. https://www.nikkei.com/article/DGXMZO87817770Y5A600C1000000/.

would help revitalization to make its way through the multiple veto points and hurdles involved in securing finance, selecting and approving viable revitalization programs, and supervising their effective implementation. All of these stages are fraught with potential difficulties that threaten to dissipate the focus and momentum of a "local revitalization" program that starts off rather ill-defined (being a work in progress) and lacking broad public support (though polls routinely indicate there is a potential popular base for action).

To be sure, some critics quickly wrote off local revitalization as "recycled" policy. Ignoring the Anthropocene and its externalities, they argued within the traditional framework of top-down versus decentralized governance. Hence, they called for a broader debate on restructuring regional governance so as to devolve most fiscal and administrative decisions away from the central government. One critic held that postwar Japan has been witness to decades of local and regional revitalization: "'Locally-focused' policies have literally been recycled in the form of industrial decentralization, nodal development, relocation of the functions of the capital, decentralization of corporate headquarters, 'hometown revitalization', and the development of 'regional core cities' and 'wide community areas.'"[18] This kind of doubt concerning the very relevance of intergovernmental policy and desire to engage in a protracted debate over a completely different strategy was indicative of the need for greater clarity in the current local revitalization program. As will be outlined below, Masuda's interest and activism on energy issues afforded one avenue that clarified flagship projects. Smart energy came to be the common theme bonding "local revitalization," "resilience," and other initiatives.

SMART ENERGY AND THE JAPANESE-STYLE *STADTWERKE*

Japan's smart energy technocrats—an expanding stream of intellectuals and bureaucrats—advocated structural reforms for diffusing a new paradigm of smart communities. The smart community paradigm is centered on distributed energy, but also encompasses other utility services, mobility, communications, governance, healthcare, and the other elements of modern urban life. One summary statement of the Japanese project was outlined (in Japanese) in *Smart Communities: A Smart Network Design for Local Government Infrastructure*. This book was organized by Japan's top mainstream energy intellectual Kashiwagi Takao and published on October 15, 2014.[19] In its initial chapter, written by Kashiwagi, the book describes how Japan's energy technocrats sought to use the feed-in tariff, *stadtwerke* (municipal business), power sector deregulation, and other key elements of Germany's green energy

18. Sasaki, N., February 23, 2015. Regional Revitalization: Another Perspective. The Japan News. http://www.yomiuri.co.jp/adv/chuo/dy/opinion/20150223.html.

19. See Kashiwagi, T., (Ed.), Smart Communities: A Smart Network Design for Local Government Infrastructure, Tokyo: Jihyosha, 2014.

transition as engines for something more ambitious. The Japanese technocrats were linking their project to fiscal and financial policy, with the potential to revitalize industry, build resilience, and bolster local democracy.

Kashiwagi's chapter depicts smart communities as the key item in Japan's Abenomics growth strategy. Kashiwagi had been making that assertion for well over a year in previous books, as well as numerous articles and smart community events.[20] It was an argument that deserves scrutiny. One reason for examining it is that Kashiwagi is an influential figure in Japanese energy policy-making circles. Kashiwagi designed Japan's first smart community, a 100% renewable microgrid project, for the 2005 Aichi World's Fair. He consistently argued that the focus of Japan's public works should shift from roads and bridges to resilient, smart energy systems that maximize efficiency and the uptake of renewable energy. After March 11, 2011, his main contribution, as a policy entrepreneur, was to help forge a broad public–private coalition that links decarbonizing smart energy systems (heat and power grids) with disaster resilience, spatial planning, and local economic development.

Kashiwagi was ideally positioned to coordinate the shift to smart communities. As an academic, Kashiwagi was professor at the Tokyo Institute of Technology from 2007, in 2012 becoming specially appointed professor and head of International Research Center of Advanced Energy Systems for Sustainability (AES Center). The AES Center was inaugurated in September of 2009, with funding from some of Japan's largest firms. Its 200 specialist researchers in environment and energy are focused on solar and fuel cells. As importantly, it also collaborates with the smart energy divisions of the leading power, gas, and other firms in Japan's energy business, along with blue-chip firms in construction, home building, engineering, automaking, and other areas. Moreover, the AES Center includes 15 local governments (prefectures and cities), many of which are exemplary smart communities.[21]

Kashiwagi was also a top-rank policy advisor. In the years preceding March 11, 2011, he had served as a member or chair of the Agency for Natural Resources and Energy's (ANRE's) main energy policy-making councils, including the Advisory Committee for Natural Resources and Energy (ACNRE), the Strategic Policy Committee of the ACNRE, and the Energy Efficiency and Renewable Energy Committee of the ACNRE, in addition to numerous other national and subnational bodies.

Kashiwagi became an even more important policy entrepreneur after March 11, 2011. From May 17, 2011 to August 1, 2011, Kashiwagi led the

20. Among myriad examples, see (in Japanese) Kashiwagi, T., August 6, 2012. Growing a JPY 500 Trillion Industry: Japan Should Now Take the Initiative to Develop a Growth Strategy Built on Smart Cities. Nikkei Smart City Consortium. http://bizgate.nikkei.co.jp/smartcity/interview/000538.html.

21. The corporate and local-government membership of the AES Center is listed (in English and Japanese) at its website. https://aes.ssr.titech.ac.jp/partner.

"Commission on Making the Effective Use of Heat Energy and Urban Planning."[22] The commission included major business leaders as well as observers from METI, MLIT, Tokyo Metropolitan Government (TMG), and other agencies. It met six times in rapid succession and plowed through an enormous amount of material on how to deploy smart heat systems to slash emissions and maximize renewables while building disaster resilience. The commission's August 1, 2011 wrap-up report is a blueprint for many of the policy reforms Kashiwagi and his collaborators have been doing since in spatial planning, national resilience, energy policy, and other areas that are now explicitly integrated (in formal administrative terms as well as via this cohort of smart energy technocrats).[23]

From November 7, 2014, Kashiwagi also chaired the research group that developed the planning and financing mechanisms to implement the MIC's "Locally Led Community Energy System Improvement Research Commission."[24] Like other commissions, this was a collaborative initiative including MIC, METI, Ministry of Environment (MoE), Ministry of Agriculture, Forestry and Fisheries (MAFF), the finance agency, and several local governments. That commission apparently remains in operation (meeting most recently on January 11 of 2017), developing a "one-stop" approach so that local communities can get all the assistance they need. They have organized a "task force" of METI, MIC, MAFF, MoE, and other agencies to go and help the locals apply for subsidies, revise their energy master plans, and secure expert advice. For well-done plans, they will help with technical advice on including the best ICT sensors, storage, and other options to maximize the effectiveness of the smart energy system. And they emphasize local government leadership as well as the role of local actors (such as Small and Medium-Sized Industries [SMEs]), as this is part of local revitalization.

Kashiwagi was also named to several major committees that are at the nexus of Japan's ambitious and energy-centered industrial policy. For example, in 2013, Kashiwagi became chair of the main working group Strategy Conference (*senryaku kyougikai*).[25] The Strategy Conference groups key scholars and business interests involved in smart energy and CC. It role is to deliberate

22. The METI website for the Commission on Making the Effective Use of Heat Energy and Urban Planning is available (in Japanese) here. http://www.meti.go.jp/committee/kenkyukai/energy_environment.html#netsu_energy.

23. The August 1, 2011 interim report by the Commission on Making the Effective Use of Heat Energy and Urban Planning is available (in Japanese) here. http://www.meti.go.jp/committee/kenkyukai/energy/nestu_energy/report01.html.

24. Details on the membership and meetings of the Locally Led Community Energy System Improvement Research Commission are available (in Japanese) at the following MIC website. http://www.soumu.go.jp/main_sosiki/kenkyu/jichitai_energy/index.html.

25. The Energy Strategy Conference's membership and details of its meeting can be found (in Japanese) at the Cabinet Office website. http://www8.cao.go.jp/cstp/tyousakai/juyoukadai/wg.html.

scientific innovation strategies. More recently, its deliberations have centered on the role of energy within Japan's overall "Society 5.0" innovation strategy. It pays increasingly close attention to smart networks and resilience. Akin to Germany's Industry 4.0 initiative,[26] Japan's Society 5.0 project seeks to harness the technologies of the fourth industrial revolution. But, incentivized by disasters, demographics, and other challenges, Japan's effort transcends Germany's smart factories and aims to deploy smart systems throughout the entire society.

Kashiwagi also chairs the working group inaugurated to achieve the goals of the National Energy and Environment Strategy for Technological Innovation Towards 2050 (NESTI 2050), in concert with top-ranking representatives from business, academe, and administration.[27] The NESTI 2050 was unveiled on April 19, 2016 by the Council for Science, Technology and Innovation. Like many of Japan's energy-related policies, the NESTI 2050 emphasizes the role of smart communities and smart energy systems. These systems are explicitly portrayed as an important basis for achieving Japan's climate goals announced at the 2015 Paris Climate Talks. The working group held its initial meeting on December 14 of 2016.

In addition to his prominence in academe and policy making, Kashiwagi is also directly involved in business circles that embody the ongoing revolution in smart energy. Kashiwagi's multiple roles in academe, policy making, and business placed him at the intersection of the Japanese public and private sectors.[28]

A second reason for scrutinizing Kashiwagi's argument is that he was using his policy-making roles quite effectively. It is difficult to exaggerate Kashiwagi's coordinating role in Japan's energy policy-making circles and policy changes. For example, the evidence indicates Kashiwagi that he helped realize Japan's June 14, 2013 New Growth Strategy's explicit commitment to ICT and energy-centered growth.[29] He also certainly had a hand in coordinating the expansion of smart community projects and the increasing streams of finance flowing from the various ministries of the central government plus their

26. On Germany's Industry 4.0 industrial policy, see "Industrie 4.0: Smart Manufacturing for the Future, Germany Trade and Invest, July 2014. https://www.gtai.de/GTAI/Content/EN/Invest/_SharedDocs/Downloads/GTAI/Brochures/Industries/industrie4.0-smart-manufacturing-for-the-future-en.pdf.

27. The working group's membership and the particulars of its meetings are available (in Japanese) at the Cabinet Office's website. http://www8.cao.go.jp/cstp/nesti/suishin/index.html.

28. On Kashiwagi's role, see DeWit, A., "Japan: Response of policy entrepreneurs to an energy crisis," in (Patrice Geoffron, Lorna Greening, Raphael Heffron Eds.) Meeting the Paris mandate: A cross-national comparison of energy policy-making. Springer-Verlag (forthcoming, 2018).

29. See Kashiwagi's arguments (in Japanese) detailed in Kashiwagi, T., May 17, 2013. Energy Policy as the Pivot for the Growth Strategy. *Nikkei Bijinesu*. http://business.nikkeibp.co.jp/article/opinion/20130513/248005/.

affiliated quangos (quasiautonomous nongovernmental organization) such as the New Energy Development Organization (NEDO).

Kashiwagi's vision, as articulated in his 2014 book chapter, puts the motive force for rolling out Japan's smart communities in a vehicle akin to the German *stadtwerke* of municipally owned utilities. Germany's 900 or so *stadtwerke* that operate in energy—out of a total of 1420 that operate in water supply, sewage, waste management, and other community functions[30]—were among the major winners from German power deregulation. They are also increasingly recognized as key to that country's ability to diffuse renewable energy, whose role in Germany's power mix has risen from about 6% in 2000 to 30% in 2014. The *stadtwerke* have helped drive this impressive progress because they have organizational, financial, and other heft together with a central role in servicing community demand for power.[31]

30. On this, see p. 5 of Julian, C., July 2014. Creating Local Energy Economies: Lessons from Germany. ResPublica. http://www.respublica.org.uk/our-work/publications/creating-local-energy-economies-lessons-germany/.

31. See Hockenos, P., September 28, 2013. Local, Decentralized, Innovative: Why Germany's Municipal Utilities are Right for the Energiewende. Energy Transition. http://energytransition.de/2013/09/local-decentralized-innovative-why-germanys-municipal-utilities-are-right-for-the-energiewende/.

Stadtwerke Japan's Local Public Corporations

Andrew DeWit
Professor, School of Economic Policy Studies, Rikkyo University, Tokyo, Japan

OVERVIEW

The earlier paper in Chapter 9 explained part of the institutional legacy that Japanese energy technocrats are working with. Their use of the *stadtwerke* model as the engine for rolling out Japanese smart communities is significant. The Japanese vision appears to be more ambitious than the German efforts: while Germany is a leader on diffusing renewable power per se, it appears to be rather a laggard on the EU rollout of smart cities, perhaps because it can export and import power.[32] The fourth-largest economy in the world, Germany's Feed-in Tariff (FIT)-driven energy shift has had a profound impact on the global public debate and policy making. Yet if Japan's incipient model finds traction, it may have an even larger impact. Japan is the third-largest economy, well over USD 1 trillion larger than Germany, and now in the midst of an unprecedentedly large economic experiment. Japan's unparalleled need for sustainable economic growth dovetails with an increasingly powerful role of innovative local governments working in tandem with smart elements of key central agencies. The *stadtwerke* approach not only puts the local community in charge of the smart community deployment; it also institutionalizes that leadership in a vehicle with the financial, administrative, and other means to take effective action. Japan's model may open an alternative to building smart communities led by behemoths of the private sector. Properly financed and incentivized local governments seem unlikely to allow large corporate actors simply to siphon income from core urban infrastructures and do as they will with data on the residents' consumption, movements, and other interactions in the smart community.

Japan's own history of local public corporations also offers a fertile basis for the insertion of energy-centered *stadtwerke*. Japan's 1700-plus local governments have long had public corporations servicing such utility functions as water. Japan's local special corporations numbered only 45 in 1953, but had grown to 3000 in the wake of the high-growth 1960s.[33] Japan has seen waves of expansion of local public corporations' roles and numbers as the need arose.

---

32. See p. 32 European Parliament Directorate General for Internal Policies, "Mapping Smart Cities in the EU," January 2014. http://energytransition.de/2014/06/remunicipalization-of-hamburg-grid/.
33. Samuels R., 1983. The Politics of Regional Policy in Japan: Localities Incorporated? Princeton University Press, Princeton, p. 38.

Their ranks swelled rapidly in the postwar years of high growth due to the imperative of diffusing such basic infrastructures as waterworks and sewerage in cities and towns undergoing what was then an unprecedented pace of urbanization. Later on, the public corporations' welfarist role grew between 1975 and 1984, and for industrial promotion between 1985 and 1995.[34]

The total number of Japan's local governments sharply declined over those decades. Amalgamations between 1953 and 1955 saw their count more than halved from 10,520 in October of 1945 to 3975 in September of 1956. They continued to merge afterward, reaching 3229 in 2000, and then dropped to 1718 in April of 2014.[35] That decline in the number of local governments, even as the number of local public corporations increased, underscores the significance of the latter's role.

As of FY 2016, Japan's 1718 local governments managed a total of 8534 public corporations, of which 3639 (42.6%) manage the sewer systems and 2041 (23.9%) the water supply. These businesses total (FY 2016) JPY 16.9 trillion in operations and are managed through special accounts that are separate from Japan's local government general budget (whose spending on sanitation, education, public works, and other categories is just under JPY 95 trillion). Of the public corporations' JPY 16.9 trillion in operations, the sewerage works represent JPY 5.4 trillion (32.3%) and the waterworks JPY 3.9 trillion (23.5%).[36] These are, in short, significant local public service business operations that contribute to the economic activity of the local area as well as to the revenue base of the local governments.

Japan's local public corporations have almost no presence in electricity and gas, representing only 0.9% of the power supply and 2.6% of gas supply. These are key infrastructures for growing the smart community, as well as diffusing economic opportunity to nearby communities that could supply larger communities with power and energy. Postwar Japanese prefectures, cities, and towns have been passive consumers of centralized privately owned power delivered by 10 monopoly firms that also dominated their respective catchment areas' political economies. Moving to smart communities and distributed power, through public agency, offered one means to bolster local communities as well as disrupt the old business model of the power utilities.

When viewed against this background, the technocrats' initiatives appear to have an unstated but potentially quite "political" dimension. As noted, the

34. Sakamoto, M., 1996. Public Corporations in Japan, with Special Emphasis on Personnel Management. In: All Farazmand (Ed.), Public Enterprise Management: International Case Studies. Greenwood Press, p 138.

35. See (in Japanese), nd. The Particulars of Meiji and Showa-era Amalgamations and Changes in the Number of Local Governments. Ministry of Internal Affairs and Communications. http://www.soumu.go.jp/gapei/gapei2.html.

36. The details on Japan's local public corporations are available (in Japanese) at the Ministry of Internal Affairs and Communications 2016 Overview of Local Public Corporations. http://www.soumu.go.jp/main_content/000510325.pdf.

power *stadtwerke* in Japan offer a mechanism that puts the incentives to champion revolutionary change, leading smart communities, into the hands of the cities and towns. The Ministry of Internal Affairs and Communications (MIC), a fortuitous blend of ICT enthusiasm coupled with responsibility for local fiscal health, set a goal of establishing no fewer than 1000 local energy firms over the 5 years from 2015. The national government would not only allow local governments to finance investments in these firms, but would pick up half the interest payments.[37] Moreover, Kashiwagi helped coordinate these initiatives as chair of the MIC "Commission for Deploying a Local-Government-Led Community Energy System." This commission was created by MIC on November 4 of 2014, and held its first meeting on November 7. It had three more meetings, seeking to devise a template for local government decentralized energy systems, through to March of 2015.[38] These and other policy initiatives suggested that smart energy bureaucrats in the MIC were acting quickly, using the Abe regime's desperation to ignite sustainable domestic growth via a focus on "local Abenomics" and "regional revitalization" since mid-2014.[39]

By the summer of 2017, these initiatives had resulted in 31 locally led energy firms, with a further 86 under consideration. These finding were the product of a 2017 survey of Japan's local governments by the Asahi newspaper and Hitotsubashi University. The survey response rate was 79% (1382 of Japan's 1788 local governments replied).[40] Importantly, many of the projects listed (such as Hamamatsu City, Miyama City, and others) were fostered by the pragmatic and coordinated, post-March 11, 2011 disaster resilience policy making described above.

Making Choices

The technocratic strategies underlying this development had a long background. Kashiwagi and his cohort had studied the German city business model for several years. They determined that German *stadtwerke* would provide a robust model for bundling postderegulation expertise and institutional clout

37. On this, see "Small-town Japan's big hopes for energy self-sufficiency," *Nikkei Asian Review*, October 28, 2014: http://asia.nikkei.com/Politics-Economy/Policy-Politics/Small-town-Japan-s-big-hopes-for-energy-self-sufficiency.
38. See (in Japanese), November 4, 2014. Opening of a Commission for Deploying a Local-Government-Led Community Energy System. Japanese Ministry of Internal Affairs and Communications (MIC). http://www.soumu.go.jp/menu_news/s-news/01gyosei05_02000053.html.
39. For a summary of the politics of "local Abenomics," see Sieg, L., Kajimoto, T., September 2, 2014. Japan's 'Abenomics' feared in trouble as challenges build. Reuters. http://www.reuters.com/article/2014/09/02/us-japan-economy-abenomics-idUSKBN0GX0VY20140902.
40. See Ishii, T., Kobori, T., (in Japanese), August 14, 2017. 31 Local Governments Doing Retail Power Businesses, Aiming at Local-Production/Local-Consumption. Asahi Shimbun. http://www.asahi.com/articles/ASK8F5FS8K8FULBJ002.html.

(e.g., for raising capital) without having to rely on unpredictable, distracted party politics or weak and fractious popular movements. But in addition, the public sector–led approach means that the Japanese smart community would not merely be a construct of Hitachi, Panasonic, and the other market players that would like to lead the smart city rather than be led.[41] Reading through the technocrats' work (as well as studies from within the MIC[42]) suggests their deliberations became increasingly detailed as they realized the post-March 11, 2011 period offered an opportunity to break through the dominance of the monopoly utilities and turn the energy political economy towards "prosumer" (power producer and consumer) cities.

Aside from the content of his recent activism and writing, what made Kashiwagi's involvement especially interesting is precisely that he is both a core member of the so-called "nuclear village" as well as an enthusiast for renewable energy.[43]

Kashiwagi straddled both Japan's deeply damaged paradigm of nuclear power as well as its rapidly emerging paradigm of distributed power and smart communities. He made sure to preface his remarks by expressing fealty to official "power mix" targets of roughly 20% nuclear by 2013, but would then spend the bulk of his attention on the role of the smart community. The diffusion of the smart community, and other forms of "local production-local consumption" (*chisan chishou*), posed a profound threat to traditional centralized models of power generation and transmission. This fact was already a point of expert debate in the global discourse.[44] In consequence, Kashiwagi was articulating a radical, disruptive program from within the core of policy-making circles. He was, however, pragmatic and careful to stress the resilience, interregional equity, and other benefits. This narrative arguably blunted the vested interests' capacity to stymie the deployment of distributed technologies that threatened their business models.

---

41. On this very important question of whether the public sector will lead, see Marshall, A., February 2014. Big Data, Big Questions," Metropolis. http://www.metropolismag.com/February-2014/Big-Data-Big-Questions/.
42. The role of the *stadtwerke* in smart communities is especially well depicted in MIC's Local Revival Group's Local Policy Division's May 13, 2014 presentation to the governing LDP. See (in Japanese) "Concerning the Project on Distributed Energy Infrastructure". https://www.jimin.jp/policy/policy_topics/pdf/pdf170_4.pdf.
43. In a March 19, 2010 presentation, for example, Kashiwagi argued that an energy system shift was underway, towards low-carbon sources including nuclear power, "clean coal," natural gas as well as renewables linked together via ICT and smart communities. See (in Japanese) Kashiwagi Takao "The Smart Community Concept and Environmental Business," Kanto METI March 19, 2010. http://www.kanto.meti.go.jp/pickup/kankyoryoku/event/20100319kaisai_gaiyou.html.
44. Hannes, B., Abbott, M., April 13, 2013. Distributed energy: Disrupting the utility business model. Bain Brief. http://www.bain.com/publications/articles/distributed-energy-disrupting-the-utility-business-model.aspx.

The severity of Japan's economic and other crises also puts a premium on making smart choices in the midst of multiple constraints. In this context, Kashiwagi opted to side with the general interest rather than the sectoral interests that dominated Japan's energy economy. This position was hardly a surprise: Kashiwagi designed Japan's first smart community, a 100% renewable microgrid project, linking NEDO and other facilities, that went functional at the 2005 Aichi World's Fair.[45] He had much intellectual capital invested in these initiatives. The power monopolies, with their focus on centralized power and control of the grid as the core of their business model, stood in the way of a nationwide diffusion of smart communities. Hence, actual deregulation of the power sector was key to Kashiwagi's argument. So too was the diffusion of distributed renewable power supported by the FIT. Kashiwagi argued that Japan's local governments stand to gain JPY 5 trillion of Japan's JPY15 - trillion power economy through distributed renewable energy supported by the FIT. He represented that opportunity as a massive boost for their finances as well as their local economies.

Importantly, one of Kashiwagi's collaborators on the community energy projects is Masuda Hiroya. We have already seen that Masuda is a major force in local revitalization. Masuda was also an advocate of renewable energy while governor of Iwate, having overseen the installation of 43 wind turbines in the prefecture's Kamaishi City while governor.[46]

Both Kashiwagi and Masuda had also served as key members of the "Plan Subcommittee" (*keikaku bukai*) of the "National Lands Deliberation Commission" (*kokudo shingikai*). The subcommittee played a central role in overseeing Japan's National Spatial Strategy (NSS), adopted in August 2015. Kashiwagi was the first energy expert to serve on Japan's spatial planning, and Masuda was alternate head of the subcommittee.[47] The 2015 NSS is a quite significant document, one the OECD *Territorial Review of Japan* (2016) deems "the most important of a number of key planning documents." The OECD added that unlike prior spatial plans, including the previous 2008 NSS,

45. On this see (in Japanese) Kashiwagi's description of the project in his article for the Ministry of Land, Infrastructure and Tourism (MLIT), "The Smart City: Achieving Both Economic Development and Environmental Measures," *MLIT Shinjidai*, Vol 71, February 2011: http://www.mlit.go.jp/kokudokeikaku/iten/service/newsletter/i_02_71_1.html.

46. Masuda gave a keynote talk on this and the need for distributed and smart-energy initiatives to a November 18, 2011 conference of the Japan Science and Technology Agency. See the attached transcript at Masuda H. (In Japanese) "Low-Carbon Society and Building Sustainable Communities," November 11, 2011, Keynote speech to Japan Science and Technology Agency conference on Towards a Low-Carbon Society Via Energy Conservation and Efficiency in Communities, Tokyo, November 11, 2011: http://www.jst.go.jp/lcs/result/agora20111118/index.html.

47. The membership list of the "Plan Subcommittee" (*keikaku bukai*) of the "National Lands Deliberation Commission" (*kokudoshingikai*) for the relevant period (2014–15) is available (in Japanese) at the MLIT website. http://www.mlit.go.jp/policy/shingikai/s103_kokudo_keikaku.html.

the 2015 NSS was a "truly horizontal initiative," one built on the basis of "an intensive exercise in inter-ministerial co-ordination and consultations extending beyond the government itself under the aegis of the National Land Council, which brings together parliamentarians, academic experts, representatives of the private sector, elected officials from the cities and regions, and others." The NSS was composed in a "whole of government" approach that brought together the MLIT and the other central agencies as well as the subnational governments. In addition, the consultation with civil society was also unprecedented in its breadth. This degree of consultation gave the NSS a legitimacy that transcended its predecessor documents. That legitimacy is backed up by legal authority because at least 20 other national laws are obligated to refer to the NSS.[48]

The 2015 NSS also paid careful attention to smart communities, renewable energy, CC, resilience, and other factors as the context for urban policy. This was a sharp contrast with the previous 2008 NSS, evident in a survey of the frequency of several keywords in the 2008 NSS with the 2015 NSS. For example, the word "energy" appeared only 54 times in the 2008 NSS, but the 2015 NSS had 207 references to "energy." Similar results were seen for "compact" (as in spatial densification), "renewable energy," "smart community," and "distributed" (in reference to distributed energy).

The comparison, coupled with a thorough reading of the two texts, shows that the 2008 NSS was concerned with disasters and the transport and other networks that are critical to economic activity and responding to crises. But the 2015 NSS displayed a far greater concern for climate and other disaster threats, as would be expected for a plan developed in the wake of the March 11, 2011 disaster. Yet the NSS 2015 also reflected the emergence of a very different, distributed network paradigm for coping with disaster threats as well as aging and other challenges noted in the introduction to this chapter.[49]

The fruits of Kashiwagi and Masuda's collaboration in such high-level planning and other initiatives was also seen in a 2015 book (in Japanese), *The Cogeneration Revolution*.[50] In the book, Kashiwagi and Masuda discussed how an energy-centered community building and local economic revitalization can help not only to build a new growth paradigm but also to alleviate the crisis of rural depopulation and the threat of "local extinction."

In *The Cogeneration Revolution*, Masuda referred to public works as having been the primary Japanese mechanism of interregional redistribution. He represented this as a contrast to what he saw as the German innovation of interregional redistribution via the feed in tariff and opportunities in energy. In

48. OECD, 2016. Territorial Reviews: Japan 2016. OECD, Paris, pp. 79–80.

49. Both the 2008 NSS and the 2015 NSS are available (in Japanese) at the following MLIT website. http://www.mlit.go.jp/kokudoseisaku/kokudokeikaku_fr3_000003.html.

50. Kashiwagi, T. (Ed.) (in Japanese), 2015. Cogeneration Revolution: A New Card for Growth in Local Revitalization. Nikkei BP Consulting.

tandem with this line of thinking, Kashiwagi emphasized the opportunity to bolster the energy system and decentralize it through the deployment of cogeneration as well as the enhancement of reliance on solar, wind, geothermal, and other local renewable energy resources. Kashiwagi portrayed smart grids and smart communities as being key networks in constructing the framework for this decentralization of energy. He also depicted both as infrastructural investment that lends itself readily to participation by the private sector.

Masuda agreed and argued that facilitating private sector investment is perhaps the most important condition. He conceded that the previous era's investments in infrastructure helped greatly to network the national economy and foster industries. But he emphasized that it is too costly to maintain the same approach in an era when the focus is shifting from supporting industry to supporting livelihoods. He also noted that the diffusion of infrastructure throughout the country had despoiled the environment and led to a debt burden that is difficult to pay down. He added that technological and other changes were leading to the need for different kinds of infrastructure networks, including ICT as well as distributed energy. He pointed out that the diffusion of distributed energy could bring the additional benefit of helping to alleviate Japan's mounting debt burden.

Kashiwagi illustrated productive energy infrastructure as waste heat pipelines linking local communities with incineration plants to make use of the waste heat. He noted that the cost of this infrastructure often makes it difficult to construct. However, he noted that the Ministry of Internal Affairs and Communications fortuitously unites the former Ministry of Posts and Telecommunications and the former Ministry of Local Autonomy. The old Posts and Telecommunications Ministry had jurisdiction over many underground infrastructure networks for communications cables. The replacement of these communication cables with fiber-optic cables had left considerable space in the communication networks. Kashiwagi argues that this empty space could be used for the deployment of heat pipelines. This approach would lower construction costs and adhere to the Japanese state's policy of maximizing the "effective use of extant infrastructure stock" (*kizon no infurasutokku no yuukoukatsuyou*), a policy that had emerged from the Koizumi-era reforms and become common sense in the 2010s.[51]

The lower costs of installation for the thermal networks arose from the fact that the thermal networks could be placed within existing utility corridors power cables, leading to a network of fiber optics, heat pipes, and wires. Kashiwagi argued that the business opportunity from distributing heat and power through these networks would allow for it to be a very viable option for

51. This policy was a guiding theme in regional policy. See, for example, (in Japanese), 2008. Strategic Infrastructure Management: Smart Use, Densification, and Reform. Ministry of Land, Infrastructure and Transport. http://www.mlit.go.jp/sogoseisaku/region/sogoseisaku_region_fr_000026.html.

productive public works. He added that the business generated would provide a viable economic base on which to build communities complete with daycare and other amenities that lead to increasing numbers of young people. Masuda agreed that the energy-centered community offered a new approach to overcoming the inertia of the past practice in regional development.

The Significance of Japan's Adaptation-Oriented Smart Cities

We have seen that the Japanese policy regime came to focus on resilience in the wake of March 11, 2011. How Japan restructures its urban communities is potentially important for a global community confronted with rapidly worsening CC and increasing resource constraints. As many expert analysts and organizations argue, these crises will have to be resolved—or at least alleviated to manageable levels—in cities. This is because cities comprise 3% of global land area but produce 50% of global waste and about 70% of global anthropogenic greenhouse gas emissions. Cities also consume nearly 75% of natural resources while generating about 80% of global GDP.[52] Over half the total global population of 7.6 billion[53] already lives in cities. The United Nations and other projections suggest 2 billion new urban residents will likely be added by 2030, during which time perhaps over 2 billion new "middle-class" consumers may increase the ranks of the present 3.2 billion. It is estimated that 65% (or just under 3.5 billion) of these middle-class consumers will be in the Asia Pacific region.[54]

Given the above, unprecedented resource efficiency and substitution is imperative. The United Nations Environment Programme (UNEP)'s June 5, 2014 report on "Decoupling 2: Technologies, Opportunities and Policy Options" warned that between 2000 and 2010, metal prices increased by 176%, rubber by 350%, and energy sources by a rough average of 260%.[55] These price signals of unsustainability were evident after just a decade and a half in which some 260 million Chinese had migrated to cities. Though prices dipped for a short period in the early 2010s, they began climbing again middecade. In its October 2017 "Commodity Markets Outlook," the World Bank noted that

52. The figures cited are available in the United Nations Environmental Programme's "Global Initiative for Resource Efficient Cities". https://resourceefficientcities.org.

53. The figure is for 2017, and cited in "World Population Prospects: Key Findings and Advance Tables, 2017 Revision," United Nations Department of Economic and Social Affairs, Population Division, 2017. https://esa.un.org/unpd/wpp/Publications/Files/WPP2017_KeyFindings.pdf.

54. See Kharas, H., February, 2017. The Unprecedented Expansion of the Global Middle Class. Global Economy and Development Working Paper 100. Brookings Institute. https://www.brookings.edu/wp-content/uploads/2017/02/global_20170228_global-middle-class.pdf.

55. See p. 23 of the report itself as well as summary "Factsheet" at the website report's UNEP website. https://www.unenvironment.org/resources/report/decoupling-2-technologies-opportunities-and-policy-options.

energy prices were set to increase and that there were substantial geopolitical, supply ,and other risks.[56]

The forces driving these price increases are myriad. Similar to the UNEP's study, the February 2015 OECD report on "Material Resources Productivity and the Environment" warned that the sheer amount of material extracted, harvested, and consumed globally had ballooned 10-fold between 1900 and 1980, reaching roughly 35 billion metric tons (Gt). Between 1980 and 1990, this volume rose yet again to reach 43 Gt. The volume then further expanded to 72 Gt between 1990 and 2010. During the same 1990 to 2010 period, the OECD share of these material flows fell from 43% overall to 27%, while the BRICS share, led by China and India, rose from 34% to 61%. Meanwhile, the non-OECD, non-BRICS, "rest of the world" share of this growing mountain actually declined slightly from 22% to 21%. But under business as usual, the non-BRICS, non-OECD share of material flows is likely to increase as the total is slated to exceed 100 Gt by 2030.[57] Indeed, the UNEP report projected that total material flows could reach 140 Gt by 2050, given business as usual.

The scale of "business as usual" was outlined succinctly by the OECD. Its report revealed that the average daily per-capita consumption of materials in OECD countries in 2011 was as follows: 10 kg of biomass, 18 kg of construction and industrial minerals, 13 kg of fossil energy carriers, and 5 kg of metals.[58] That volume of material consumption is one of the reasons for the onset of the Anthropocene era.[59]

More striking is the calculation of human impact over time. Vaclav Smil, the noted student of energy and materials, calculated the increase of global "anthropomass" (the sheer mass of homo sapiens) between 1900 and 2000. Smil's calculations found that the anthropomass increased from 13 megatons of carbon (Mt C) in 1900 to 55 Mt C in 2000. Over the same period, the stock of wild terrestrial mammals halved, from 10 Mg C to 5 Mt C, while domesticated animals (especially cows and pigs) rose from 58 Mt C to about 200 Mt C.[60]

---

56. See World Bank, October 2017. *Commodity Markets Outlook*, International Bank for Reconstruction and Development/World Bank. http://pubdocs.worldbank.org/en/743431507927822505/CMO-October-2017-Full-Report.pdf.
57. The report's overview and relevant data can be found at OECD, Material Resources, Productivity and the Environment, Paris: OECD, February, 2015. http://www.oecd.org/env/waste/material-resources-productivity-and-environment.htm.
58. See OECD, February, 2015. Material Resources, Productivity and the Environment. OECD, Paris, p. 9.
59. See Krausmann, F., et al., October 2017. Material Flow Accounting: Measuring Global Material Use for Sustainable Development. Annual Review of Environment and Resources 42. http://www.annualreviews.org/doi/full/10.1146/annurev-environ-102016-060726.
60. See Smil, V., December 2011. Harvesting the Biosphere: The Human Impact. Population and Development Review 37, 4. http://vaclavsmil.com/wp-content/uploads/PDR37-4.Smil_.pgs613-636.pdf.

Moreover, this amount of material consumption omits the much larger and increasing "overburden" and other waste removed to extract useful resources. These materials do not enter the flow of consumable resources, yet also cost more and more energy as well as equipment and other inputs to dig up, separate, and throw away. There is as yet no consistent international accounting for this overburden and other waste. But in the United States alone, these "extractive wastes" are estimated to total roughly 10 Gt per year.[61]

"Business as usual" would mean a 140-Gt flow of resources in 2050, thrice the gargantuan figure for 2000. The UNEP argues that it would probably exceed the limit of all available resources as well as this planet's capacity to absorb the massive ecosystem damage of extracting it. So consider again the anticipated 3 billion more middle-class consumers, a near doubling of the present 1.8 billion who collectively account for 80% of current material consumption. Math, physics, and other realities indicate that urbanization and middle-class lifestyles simply cannot follow the resource- and energy-intensive pattern common to the developed world and rapidly diffusing throughout the BRICS. Without radical change, resources and energy prices seem likely to increase significantly, especially due to interactions with water and other constraints. Moreover, accelerating CC will—as even the US military has warned—become increasingly marked by extreme weather, declining crop yields, expanding disease outbreaks, and resource wars.[62]

Done collaboratively, democratically, and quickly, the smart city paradigm may help steer humanity away from—or at least minimize—the fraught outcomes detailed by the UNEP, the OECD, the US military, and a host of other agencies. The smart initiatives build on past programs for resource efficiency, especially in urbanization, by putting cutting-edge monitoring and management technology into the mix. In Japan, as elsewhere, the smart city essentially blends the ICT in, to take a trivial example, a smart phone with such large-scale and resource-intensive urban infrastructures as power and energy systems, waterworks, transportation, waste treatment, street lighting, administration, and healthcare. Measuring and monitoring these massive systems, and helping urban citizens and businesses avoid waste, opens the door to radical efficiencies. This is neither the stuff of fantasy nor ICT firms' PR materials. Thanks to the rapid cheapening and miniaturization of the GPS, thermometers, accelerometers, hygrometers, ambient light monitors, and other sensors "stuffed" in the smart phone,[63] the power of ICT

61. See Hill, E., January 4, 2016. Material Wastes. In: Materials, Society and the Environment. http://www.trunity.net/sam2/view/article/51cbf4897896bb431f6afcf7/.

62. See the US Center for Naval Analysis Military Advisory Board, May 2014. National Security and the Accelerating Risks of Climate Change," CAN. http://www.cna.org/sites/default/files/MAB_2014.pdf.

63. On the cheapening of sensors, see Jesse Berst, April 25, 2014. A 'perfect storm' that's good news. (Falling prices on comms and sensing). SmartGridNews.com. http://www.smartgridnews.com/artman/publish/Business_Markets_Pricing/A-perfect-storm-that-s-good-news-Falling-prices-on-comms-and-sensing-6487.html.

is already enabling communities to realize significant gains in efficiency, resilience, responsiveness, and creativity.

Nor are rising resource costs the only driver: the development of smart cities, an inherently decentralized model, is also being accelerated by the patent vulnerability of large-scale and centralized power and other systems to increasingly frequent and intense natural and human disasters. Not all smart systems are inherently resilient to natural disasters and cyberattack, which is one reason for the rise of "network science."[64] As Anthony Townsend, research director at the Institute for the Future and Senior Research Fellow at New York University's Rudin Center for Transportation, illustrates very well,[65] these profound trends are already reshaping urbanization.

POLICY INTEGRATION, CITIZENS, AND RESILIENCE

As described earlier, Japan's technocrats have built an array of collaborative mechanisms, some described in the OECD "Territorial Review of Japan" (2016). These institutions and organizations link together all levels of government, business (emphasizing what might be deemed "fiscal affirmative action"[66] opportunity for SMEs), civil society, and academe in addressing the challenges (e.g., aging, depopulation, disaster hazards, energy risks) and diffusing solutions (compact cities, efficiency, renewables, microgrids) in ways that suit local circumstances.

It is indeed indubitable that these approaches emphasize local circumstances, rather than just impose the centralized models that characterized much of Japan's center—local relations during the postwar years. To confirm that point, one has only to read the local versions of renewable energy, compact city, disaster resilience, and related strategies. Perusing them confirms that the plans discuss—in extensive detail—local disaster risks and resource endowments (e.g., biomass, geothermal). And searching for regular updates reveals that the plans are not only being implemented but are being revised in light of changing conditions.[67]

One standout case in this regard is the city of Higashi Matsushima, one of the Future Cities noted. The Higashi-Matsushima City Smart Disaster Prevention Eco-Town offers a fascinating account of the institutional context that has fostered it and other rapidly proliferating cases.

64. On this, see Marchese, D., Linkov, I., May 10, 2017. Can You Be Smart and Resilient at the Same Time? Environmental Science and Technology 51. http://pubs.acs.org/doi/pdfplus/10.1021/acs.est.7b01912?src=recsys.

65. See Townsend, A., 2013. Smart Cities: Big Data, Civic Hackers, and the Quest for a new Utopia, New York: W.W. Norton.

66. This is because the subsidies and other supports are more generous for smaller businesses.

67. The plans are also made public. They can all be accessed (in Japanese) via the Cabinet Secretariat's National Resilience website, which links to them. https://www.cas.go.jp/jp/seisaku/kokudo_kyoujinka/tiiki.html.

The population of Higashi Matsushima prior to March 11, 2011 was 43,142. Over 1130 residents died in the disaster, which saw 65% of the city inundated by the sea. As noted earlier, the disaster led to the official opening of Japan's first microgrid-based smart community on June 12, 2016. The microgrid allows the community's power network to "island" from the regional power grid in the event of a disaster. Higashi-Matsushima eco-town's renewable generation assets, complete with battery storage, are linked through the microgrid. Even in a protracted disaster, this system will provide power to the municipal hall and critical care facilities.

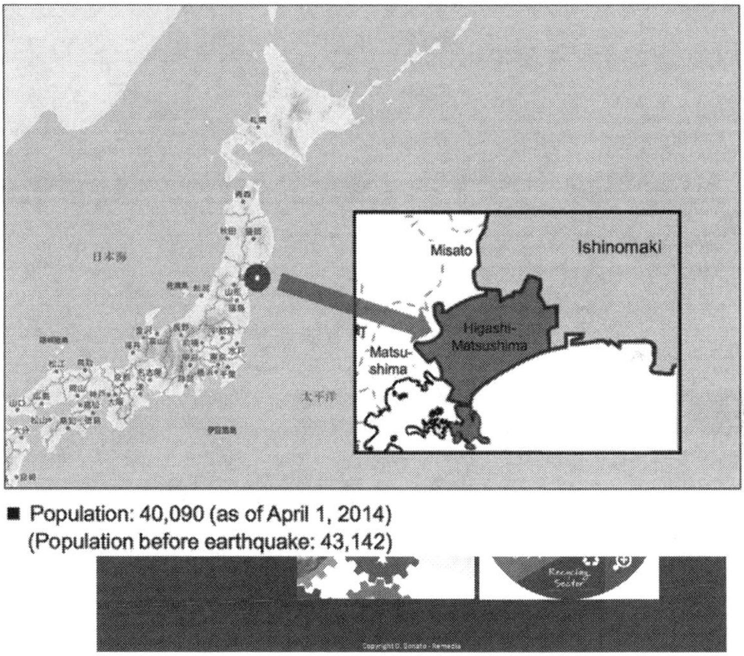

Higashi Matsushima's project is not the product of top-down, cookie-cutter policy making. Higashi Matsushima Mayor Hideo Abe was a practitioner of citizen-centered city management before March 11, 2011. He led "citizen-collaborative city planning" from 2005, allowing citizens into the center of policy making.[68] For example, Mayor Abe put community groups in charge of local assets, rebranding them as citizen centers (*shimin sentaa*) empowered to make decisions on public safety, aging, and other issues, well before

68. On Higashi-Matsushima's "citizen-collaborative city planning," see (in Japanese) the following Higashi Matsushima City website. http://www.city.higashimatsushima.miyagi.jp/city/kyoudou/index.html.

March 11, 2011. The Japanese Association of City Mayors certainly recognized how important Mayor Abe's innovation were. On October 8, 2014, the association asked his advice on "local revitalization" policies, and how he and his city officials undertook countless workshops with residents to build trust has led a bottom-up rebuild with the ambition of energy autonomy.[69]

Mayor Abe personally lost relatives in March 11, 2011 and hence led the post-March 11, 2011 rebuild of Higashi Matsushima under the slogan of "Never Forget That Day" and with the ambition of energy self-sufficiency. The city's residents were fully in support because virtually all knew of residents who had died in the open, due to the cold, or in hospital because there was insufficient backup power to run critical services.

The residents and local businesses are organized in the "Higashi Matsushima Organization for Progress and Economy, Education and Energy (HOPE)." HOPE was set up as a company on October 1, 2012 and includes former prime minister Kan Naoto and environmentalist CW Nicol on its advisory board. It also lists Sekisui House, NEC, Panasonic, Nomura, Fujitsu, and other firms among its members (Fig. 1).[70]

Interestingly, Japan's resolutely antinuclear Institute for Sustainable Energy Policies (ISEP) was listed as an advisor organization on October 24, 2011, when Higashi Matsushima applied for "Environmental Future City" recognition and finance.[71] That was not surprising because ISEP works with Denmark,[72] and Higashi Matsushima was visited by the crown prince of Denmark on June 14, 2011. Among other things, the crown prince participated in a seminar on renewables and reconstruction. Higashi Matsushima also received USD 1 million from Maersk, the Denmark-based shipping conglomerate. In fact, the city became the focus of Danish March 11, 2011 relief efforts in Japan largely because they knew a contact at Tohoku University (a Japanese assistant professor of agriculture and with links to Denmark).

---

69. On Mayor Abe's comments to the Japan Association of City Mayors, see (in Japanese) Higashi Matsushima City Mayor Abe Attends October 8, 2014 5th Hearing of Basic Policy Deliberation Team on Local Revitalization," Japan Association of City Mayors, October 8, 2014: http://www.mayors.or.jp/p_action/a_mainaction/2014/10/261008kihonseisaku.php.
70. The membership list of Higashi-Matsushima Organization for Progress and Economy, Education and Energy (HOPE) is available (in Japanese), current to April of 2017, at the following website. http://hm-hope.org/organization/.
71. See p. 50 of the Higashi Matsushima City October 24, 2011 application (in Japanese) for "Environmental Future City" support. http://www.kantei.go.jp/jp/singi/tiiki/kankyo/teiansyo/higashimatsushima1.pdf.
72. ISEP apparently fell from favour after the DPJ lost government in December of 2012 to the LDP. Perhaps that explains why ISEP does not (evidently) discuss HOPE and Higashi Matsushima as examples of community power. But the eclipse of ISEP and return of the LDP certainly did not mean that Higashi Matsushima City's smart local leadership and community trust went away.

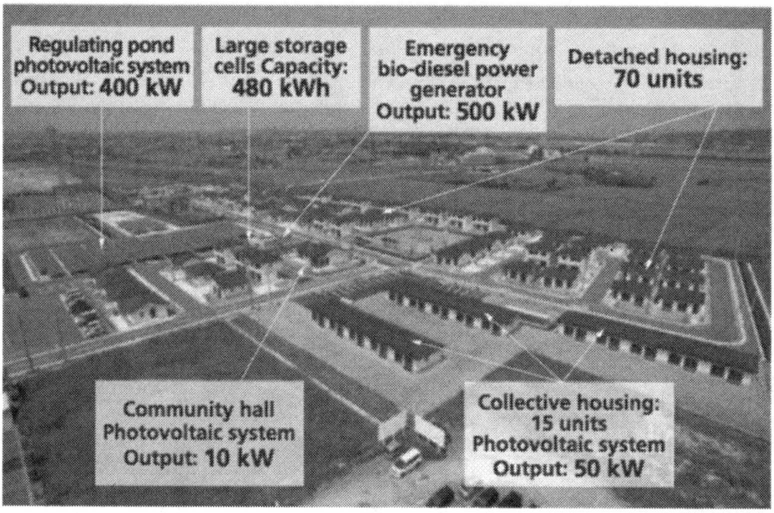

FIGURE 1 Higashi-Matsushima Disaster-Ready Smart Eco-Town. Note: This project, subsidized by the Ministry of the Environment, is an "independent and distributed low carbon energy society creation" initiative conducted by the Low Carbon Society Promotion Association. *Source: Sekisui House, 2016, p. 25.*

Moreover, unlike many overburdened local governments in the disaster area, they were able to answer the phone on March 30, 2011 when the Danish Ambassador was seeking a March 11, 2011 site to visit.[73] In short, the city was lucky. But Higashi Matsushima was also able to seize the opportunity because it already had smart local leadership and community trust.

Indeed, Higashi Matsushima City undertook a very citizen-centered project on relocating the 1395 households whose domiciles were destroyed. The effort was the subject of several major conferences, including a July 25, 2013 symposium.[74] Over 2000 city residents were involved in the deliberations, and the plans met with an 80.5% approval rate.[75] The Ministry of Environment understood the depth of Higashi Matsushima's local capacity when, in the

73. On this and other background facts concerning Higashi Matsushima, see (in Japanese) Kono, H., 2017. Community Energy, Tokyo: Chuo Koron Shinsha, p. 84.

74. The symposium was featured in a Nikkei Smart City Consortium article (in Japanese), July 25, 2013. The Mayor of Higashi Matsushima on Compact Community Building Through Group Relocation. Nikkei BP. http://bizgate.nikkei.co.jp/smartcity/symposium/symposium5/001299.html.

75. On these numbers, see p. 7 of (in Japanese) Furuyama, M., October 12, 2017. Steps in Higashi Matsushima's Reconstruction and Building Disaster Resilient Community. Presentation to Fukuoka Prefecture Seminar on Local Approaches to Promoting Renewable Energy and Energy Conservation, Kitakyushu City. http://www.pref.fukuoka.lg.jp/uploaded/attachment/33818.pdf.

spring of 2014, they asked the city officers to consider applying for an "autonomous and distributed power supply test-bed project." The city secured the funds and was able to use them (along with other monies) to cover three-quarters of the JPY 500 billion cost of the entire disaster-resilient system. These expenditures included JPY 160 billion for solar capacity, JPY 95 million for the microgrid, JPY 60 million for a 500-kW battery system, JPY 50 million for a 500-kW biodiesel generator, along with costs as transformers, energy management systems, and project design.[76]

Higashi-Matsushima City Eco-Town's 460 kW of solar is just part of the story. The citywide deployment of renewable energy multiplied by nearly 20 times between 2011 and 2015, rising to 35% of the city's power consumption. The city aims at a lofty 120% target by 2026. Moreover, from April of 2016, the community started a local power company, HOPE Electricity. As of October 2016, HOPE Electricity had a client base of 34 city and 98 private-sector facilities, delivering just over 8 megawatts (MW) of power at 1.5% below the cost of Tohoku Power, the regional major. HOPE Electricity's power sales are expected to bring in JPY 10 million of profit to be reinvested in the community. HOPE also aims to increase its power sales to over 20 MW, which seems achievable given the attraction of being local and the lack of competitors in the region.[77]

Minamisoma City

A further example of local resilience and renewable energy is seen in Fukushima prefecture's Minamisoma City. The city's location is outlined in Fig. 2.

Minamisoma City's population was 55,359 as of November 1, 2017.[78] The city's smart community plan was updated to August 9, 2016. The city is very close to Fukushima, the epicenter of the March 11, 2011 nuclear disaster, so it naturally emphasizes renewable energy and efficiency along with EVs and other green technology. The city renewable assets in 2011 were a mere 4% of their power mix. But the goals are to achieve 65% by 2020 and then 152% by 2050.[79] That would lead the city to become a regional exporter of power.

76. The particulars are described (in Japanese) in Kono, H., 2017. Community Energy, Tokyo: Chuo Koron Shinsha, pp. 60−62.
77. The particulars regarding HOPE Electricity are available (in Japanese) at the firm's website. http://hm-hope.org/denki/. See also (in Japanese) Kono, H., 2017. Community Energy, Tokyo: Chuo Koron Shinsha, pp. 85−87.
78. See (in Japanese) the Minamisoma website for details on population and other demographic information. http://www.city.minamisoma.lg.jp/index.cfm/8,2740,44,html.
79. Minamisoma's summary smart community plan is available (in Japanese) at the following internet website. http://www.city.minamisoma.lg.jp/index.cfm/8,4168,c,html/4168/vision_01.pdf. The details concerning the energy plan are available (in Japanese) at Table 4.1 of the following file. http://www.city.minamisoma.lg.jp/index.cfm/8,4168,c,html/4168/vision_02-4.pdf.

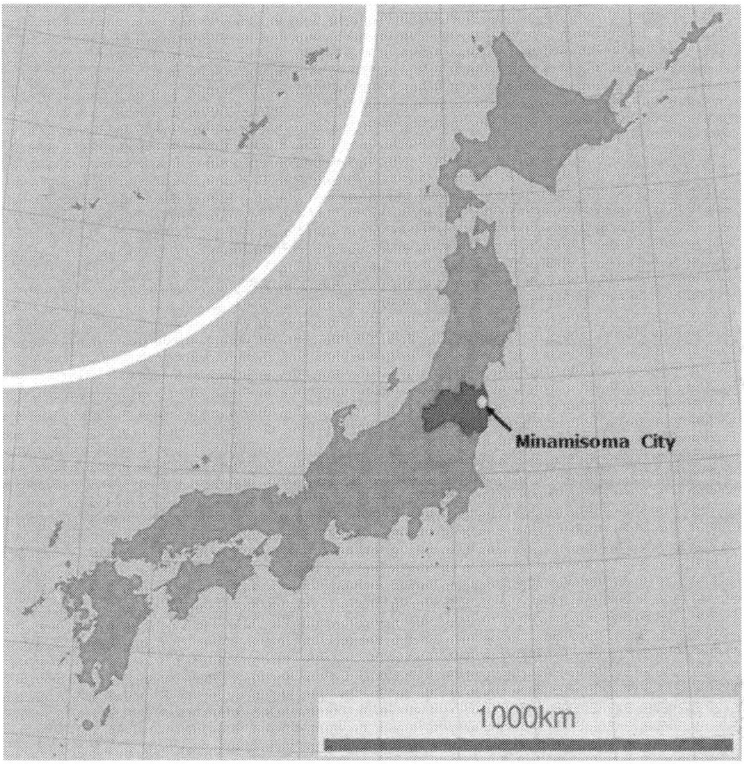

FIGURE 2 Fukushima Prefecture's Minamisoma City. *Minamisoma Smart Community.*

The city keeps track of its progress and publishes the details, albeit not as timely as it might. The most recent assessment shows that renewable deployment went from 4% of power to 8.2% in 2014. This number was ahead of Minamisoma's planned pace of deployment. The city's progress on efficiency continues through the deployment of Home Energy Management System (HEMS), Business Energy Management System (BEMS), and other energy management systems. Minamisoma has a 40-MHz battery system in place (in collaboration with Tohoku Electric Power Company (EPCO) and New Energy Promotion Council (NEPC)).[80]

Fig. 3 illustrates the model that guides the city's smart energy investments and their location. Key to this example and myriad others is the resilience framework developed by the policy integration outlined earlier. As can be seen from the outline and its collaboration, the efforts have built an increasingly

80. A February 2017 report on the battery system is available (in Japanese) at the NEPC website. http://nepc.or.jp/topics/pdf/170331/170331_1.pdf.

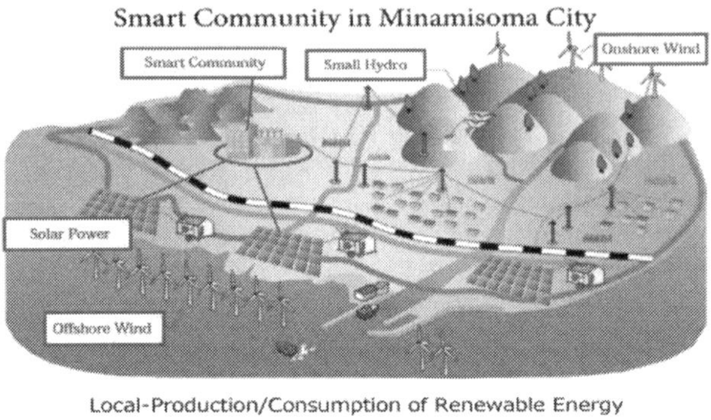

FIGURE 3 Minamisoma's illustration of the smart community. *Source: Minamisoma City.*[81]

integrated, industrial policy paradigm. It deserves further scrutiny to see whether it can evolve vested interests (such as Tohoku EPCO) as well as normalize renewable energy among the Not in My Back Yard (NIMBY) and distracted mainstream as resilience against myriad hazards.

Hirosaki City

Another example of a Japanese smart city is Hirosaki City in Aomori prefecture. The city's geographical location is portrayed in Fig. 4.

Hirosaki City's population was 174,287 as of October 1, 2017. Hirosaki Smart City's updated plan was open to public comment between March 8 and March 23 of 2017 and then adopted. Fig. 5 outlines the community energy system, which links up heat, power, and information networks as well as an array of decarbonizing inputs. In 2017, the city's initial smart city plan became 4 years old and required updating because (as its authors emphasize) of massive and rapid technological changes and the fact that its phase 2 is about to start.

Phase 1 of Hirosaki City's smart city project ran between FY 2013 and FY 2016. The plan emphasized the deployment of extant technologies and disaster resilience. The city installed significant amounts of LED lighting, some solar, energy management, advanced waste treatment, and other technologies. The next phase of the smart city plan is to deploy the "community energy system" outlined in Fig. 5.

Fig. 6 also outlines the steps in Hirosaki City's aim to produce and use hydrogen. As can be seen from the figure, the plans focus on the use of local resource endowments, including wind and biomass, to produce the hydrogen.

81. The figure is a translation of the graphic posted (in Japanese) by Minamisoma City at the following website. http://www.city.minamisoma.lg.jp/index.cfm/8,4168,75,html.

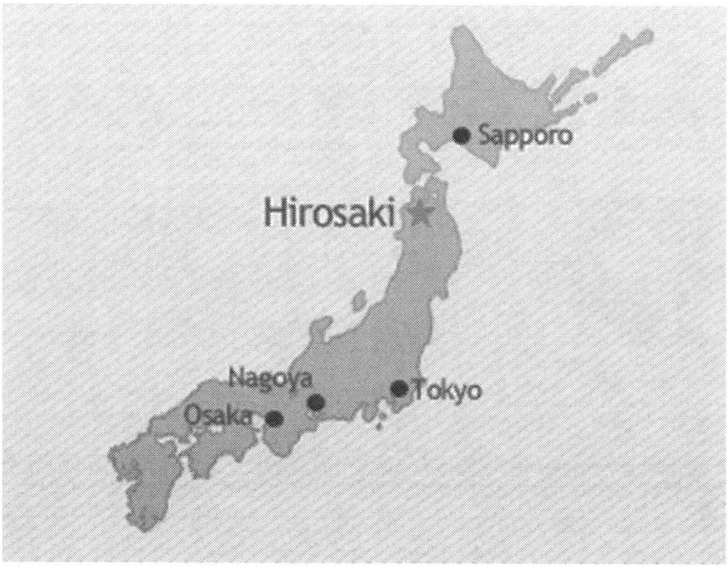

FIGURE 4 Aomori Prefecture, Hirosaki City. *Source: Hirosaki City.*

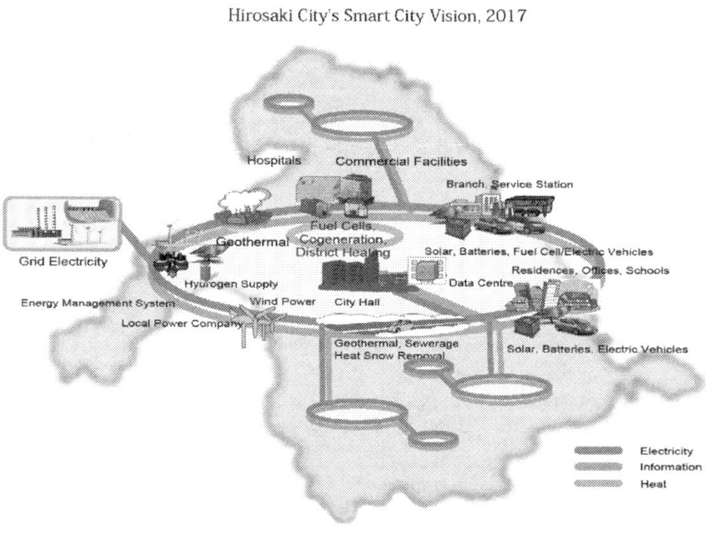

FIGURE 5 Hirosaki smart city concept. *Source: Hirosaki City.*

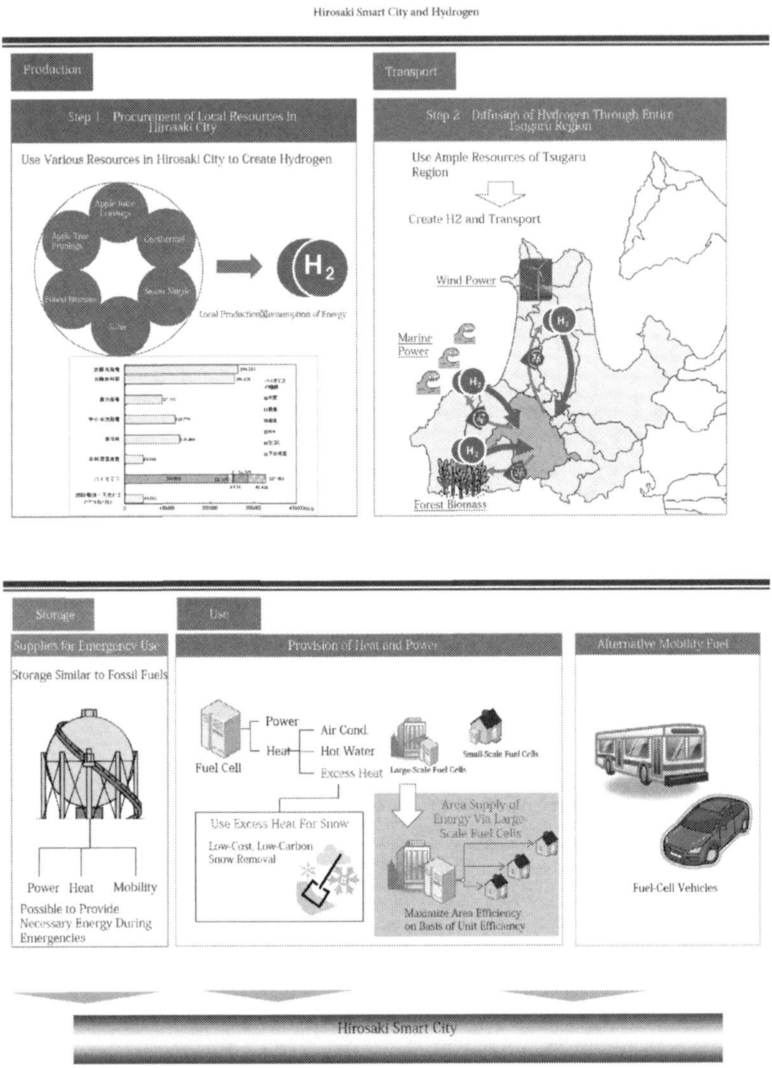

FIGURE 6 Hirosaki City's hydrogen plans. *Source: JRI.*[82]

The fuel will then be used in cogeneration systems as well as in local mobility. The promotion of hydrogen as an alternative fuel has become a major initiative for the Japanese government.

82. The Japan Research Institute file is a posting of Hirosaki City's 2012 plan for hydrogen diffusion. See the item (in Japanese) at the following website. https://www.jri.co.jp/MediaLibrary/file/pdf/company/release/2013/130329/130329-3.pdf.

FIGURE 7 Toyama Prefecture's Toyama City. *Source: Toyama City.*

Toyama City

Toyama Prefecture's Toyama City is a particularly interesting project. The city's location is portrayed here (Fig. 7).

Toyama City's population in October of 2017 was 417,966. Toyama was selected to be an "environmental model city" in 2008, one in which revitalization of public transport was made the core element in shifting the city towards a more compact and sustainable spatial profile.[83] The project centered on planning, with improved transport as the focus. Toyama City was also selected as an "environmental future city" in 2011.[84] And in June of 2012, the

83. See (in Japanese) "Concerning the Selection as an Environmental Model City," Toyama City. http://www.city.toyama.toyama.jp/kankyobu/kankyoseisakuka/ondankataisakukikaku/kankyomoderutoshi.html.

84. See (in Japanese) "Concerning the Environmental Future City Concept," Toyama City. http://www.city.toyama.toyama.jp/kankyobu/kankyoseisakuka/ondankataisakukikaku/kankyomiraitoshi-1.html.

OECD chose Toyama City as an exemplary case of "compact city development."[85]

In addition, Toyama was also chosen to be one of the Rockefeller 100 Resilient Cities[86] in December 2014. Toyama was the first Japanese city to be selected as one of these 100 resilient cities (Kyoto was added later). Since December of 2013, the Rockefeller Foundation has used the resilient cities program to disseminate awareness of CC and other challenges to cities. The foundation makes its selections based on applications from the cities themselves. One major goal is to foster an international cohort of chief resilience officers, each working in his or her designated city and funded by the foundation. Ideally, these chief resilience officers are empowered to build resilience within their own urban and regional contexts, borrowing from international best practice as well as contributing to it. Another resilient city goal is to maximize the "resilience dividend" by deploying infrastructures and practices that pay off in greater resilience against disaster as well as during routine operations.[87] These are the mitigation and adaptation synergies illustrated in Fig. 1 of this paper.

Toyama City is thus a notable case internationally as well as in terms of multiple overlapping policy lines. These policies and their main sponsoring agencies (in brackets) include compact city (MLIT), smart community (METI), distributed energy (MOE, METI); hydrogen generation and distribution (METI), heat recovery in sewers (MLIT), ICT-enabled efficiencies (MIC), and new-model mobility, such as car-sharing, small electric, and fuel-cell vehicles, buses, and light rail (MLIT, METI). As noted earlier, the central element of Toyama's approach is to use revitalized public transport to focus urban functions on the downtown area as well as along urban transit lines. The city also aims to revitalize its downtown area, enhancing walkability and thus bolstering its social capital.

The overall image of what Toyama aims at is seen in Fig. 8. As is evident in the figure, the compact city approach seeks to counter sprawl and motorization through plans that relocate core urban services (medical, welfare) in downtown areas and link them.

Toyama has also sought to increase the use of renewable energy. Toyama's renewable energy and related projects include the deployment of solar and other power generation in the downtown area. The city also aimed at building a

85. See June 13, 2012. Governance: Compact Cities: the way of the future. OECD. http://www.oecd.org/newsroom/governancecompactcitiesthewayofthefuture.htm.

86. The website for the Rockefeller 100 Resilient Cities is available at the following URL. http://www.100resilientcities.org.

87. Concerning the Rockefeller Foundation's 100 Resilience Cities initiative, see: https://www.rockefellerfoundation.org/our-work/initiatives/100-resilient-cities/.

88. The figure is available at "G7 Environment Ministers' Meeting to be Held in Toyama, an Eco-City Japan Boasts to the World," *Japan Environment Quarterly*, Ministry of the Environment, Volume 13, March 2016: http://www.env.go.jp/en/focus/jeq/issue/vol13/fea01.html.

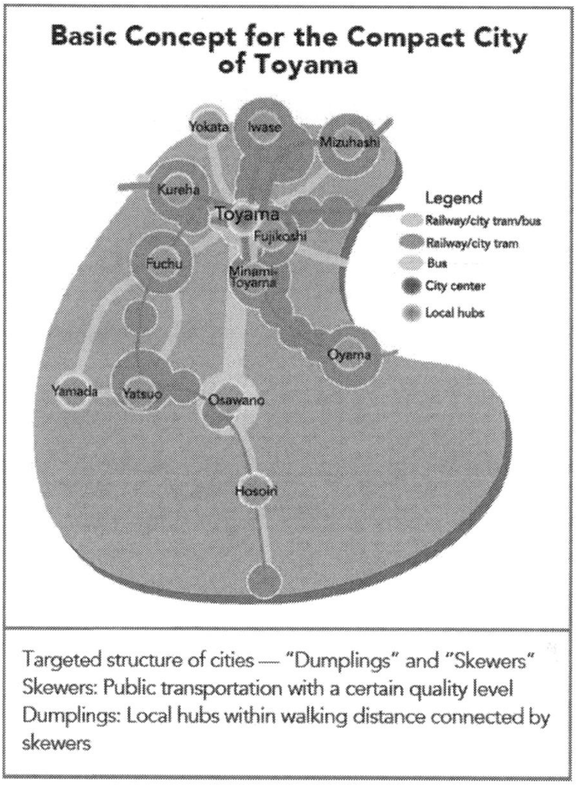

FIGURE 8 The compact city of Toyama. *Source: MOE.*[88]

distributed energy and materials circulation system that includes biomass, especially marine biomass as well as a biogas network that uses leftover foodstuffs as the input resource. Toyama City's outlying regions were also to be stimulated by a program that stresses the revitalization of farms via small hydro and other renewable energy projects. The city was also to build a factory for the cultivation of plants related to the pharmaceutical industry, one that uses geothermal, small hydro, solar, and other energy sources. The city also aimed at building an "eco-forest Toyama," a project that aimed to increase forested area from 0 ha in 2010 to 500 ha by 2016. The amount of forest-origin biomass was also to be increased from 370 tons in 2010 to 2000 tons in 2016.[89]

89. On this, see (in Japanese), 2013. Toyama City Environmental Future City Plan. Toyama City. http://www.city.toyama.toyama.jp/data/open/cnt/3/9959/1/05siryou2.pdf.

FIGURE 9 Toyama City safe and smart model block. *Source: Toyama City.*[92]

Toyama City's most recent reports on its progress indicate that its forested area increased to 699 ha.[90] The city had also subsidized the installation of 156 energy management systems by 2016.[91]

More recently, Toyama opened a newly renovated downtown area as a model smart and compact community, as shown in Fig. 9. The project is 8500 m^2 and contains 21 housing units and city facilities. All of the housing units include solar power, battery backup, and household-size fuel cell cogeneration. There is also a community solar project of 11 kW used to deliver power to a community battery. This arrangement is to provide power supplies in emergencies to the community park, which is designated a disaster shelter for emergencies. The public facilities became operational in October of 2017.[93]

Hioki City

Hioki City (population: 49,056) is yet another example of policy collaboration. It is the seventh-largest city in Kagoshima prefecture (population: 1.64 million), the southernmost prefecture on Japan's southern island of Kyushu. Hioki City borders the west side of the prefectural capital, Kagoshima City (population: 598,936), with Satsuma Sendai City (population: 95,513) to its north, and lies midway along the west coast of Satsuma Peninsula. Hioki City is especially instructive precisely because it is not a special case. Save for the fact that it is close to Japan's two operating nuclear reactors at the Sendai

90. See the data in the table on p. 2 of (in Japanese), March 2017. Toyama City Forestation Plan. Toyama City. http://www.city.toyama.toyama.jp/data/open/cnt/3/5391/1/kanseipuran.pdf.
91. See p. 120 of (in Japanese), 2017. Toyama City's Environment, Toyama City. http://www.city.toyama.toyama.jp/data/open/cnt/3/3487/1/H29toyamashinokankyou.pdf.
92. The figure was provided by the Toyama City Environmental Division's Office of Environmental Policy.
93. On the details, see (in Japanese) Kodo, S., October 30, 2017. Zero Energy Block in Toyama City: Solar, Batteries and Fuel Cells. Nihon Keizai Shimbun. https://www.nikkei.com/article/DGXMZO22873290Q7A031C1000000/.

plant in Satsuma Sendai, Hioki is generally representative of Japan's midsized local cities.

Like most of Japan's regional urban centers, Hioki's economic base and population have declined over the past several years and—barring smart policy intervention—the bad news is expected to continue. Hioki's circumstances have been aggravated by the 2011 reduction in operations at the local Panasonic semiconductor factory and other setbacks. The demographics are also daunting. In 2005, Hioki's population was 52,411, a number that had declined to 49,056 by 2016. The drop is projected to continue, leading to 47,261 residents in 2020, 37,866 in 2040, and about 29,000 in 2060. This population decline would represent a plunge of 43% between 2010 and 2060. In tandem with population decline, the percentage of elderly is increasing. Hioki's over-65 population was 27.9% in 2005, and this had risen to 31.1% in 2015, versus a national average of 26.7%. The share of elderly in Hioki's demographic profile is expected to increase to 33.7% in 2020 and perhaps 39.7% in 2060. Moreover, the single-household rate among the elderly population is also relatively high, at 28.7%. Left unchecked by smart policy, this shrinking and aging population means further constriction of local business and the local tax base. Moreover, increased numbers of single-person, elderly households leads to greater energy consumption and reduced resilience in the face of natural disasters, with elderly women being the most vulnerable.

Like many other Japanese communities, one of Hioki City's strategies to confront this decline has centered on local energy projects. The March 11, 2011 disasters were a spur, later amplified by the nationwide debate on revitalizing depleting regions as well as the evidence that distributed energy offers multiple payoffs for local resilience. Surveys of Hioki's local renewable energy resource endowment indicated that the city and its immediate area possess significant potential in solar, wind, small hydro, and other resources. In December of 2013, Hioki City itself, along with area businesses, collaborated by investing in the "Hioki Wind Power Corporation." This wind power project is worth JPY 2.4 billion and, as of September 2015, operates an installed capacity of 6.9 MW of wind generation, via three 2.3 MW turbines. In addition, in 2014, the Ministry of Agriculture, Forestry and Fisheries provided a grant to support the promotion of renewable energy businesses. This funding has been used for promoting small hydro projects. Via the Hioki Hydro Power Promotion Association, the city mayor leads local business firms and financial institutions in surveying and exploiting local small hydro assets.

Hioki's Smart Community Plan

These public—private collaborations gave rise to increasingly ambitious aims, particularly in light of the opportunities offered by the ongoing deregulation of

Japan's power markets in tandem with expanding central government support for heat and power grids as well as other ancillary infrastructure. Hence, in 2015, Hioki City and its business partners formalized their collaboration in an association. Their aim was to investigate how to promote the development of a smart community. To this end, they deliberated on the available options while assessing the energy demand profiles of public and private facilities to be included in the project.

The city and its partners took was to launch the "Hioki Local Energy Corporation" in November of 2015 as a vehicle for undertaking power supply operations. Japan's staggered deregulation of its retail power markets took a major step on April 1 of 2016 via deregulation of the retail power sector. Taking advantage of the opportunity, Hioki Local Energy Corporation announced that it will commence operations as a new power company from August of 2016. It plans to offer electricity at a 3% discounted rate, compared to the power prices of Kyushu Electric, the regional utility. Hioki Local Energy Corporation also commits itself to devote 1% of the proceeds of power sales to local revitalization. It plans to deposit the monies into a "Hioki Future Fund" and then donate to local agricultural and other projects.

Moreover, the city and local business collaboration also developed an energy-centered smart city plan, working in tandem with Hitachi Power Solutions Corporation. This project was first explicitly declared an aim by Hioki City mayor Miyaji Takamitsu in the wake of winning the May 19, 2013 municipal election. The project construction began in August of 2018 and is to be completed by February of 2019. As illustrated in Fig. 10, the city and its collaborators' design envisions an energy system that links various city and business facilities as well as heat and power sources via three "compact grids" and a common power grid, or "compact energy network." Among the local buildings included in the project are the city offices, sports facilities, health-care facilities, local factories, and other sources of heat and power demand. The network infrastructure includes a common grid network to link three compact grids, each deployed where there are clusters of public buildings and factories. Cogeneration, solar, and other energy inputs will be used, and the system's inputs and efficiencies will be managed via the cloud (the energy management system at its core). Efficiencies gained via energy management and other means are expected to reduce power prices by at least 10% while securing energy supplies in the event of disaster, and at the same time, significantly reducing emissions of greenhouse gases.

Compact grid number one, which is at the top of the figure, will link five of the city's facilities (the city hall, waterworks bureau, main community center, cultural center, and gymnasium). About 200 kW of solar power generation will be worked into this compact grid. One primary goal is to use the solar capacity to reduce the peak power demand of the cultural center during events. Moreover, during natural or other disasters, this power could be used to supply

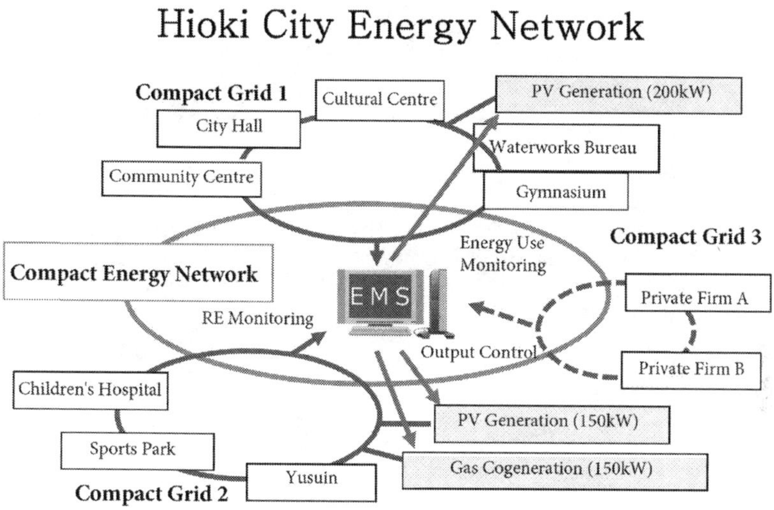

FIGURE 10 Hioki City's energy network. *Source: Hioki City, 2017.*[94]

the city hall and the main community center, with the latter acting as an evacuation shelter.

Compact grid number two, on the figure's left-hand side, would link the city's hot springs facilities (Yusuin) with a children's hospital and a comprehensive sports park. The hot springs would become the site for 100 kW (4 units of 25 kW) of cogeneration capacity. This cogeneration capacity would be linked to the cluster's other facilities by power and heat network infrastructure. The grid would provide heat and power under normal circumstances and act as a power source in emergencies.

Compact grid number three, on the figure's right-hand side, is yet to be funded. If undertaken, it is likely to see 1000 kW of cogeneration capacity installed in the city's "Kaida Factory Cluster" and heat and power links established among three site factories.

COLLABORATION ON BUILDING RESILIENCE

We have seen that many Japanese cities are sites of resilience collaboration that developed within and among the central agencies. Indeed, there is an impressive degree of collaboration among Japanese center–local governments, academe, business, NGOs, and other actors on national resilience.

94. Author's translation of Hioki City (in Japanese), December 28, 2017. Local Production/Local Consumption Energy Network Project. https://www.city.hioki.kagoshima.jp/kikaku/shisejoho/energynetwrok.html.

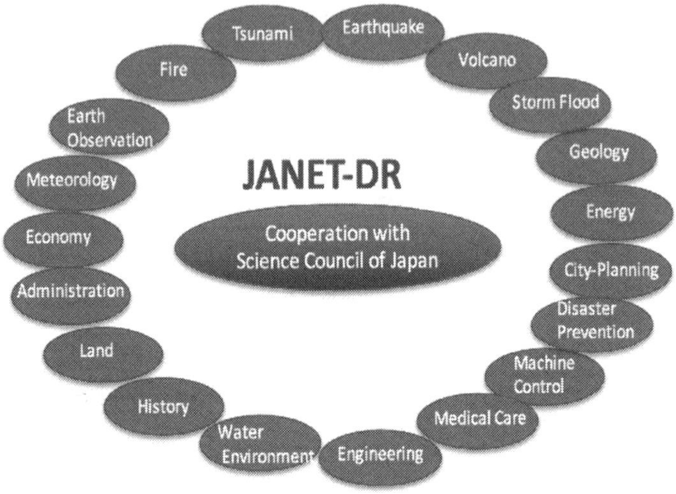

FIGURE 11 The Japan Academic Network for Disaster Reduction (JANET-DR)'s scope. *Source: JANET-DR.*[95]

For example, National Residence and related policy initiatives benefit from the Japan Academic Network for Disaster Reduction (JANET-DR).[96] This network links 55 academic associations as of November 1, 2017. As Fig. 11 shows, these associations cross a lot of relevant disciplinary boundaries. These disciplines include energy, disaster prevention, city planning, hydrology, and several other areas especially relevant to coping with the Anthropocene. The JANET-DR also cooperates with the Science Council of Japan, whose role is akin to the US National Academy of Sciences.

The JANET-DR was formalized on January 9, 2016. It emerged from a 30-association liaison that followed March 11, 2011 and played a large role in shaping the resilience debate through numerous major events and several publications. Since its formal founding in January of 2016, the JANET-DR has held multiple disaster resilience symposia, analyzing the worsening threat of typhoons and intense rain. Their work is very impressive. They JANET-DR have been important in making sure that National Resilience is world-class in how it assesses the full range of hazards and designs countermeasures. They also push to have energy, biodiversity, and other aspects central to the paradigm.

Fig. 12 also shows that the JANET-DR collaborates strongly with a variety of stakeholders, including local governments and citizen groups.

95. The figure is taken for the JANET-DR's updated (to November 1, 2017) English-language brochure on its organization and activities: http://janet-dr.com/02_about/janetdr_A4_eng.pdf.
96. The website for the JANET-DR is here: http://janet-dr.com.

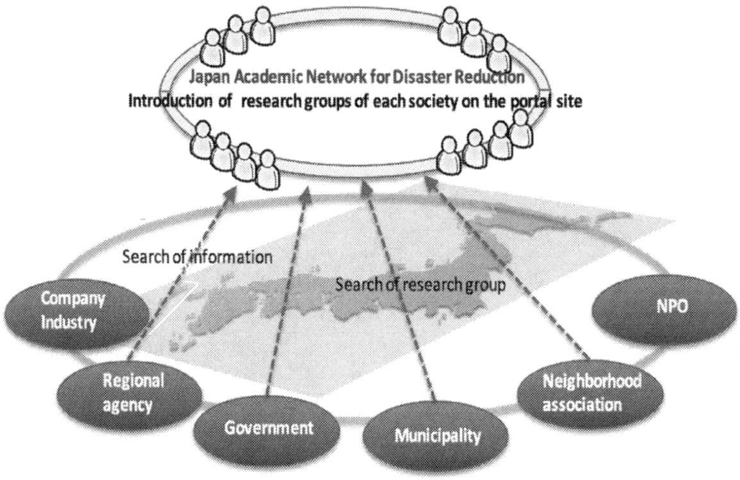

FIGURE 12 The Japan Academic Network for Disaster Reduction (JANET-DR)'s broader collaboration. *Source: JANET-DR.*[97]

Building Resilience by Policy Integration

Japan's policy integration for coping with the Anthropocene is quite extensive. As discussed earlier, Japanese technocrats have worked the smart community paradigm into energy policy, spatial planning, disaster policy, demographic policy, ICT policy, and other areas. That diversity of goals has made it hard to follow the smart community paradigm in Japan because it evolves through linked policies and budgets rather than as a specific policy regime per se. Yet the diversity of goals, integrated in "whole of government" collaboration, has perhaps ensured that the paradigm does not get cut back to minimalism by the MOF. The latter do not have the time to understand industrial policy.

But a lot of Japan's resilience work is clearly industrial policy. For example, the most advanced radar systems are included. These "dual polarized" radar systems are able to monitor weather systems and predict the likely intensity of precipitation. The radar is also networked to supercomputers to enhance the capacity to handle voluminous data. Moreover, the Dam Revival Vision depends on this system integration to help adjust the overall water network's capability to maximize flood control as well as boost hydropower generation.

Indeed, it is no exaggeration to say that after March 11, 2011, Japan has seen the rise of a new kind of strategic industrial policy. This policy is rooted

97. The figure is taken from the JANET-DR's updated (to November 1, 2017) English-language brochure on its organization and activities: http://janet-dr.com/02_about/janetdr_A4_eng.pdf.

in distributed energy and smart energy infrastructure as the core of disaster-resilient smart communities. After March 11, 2011, the technocrats ramped up their test bedding of smart energy systems, revised spatial planning to foster density (the key to making distributed energy systems work), and put disaster resilience on par with mitigation.

Japan's national resilience programs (and their prefectural and local plans) also include policies for restructuring the local communities, moving hospitals, schools, daycare, community centers, and other facilities into downtown districts. This relocation is specifically designed to shift residences and other facilities away from areas with high flood potential.

Japanese can undertake this relocation because aging and rural depopulation compel policy makers to act. Another reason is there is a fair degree of public ownership of land, healthcare, and other facilities, giving planners a large lever to use. Moreover, Japan is also a unitary state with powerful interregional redistribution. The Local Allocation Tax and other fiscal mechanisms are being used to cover many of the costs of this spatial restructuring.

And much of it appears to be good investment. The spatial restructuring—the centralization of previously dispersed daycares, hospitals, libraries, community centers, and other functions—reduces the costs of maintenance and heating, cooling, lighting, water, and other services. Moreover, while relocating or refurbishing, systems can be upgraded to state of the art. In addition, the relocation leads to increased spatial densification of demand for energy-related services. This densification of public building makes it more cost-effective and politically easier (because they are public facilities) to install highly efficient district heating and cooling and other elements of the smart community models that Japanese specialists have been at work on since March 11, 2011.

On top of all that, the densification leads to more interaction among the elderly and children (as daycare centers are often in multifunction buildings that include old-age facilities), reducing the risk of dementia. Moreover, the denser and more interesting the downtown area, the more residents walk around, helping them maintain mental and physical fitness. Walkability thus helps cut healthcare costs as well as energy use.

All this smart densification, if done well (emphasizing walkability and green space), leads to increased property values in the downtown cores as well as increased traffic for downtown shops, restaurants, and other services. These results are powerful incentives for city managers, as they see increased local revenues, a tangible reward for their opting to work with central agencies in designing and implementing the compact, resilient, smart city paradigm.

And Japan's emphasis on disaster resilience is rooted in its prominent role, since at least 1994, in international work on disaster risk reduction. The first framework for disaster risk reduction is the Yokohama Strategy (1994), the second the Hyogo Framework (2005–2015), and the third the Sendai Framework (2015–2030).[99]

Japan was the host country for all three of the World Conferences on Disaster Risk Reduction. For the Sendai Conference (March 14–18, 2015), Japan advocated the need for "prior investment" so as to build resilience in the face of multiple hazards (including CC) and reduce their impact. Another of Japan's proposals was to "build back better" (as opposed to merely rebuilding) through such initiatives as relocating communities threatened by tsunami, floods, or other hazards. Japan also emphasized the need for "mainstreaming disaster risk reduction," calling for a whole-of-government approach that makes coping with hazards a priority in all planning initiatives. These approaches are core to Japan's National Resilience and related policies.

In short, Japan is making significant progress in reaping the synergies in the overlap between adaptation and mitigation. What the Japanese state is doing with its smart communities has not received much international recognition. But the way Japanese governance and infrastructure intersect deserves a lot more attention.

99. On the frameworks, see the website of the United Nations Office for Disaster Risk Reduction: http://www.unisdr.org.

Japan Cases: Integrated Green Smart Policies and Finance for Communities

Andrew DeWit

Professor, School of Economic Policy Studies, Rikkyo University, Tokyo, Japan

INTRODUCTION

Japanese industrial policy emphasizes smart communities as the locus for pursuing decarbonizing via a combination of mitigation- and adaptation-led approaches. Japan's smart communities reflect the fact that "industrial policy is back," and for good reasons.[1] In Japan, escalating damage from extreme weather and other disaster events have led to the recognition that inclusive, "whole of government" planning and investment in resilience and adaptation is imperative.[2] Scientific evidence strongly indicates the global climate system has transitioned from the relatively placid Holocene era to an increasingly uncertain Anthropocene.[3] This change poses worsening hazards for human communities, their critical infrastructure,[4] agriculture, and other crucial elements of the built and natural environments. It also poses increasing threats to national security, and hence transcends partisan politics.[5] One particularly severe threat is the "end of stationarity" in the hydrologic cycle,[6] which means past patterns of precipitation are no longer a reliable indicator of the future. Many of the challenges posed by the Anthropocene are recognized in the 2030

1. See e.g., Dietsche, E. "New industrial policy and the extractive industries," WIDER Working Paper 2017/161, United Nations University World Institute for Development Economics Research, August 2017.
2. DeWit, A. "Japan: Response of policy entrepreneurs to an energy crisis," in (Patrice Geoffron, Lorna Greening, Raphael Heffron eds.) *Meeting the Paris mandate: A cross-national comparison of energy policy-making*. Springer-Verlag (forthcoming, 2018).
3. The science of the Anthropocene is described in detail in Waters C. N., et al. "The Anthropocene is functionally and stratigraphically distinct from the Holocene," *Science*, Vol. 351, January 2016. See also Waters, C.N. J.A. Zalasiewicz, M. Williams, M.A. Ellis and A.M. Snelling (eds.). *A Stratigraphical Basis for the Anthropocene*, Geological Society Special Publications, Special Publication 395, June 4, 2014. A more general treatment of the Anthropocene and its implications for other academic disciplines and general intellectual debate is found in Hamilton C. *Defiant Earth: The Fate of Humans in the Anthropocene*, Polity Press, 2017.
4. Critical infrastructure includes power, communications, water, transport, and other systems essential to human communities.
5. On the multiplicity of threats to national security, see Werrell, C.E. and Femia F. (eds.) *Epicenters of Climate and Security: The New Geostrategic Landscape of the Anthropocene*, The Center for Climate and Security, June 2017: https://climateandsecurity.files.wordpress.com/2017/06/epicenters-of-climate-and-security_the-new-geostrategic-landscape-of-the-anthropocene_2017_06_091.pdf.
6. Milly P.C.D. et al. "Stationarity Is Dead: Whither Water Management?" *Science*, Vol. 319, February 2008.

Agenda for Sustainable Development and Sendai Framework for Disaster Risk Reduction 2015–30.[7]

Building resilience and sustainability in the Anthropocene entails policy integration to cope with multiple externalities. The existence of significant externalities places a premium of collaborative public-sector leadership.[8] Yet there is scant comparative literature on policy integration and fiscal flows linking the often-siloed domains of disaster resilience, energy policy, environmental policy, urban planning, and other areas relevant to mitigating and adapting to the Anthropocene.

Indeed, most case studies of mitigation and/or adaptation are microlevel, examining civil society's "soft resilience" or single cities. Or they merely compare national targets on emissions reductions and the deployment of renewable energy. Many of these studies are at best minimally engaged with the larger governance challenges and the interaction between contemporary energy and material flows as well as accelerating climate hazards. However, the challenges are global, dynamic, and interrelated. These facts were underscored on Nov. 13, 2017 by the Union of Concerned Scientists and 1700 other independent scientists. Together, these experts published a dire, evidence-based warning concerning the increasingly dynamic variables on hazards.[9] In this context, some social scientists are refining a "complexity framework," which specifically addresses the Anthropocene, to use in work on sustainability.[10]

This approach may be especially important for the Asia–Pacific region because by 2018, over half of the Asia–Pacific population will have been living in cities and requiring over USD 1.7 trillion in annual climate-resilient infrastructure investments in the face of mounting disasters.[11]

Abundant evidence indicates the potentially decarbonizing impact of state-fostered resilient adaptation, coordinated at the national level. For example, compact cities and green infrastructure are inherently mitigating as well as robust in the face of mounting risks of flood, fire, and other disasters.[12] So, too,

7. On the 2030 Agenda and Sendai Framework, see: https://www.unece.org/environmental-policy/conventions/industrial-accidents/activities/sdgs-and-sendai.html.
8. See e.g., Helbling T. "Externalities: Prices Do Not Capture All Costs," International Monetary Fund, July 2017: http://www.imf.org/external/pubs/ft/fandd/basics/external.htm.
9. On this, see Ripple W.J. et al. "World Scientists' Warning to Humanity: A Second Notice" *BioScience*, bix125, November 13, 2017: https://academic.oup.com/bioscience/advance-article/doi/10.1093/biosci/bix125/4605229.
10. Note, for example, Smith-Nononi, S. "Making Complexity Your Friend: Reframing Social Theory for the Anthropocene," *American Meteorological Society*, August 23, 2017: http://journals.ametsoc.org/doi/10.1175/WCAS-D-16-0124.1.
11. The number is taken from *Meeting Asia's Infrastructure Needs*, Asian Development Bank, 2017: http://www.preventionweb.net/files/53939_specialreportinfrastructure.pdf.
12. See e.g., Bay J. and S. Lehmann (eds.), *Growing Compact: Urban Form, Density and Sustainability*, Routledge, 2017.

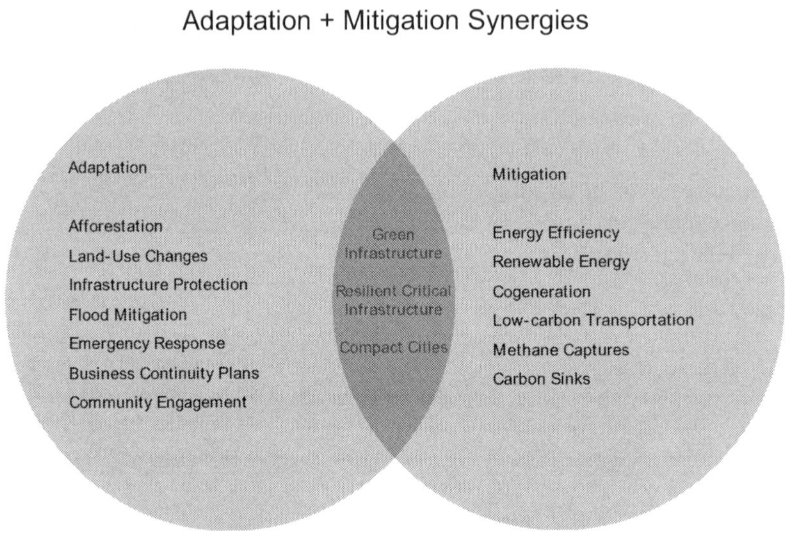

FIGURE 9.1 The intersection of adaptation and mitigation strategies. *Adapted from Winkelman.*[15]

are smart distributed energy networks that are growing from within vulnerable, carbon-intensive conventional power and heating systems.[13] Hence, this overview of what Japan is doing focuses on its integrated planning to promote resilient, compact, and decarbonizing communities and critical infrastructure.[14]

This section asserts that it is possible to achieve synergies by smart adaptation and mitigation. This claim is not original, because there have been efforts to enhance such synergies by breaking down siloes.[16] Fig. 9.1 illustrates

13. This development is outlined in *The Utility of the Future*, MIT Energy Initiative, 2016: https://energy.mit.edu/wp-content/uploads/2016/12/Utility-of-the-Future-Full-Report.pdf.
14. Similar work includes Kwok, S and Watanabe, C. *From Compact City to Smart City: A Sustainability Science & Synergy Perspective*, Journal of Environmental Science and Engineering A 6, 2017: http://www.davidpublisher.org/Public/uploads/Contribute/5962f22d0f424.pdf.
15. Winkelman, S. "Green Resilience: Adaptation + Mitigation Synergies," Center for Clean Air Policy, December 21, 2016: https://www.ctc-n.org/sites/www.ctc-n.org/files/ctcn_ccap_webinar_green_resilience_adaptation_mitigation_synergies.pdf.
16. For example, Laurikka, H. Synergies Between Mitigation and Adaptation Exist in Several Sectors, SDG Knowledge Hub, March 12, 2013: http://sdg.iisd.org/commentary/guest-articles/synergies-between-mitigation-and-adaptation-exist-in-several-sectors/, Rizvi, A.R., Baig, S., Barrow, E., Kumar, C. (2015). *Synergies between Climate Mitigation and Adaptation in Forest Landscape Restoration*. Gland, Switzerland: IUCN, 2015: https://portals.iucn.org/library/sites/library/files/documents/2015-013.pdf, Locatelli, B. Fedele, G., Fayolle, V., Baglee, A., "Synergies between adaptation and mitigation in climate change finance," *International Journal of Climate Change Strategies and Management*, Vol. 8 Issue: 1 2016: https://doi.org/10.1108/IJCCSM-07-2014-0088.

several of these synergistic outcomes. The figure indicates that the nexus between conventional adaptation (e.g., afforestation, flood countermeasures, emergency response) and mitigation (e.g., energy efficiency, renewable energy deployments, modal shifts in transportation) strategies can be synergistic. These synergistic outcomes include, but are not limited to, the green infrastructure, resilient critical infrastructure, and compact city initiatives illustrated at the intersection of the Venn diagram.

All countries have multiple reasons for mitigating climate change as well as adapting to the accelerating hazards that appear unavoidable in the coming decades of the Anthropocene. These latter include risks of global food supply shocks and military conflict. The "threat multiplier" effect of climate change is already evident and hence recognized by the US military and the North Atlantic Treaty Organization.[17]

Therefore, Japan is not alone in possessing ample reasons to address climate change. Nevertheless, this section argues that Japan has an especially potent set of incentives to do so, and that these incentives dovetail with the country's sobering demographic, fiscal, energy insecurity, and other challenges. Driven by these incentives, coupled with period crises (such as the Mar. 11, 2011 natural and nuclear disasters in Tohoku; hereafter 3−11), Japan is indeed taking action.

Fig. 9.2 is one means to visualize what Japan is doing. Japan has built a comparatively well-integrated coordination among state agencies, market players, and civil society. This coordination is especially pronounced in policies relating to smart communities, compact cities, and national resilience.

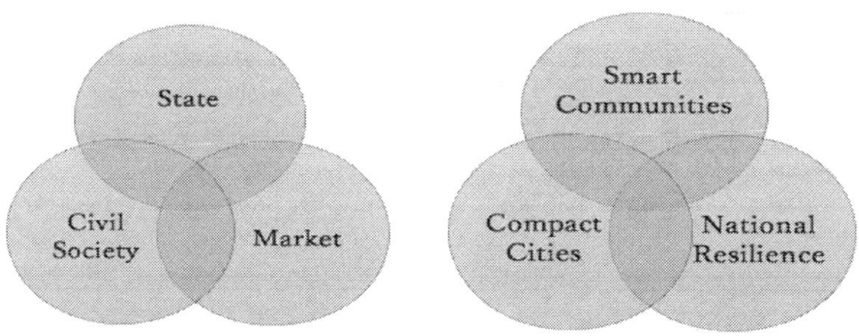

FIGURE 9.2 Agency and policy regimes in building Japanese resilience. *Provided by the author.*

17. On these agencies' assessments, see Nuccitelli D. "NATO joins the Pentagon in deeming climate change a threat multiplier," *Bulletin of the Atomic Scientists*, May 25, 2017: https://thebulletin.org/nato-joins-pentagon-deeming-climate-change-threat-multiplier10790. See also Causevic, A. "Climate Change and NATO: A New Study," The Center for Climate & Security, October 4, 2017: https://climateandsecurity.org/2017/10/24/climate-change-and-nato-a-new-study/.

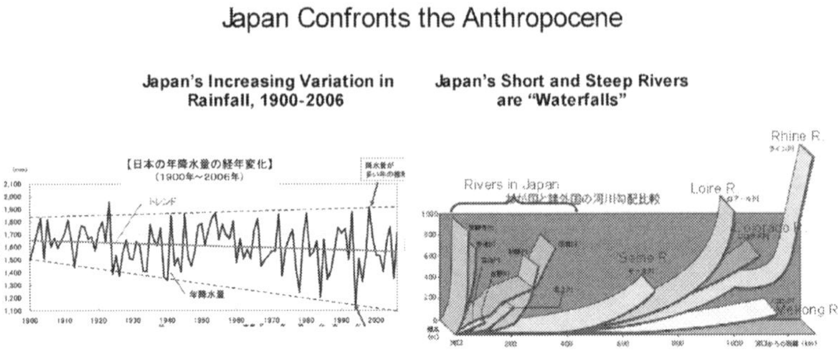

FIGURE 9.3 The end of stationarity in Japan. *Adapted from MLIT, 2009 "River management in Japan," River Bureau, Ministry of Land, Infrastructure, Transport and Tourism (MLIT), Japan, January 2009: https://www.mlit.go.jp/river/basic_info/english/pdf/conf_05.pdf.*[18]

The most crucial area of overlap among these policies regimes is critical infrastructure, especially energy-related infrastructure.

As noted, Japan has ample incentives to implement policy coordination. Fig. 9.3 illustrates a potent and worsening threat that Japanese policy makers simply cannot ignore. This figure is one indicator of what the end of the Holocene and onset of the Anthropocene means for Japan's built and natural environments. It also shows why integrated, robust industrial policy is required to confront climate change. The figure highlights the increasing variation in Japan's annual rainfall from 1900 to 2006, which has continued to accelerate in recent years. Note Japan's short and steep rivers, compared with the Seine, Rhine, Mekong, and other major global rivers in the right-hand section of the figure. Japanese rivers' short length and steep pitch derive from the fact that the country is narrow (no point in Japan is more than 150 km from the sea) and mountainous. Moreover, Japan receives roughly twice the global annual average of rainfall, all of which make Japanese rivers more like waterfalls during the country's seasonal bouts of concentrated rainfall.

Keihanna and Toyota City were slated for accelerated deployment domestically and overseas. Based on research by Nikkei BP Cleantech Institute, in 2010 Japanese experts estimated that the smart-city market would reach at least a cumulative JPY 5000 trillion in value between 2011 and 2030, when smart infrastructure such as water and housing are included along with smart grids, renewable energy, battery storage, and other core elements of the smart market.[19] That 2010 estimate of a cumulative JPY 5000 trillion smart-

18. See "River management in Japan," River Bureau, Ministry of Land, Infrastructure, Transport and Tourism, Japan, January 2009: https://www.mlit.go.jp/river/basic_info/english/pdf/conf_05.pdf.
19. See (in Japanese) "The Global Smart City Market Will Be 3100 trillion," *Nihon Keizai Shimbun*, September 9, 2010: https://www.nikkei.com/article/DGXNASFK2401D_U0A920C1000000/.

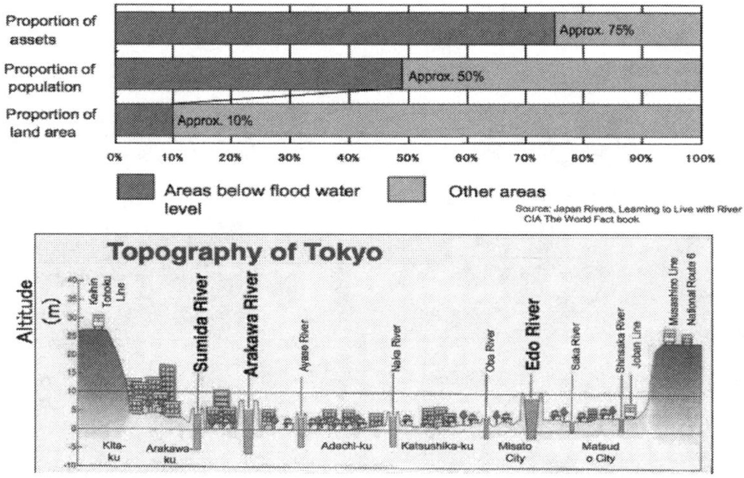

FIGURE 9.4 Tokyo's flood risk. *Adapted from MLIT, 2009.*[21]

city market was roughly 10 times the size of the Japanese economy. It is thus understandable that Japanese policy makers began devoting considerable fiscal, administrative, and other resources to the effort.

In addition, the flood risk, which is always present in Japan, is also rapidly worsening. It is useful to recall that the Meiji-era state-building elite brought over Dutch hydrologist Johannes de Rijke in 1873. Upon seeing the Joganji River in Toyama Prefecture, de Rijke reportedly exclaimed that "rivers in Japan are like waterfalls."[20] In geologic terms, that remark was made back in the Holocene era. The dramatically increasing variation in annual rainfall displayed in Fig. 9.3 indicates that Japan's hydrologic challenges have worsened greatly with the onset of the Anthropocene.

Fig. 9.4 places these hydrological facts in perspective. As is well-known, roughly 70% of Japan is mountainous, with half the population and three-quarters of the assets concentrated into the 10% of land that is coastal flood plains. The topography of Tokyo is particularly fraught. The figure shows that the levels of the Sumida, Ara (in the figure, "Arakawa"), Edo, and other rivers that run through Tokyo are already considerably higher than their surrounding districts. As sea levels continue to rise and bouts of intense rain become more common, the threat of catastrophic flooding correspondingly increases.

20. On Johannes de Rijke's comment, see Asian Disaster Risk Reduction Center (ADRC), "Rivers in Japan 1998": http://www.adrc.asia/management/JPN/RIVERS%20IN%20JAPAN%201998.html.

21. See "River management in Japan," River Bureau, Ministry of Land, Infrastructure, Transport and Tourism, Japan, January 2009: https://www.mlit.go.jp/river/basic_info/english/pdf/conf_05.pdf.

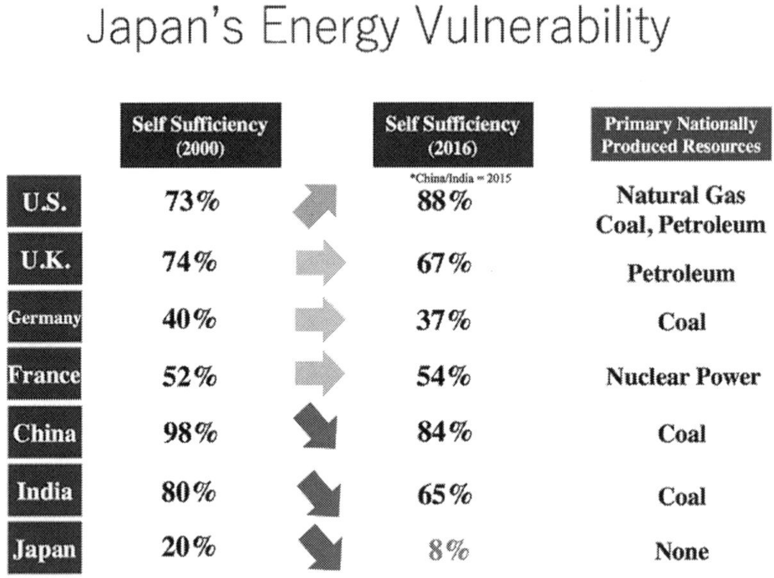

FIGURE 9.5 Japan's energy self-sufficiency in comparative perspective, 2000–16. *From ANRE, 2017.*[23]

Another reason Japan has to act in the Anthropocene is that it is dangerously dependent on fossil fuels. The consumption of these energy resources is not only the primary driver for the end of the Holocene era,[22] it also poses multiple other risks for resource-poor Japan.

Fig. 9.5 compares the change in Japan's energy self-sufficiency for the years 2000 and 2016 relative to several other large economies in Asia, North America, and Europe. These data measure primary energy as opposed to merely electricity, which is generally only 30% of all energy inputs. Most transport and industrial processes and the like rely on liquid or solid fuels rather than electricity, which is thus a secondary energy source (or "energy

22. On the role of fossil fuels, see Zalasiewics, J. et al., "The New World of the Anthopocene," in *Environ. Sci. Technol.* 2010, 44: http://pubs.acs.org/doi/pdfplus/10.1021/es903118j. A recent book-length treatment of fossil-fuels' role is available in Austin G (ed) *Economic Development and Environmental History in the Anthropocene: Perspectives on Asia and Africa.* New York, Bloomsbury Academic, 2017.
23. ANRE, "Resources, Geopolitics, and National Strategies," Agency for Natural Resources and Energy Ministry of Economy, Trade and Industry, September 29th, 2017: http://www.enecho.meti.go.jp/committee/studygroup/ene_situation/002/pdf/002_008_01.pdf.

carrier") because it is a means of moving energy in a useable form from one point (e.g., the generation facility) to another (e.g., the site of consumption).[24] In Japanese use, energy self-sufficiency measures the share of domestically produced energy, including hydro and other renewable energy along with nuclear power.[25] As is evident from the figure, in 2000, Japan's self-sufficiency was lowest among its peer countries in the industrialized world as well as its increasingly large neighbors, China and India. Moreover, in 2016, Japan's self-sufficiency had declined to the striking low level of only 8%. This number was far less than the next lowest level of 37% self-sufficiency in Germany.

The right-hand column of Fig. 9.5 also indicates that Japan lacks the relatively abundant domestic energy resources that underlie its peers' comparatively high levels of self-sufficiency. The implications of this are clear: Japan is extremely vulnerable to supply, price, and other shocks in energy. Although conventional economics generally treats energy as just another commodity, it is in fact the "master resource."[26] Energy is essential for every other sphere of economic activity, including mobility, resource extraction, material production, and virtually all household consumption. Japan's low self-sufficiency in energy renders it uniquely exposed to disruptive events.

Fig. 9.6 further details Japan's vulnerability. The figure shows that Japan's import reliance on the two most critical fossil fuels, petroleum and gas, is generally much higher than its peer countries. Japan imports 99% of its oil and 98% of its gas (meaning natural gas in the form of liquefied natural gas). In addition, Japan is dependent on the increasingly unstable Middle East for supplies of these fuels. Japan's high reliance on Saudi Arabia in particular is a matter of concern, because in the latter part of 2017 the Saudi state became increasingly unstable.[27]

The right-hand columns in Fig. 9.6 also show that Japan's imports are via tanker transport. The Japanese archipelago lacks pipeline connections to neighboring countries (it also lacks external power grid connections). Such infrastructure networks are not a guarantee of security, because they can be

24. On this crucial difference between primary and secondary energy, see US EIA, "What are secondary energy sources?," US Energy Information Administration (no date): https://www.eia.gov/energyexplained/?page=secondary_home.

25. Whether Japan's nuclear power is in fact a domestic source is open to question. The uranium fuel (and its by-products) used in Japan's nuclear reactors is ultimately an imported commodity. However, the Japanese energy authorities deem nuclear energy production a domestic energy source. This chapter draws on their data without necessarily agreeing that nuclear power is a completely domestic resource.

26. On the critical role of energy, see Day, JW and C. Hall "Energy: The Master Resource," in Day, JW and C. Hall, *America's Most Sustainable Cities and Regions. Surviving the 21st Century Megatrends*, Springer, 2016, pp. 167–216.

27. On Saudi Arabia's increasing turmoil, see Lippman, T. W., "The End of Saudi-Style Stability," Middle East Institute, November 9, 2017: http://www.mei.edu/content/end-saudi-style-stability.

	Petroleum			Gas		
	Import Reliance	% Middle East	Largest Importer	Import Reliance	% Middle East	Largest Importer
U.S.	41%	8%	15% Canada – Connected via Pipeline	3%	0%	3% Canada – Connected via Pipeline
U.K.	22%	1%	12% Norway – Connected via Pipeline	46%	10%	32% Norway – Connected via Pipeline
Germany	96%	4%	37% Russia – Connected via Pipeline	90%	0%	44% Russia – Connected via Pipeline
France	97%	25%	15% Saudi Arabia – Tanker Transport *Connected via European Pipeline	99%	2%	40% Norway – Connected via Pipeline
China	61%	31%	9% Saudi Arabia – Tanker Transport *Connected via pipeline to Russia etc.	29%	4%	15% Turkmenistan – Connected via Pipeline
India	83%	46%	15% Saudi Arabia – Tanker Transport *No pipeline	40%	25%	22% Qatar – Tanker Transport *No pipeline
Japan	99%	85%	37% Saudi Arabia – Tanker Transport *No pipeline	98%	23%	28% Australia – Tanker Transport *No pipeline

FIGURE 9.6 Japan's import reliance in comparative perspective (2016). *From ANRE, 2017.*[31]

attacked by enemies or used to apply diplomatic pressure.[28] Yet external energy networks can also alleviate isolation and foster supply sharing. The capacity to trade power flows is especially important for increasing the penetration of intermittent renewables and meeting renewable energy targets.[29] External networks also reduce the geopolitical risk of shipping crucial energy commodities through dangerous "choke points" such as the straits of Hormuz and Malacca.[30]

Fig. 9.7 indicates Japanese policy makers' concern regarding their country's energy security. The figure is taken from the Ministry of Economy, Trade, and Industry (METI) 2017 White Paper on Energy and shows the steep decline in global investment in exploration and production. This decline is expected to

28. On these matters, see Vladimirov M.and De Jong S. Deciphering Gazprom's Pipeline Agenda in Europe, The Atlantic Council, March 14, 2017: http://www.atlanticcouncil.org/blogs/new-atlanticist/deciphering-gazprom-s-pipeline-agenda-in-europe.
29. The latter fact is evident, for example, in the Northern Pass Transmission Line project that will transmit hydropower from Quebec to Massachusetts. See Petrova, V. Northern Pass wins bid to deliver hydropower to New England, Renewables Now, January 26, 2018: https://renewablesnow.com/news/northern-pass-wins-bid-to-deliver-hydropower-to-new-england-599544/.
30. A detailed discussion of choke points and their risks can be found in US EIA, "World Oil Transit Chokepoints," US Energy Information Administration, July 25, 2017: https://www.eia.gov/beta/international/regions-topics.cfm?RegionTopicID=WOTC.
31. ANRE, "Resources, Geopolitics, and National Strategies," Agency for Natural Resources and Energy Ministry of Economy, Trade and Industry, September 29, 2017: http://www.enecho.meti.go.jp/committee/studygroup/ene_situation/002/pdf/002_008_01.pdf.

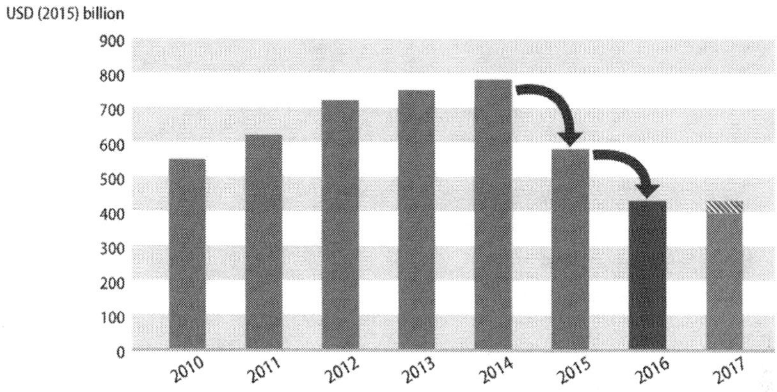

FIGURE 9.7 Global oil and gas investment, 2010–17. *From METI, 2017* [33]

cause uncertainty and perhaps havoc, in global oil and gas supply and prices over coming years. There is, of course, a great deal of "peak demand" optimism that the global economy is abandoning conventional energy in favor of renewables. Seen from this perspective, declining investment in conventional production is to be welcomed. However, there are also credible arguments to the contrary.[32] Given Japan's extreme reliance on conventional energy imports, the risk of price and supply shocks cannot be ignored. If one is to understand what Japanese policy makers are doing, and why, one has to see energy and other risks from their perspective.

Japan has responded aggressively to these risks. Indeed, analyses of Japan's primary energy supply and its component shares for the period 2010–16 show that the primary energy supply (measured in millions of tons of oil equivalent [Mtoe]) fell from 514.7 Mtoe in FY 2010 (meaning Apr. 1 to Mar. 31 of 2010–11) to 465.6 in FY 2016. This decline was expected to continue through FY 2018, primarily because of increasing efficiency.[34] Even so, the lion's share of primary energy remains the fossil fuels of oil, gas, and coal. In FY 2014, these three fuels composed fully 94.7% of Japan's primary energy, much

32. Some of these arguments are detailed by energy analysts Ellen R. Wald. See Wald, E. "The Problematic Truth About U.S. Shale Oil Production," Investing, January 10, 2018: https://www.investing.com/analysis/the-truth-about-american-shale-oil-production-200279008.

33. See Japanese Ministry of Economy, Trade and Industry (METI), Japan's Energy White Paper 2017, p. 3: http://www.meti.go.jp/english/report/downloadfiles/energy_hakusho_201711.pdf.

34. See IEEJ, "Economic and Energy Outlook of Japan through FY2018," The Institute of Energy Economics Japan, July 25, 2017, Table 1, p. 4: http://eneken.ieej.or.jp/en/press/press170725.pdf.

higher than the 87.5% level in China and 82.9% in the United States for the same year.[35]

Conventional analyses made a number of assumptions about Japan's energy future. One is that the number of nuclear reactors in operation would roughly double, from 5 (in Nov. 2017) to 10 in FY 2018. The Institute of Energy Economics Japan (IEEJ), for example, assumed that this increase in the role of nuclear energy would have a large role in raising the country's rate of self-sufficiency to just over 10%.[36]

It is true that 12 reactors (at six separate plants) have passed safety inspections, including two at the Kashiwazaki Kariwa plant (the world's largest). Yet, putting them back online is fraught with political and legal hurdles. Moreover, even the Nuclear Regulation Authority's (NRA's) chairman, Fuketa Toyoshi, made it clear that the slow pace of inspections will not be accelerated. Fuketa concedes that the NRA has considerable accumulated experience in conducting safety checks, but points out that seismic and other risks require thorough inspections. He made it clear to the press that he simply could not forecast the number of reactors that would be approved over the coming 5 years from 2017 through to 2022.[37] Hence, the IEEJ's estimate that 10 reactors would be in operation by 2018 was simply a guess.

The IEEJ also assumed that non—hydro renewable energy would only double in output between 2010 and 2018, increasing from 63.9 to 127.3 TWh. It also estimated that the share of hydro would decline over the same period, dropping from 84.3 TWh in 2010 to 80.1 TWh in 2018. Given the geopolitical and other risks of relying on fossil fuels, these numbers indicate either inertia in the energy economy and its governance or difficulty in assessing what Japan is doing.

It is certainly the case that Japanese deployments of alternative energy have been less than optimistic forecasts that followed the 3—11 disaster. Fig. 9.8 shows that Japan's deployment of renewable energy accelerated to an annual average growth rate of 29% between 2012 and 2015. However, that increase started from a very low base. Moreover, the International Energy Agency's (IEA's) Oct. 2017 report, *Renewables 2017: Analysis and Forecasts to 2022*, is

35. For these data, see the relevant figure on comparative reliance on fossil fuels in ANRE (in Japanese), "The Energy Problems Confronting Japan," Agency for Natural Resources and Energy Ministry of Economy, Trade and Industry, July 21, 2017: http://www.enecho.meti.go.jp/about/special/johoteikyo/energyissue.html.

36. See IEEJ, "Economic and Energy Outlook of Japan through FY2018," The Institute of Energy Economics Japan, July 25, 2017, Table 1, p. 4: http://eneken.ieej.or.jp/en/press/press170725.pdf.

37. On this, see Tsukimori O. and A. Sheldrick, "Japan nuclear regulator says restart approval pace unlikely to speed up," *Reuters News*, November 7, 2017.

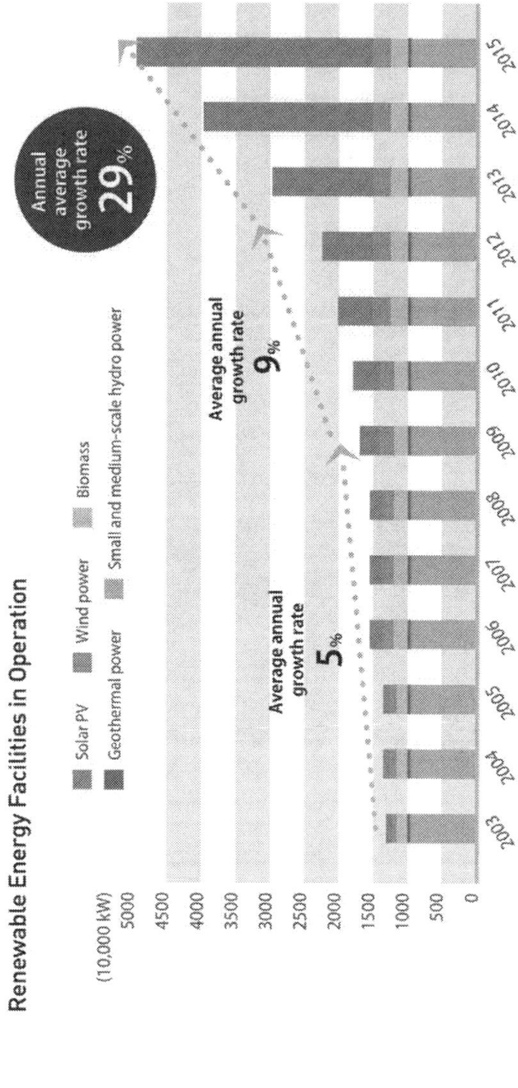

FIGURE 9.8 Renewable energy growth rate in Japan. *From METI, 2017.*[39]

pessimistic about Japan's present and future role as a site for non−hydro renewable energy as well as hydro.[38]

The IEA's data show that Japan's added capacity had slowed. Japan's capacity additions for 2015 and 2016 were almost entirely solar power and declined from just over 10 GW in 2015 to roughly 8 GW in 2016.

Moreover, capacity additions do not equate with generation. The reason is that solar and wind power generation totals in Japan are at best, 12% and 20%, respectively, of their rated capacity. As in all countries, solar panels do not generate power at night or inclement weather. Moreover, wind turbines often do not generate energy owing to insufficient wind speeds.

The IEA undertakes comparative studies on average generation capacity factors (meaning the percentage of energy output versus rated capacity). Its survey of output between 2008 and 12 indicates that Japan's aggregate figure for both solar and wind is 15%, far less than the 27% recorded for the United States, the 26% seen in Canada, and the 18% figure for China.[40] Japan's performance in this regard is not especially poor, but rather about average. Better technology, siting, and other factors may raise Japan's average generation capacity factors in the future, but at present the difference is great between the gross figures on capacity additions and actual output.

The IEA's projections for the future of Japanese renewables is also not optimistic. The pessimism holds compared with Japan's peer countries in addition to Japan's desperate need for alternatives to fossil fuels and nuclear.

The IEA also projects that between 2017 and 2022, Japan will rank low in the deployment of non−hydro renewable and hydro. In solar, Japan is slated to only fourth (20.5 GW), and in bioenergy (2.6 GW) only third. In no other technology is Japan forecast to rank among the global top five.

Japan's solar and wind provided about 5% of total power generation in 2016, higher than India and not far from the total recorded for China, the global leader in renewables. However, by 2022, Japan's share of solar and wind power generation (i.e., not just capacity, but actual output) is assessed as remaining well below 10% of total power generation. By contrast, all other countries in the figure are expected either to meet or greatly exceed a 10% share.

The IEA concedes that Japan's capacity additions after the 3−11 disaster indeed exceeded numerous other countries and regions. That is only to be expected, however, considering the scale of Japan's economy, the extent of the disaster's effects on the energy economy, and the need for alternatives. What is surprising is that Japan's capacity additions slow down so much when most

38. IEA, *Renewables 2017: Analysis and Forecasts to 2022*, International Energy Agency, October 2017.
39. See Japanese Ministry of Economy, Trade and Industry (METI), Japan's Energy White Paper 2017, p. 3: http://www.meti.go.jp/english/report/downloadfiles/energy_hakusho_201711.pdf.
40. IEA, "Electric generator capacity factors vary widely across the world," International Energy Agency, September 8, 2015: https://www.eia.gov/todayinenergy/detail.php?id=22832.

other countries and regions in the figure are expected to add significantly to their stock of renewables.

One problem hindering Japan's deployment of renewable energy is uncertainty. The IEA analysis shows that the bulk of Japan's renewable deployment is utility and residential solar projects. Utility-scale projects are expected to comprise nearly half of all of Japan's 9.6 GW in solar growth between 2017 and 2022. However, ongoing policy changes have created a great deal of uncertainty for solar power. These policy changes include the introduction of an auction scheme in 2017, to replace the overly generous feed-in tariff that had been in operation from Jul. 2012.[41] Japan has approved ("certificated") well over 80 GW of solar projects. However, the IEA expects about 48 GW of these projects to be canceled owing to a lack of grid agreement and other reasons.[42]

The risk of curtailment is indeed a major source of uncertainty. Although the Japanese electricity market is being deregulated and aims at separating generation from transmission, the legacy utilities still possess significant influence. One result of this influence has been the expansion of their regulated authority to deny renewable projects' access to the grid, citing the lack of transmission and other capacity to cope with intermittent power. The IEA emphasizes that curtailment has become a significant business risk in areas where renewable generation exceeds local demand.[43]

The IEA expects appreciable increases in Japanese renewable deployment if grid curtailment and other barriers are reduced. The IEA poses two different scenarios for 2022: a "main" scenario in which policy does not change greatly and an accelerated scenario in which significant favorable policy changes are made. An accelerated scenario results in roughly 17 GW of additional PV capacity compared with the main scenario. The IEA also projects a significant change for bioenergy and onshore wind deployments, whereas hydro barely changes. The result is that under an accelerated policy regime, Japan's renewable deployment leads to 144.9 GW of additional capacity by 2022 versus the 125 GW of capacity addition forecast under continuation of the current policy.

In short, the most authoritative forecasts see renewable energy as, at best, achieving a comparatively moderate role in Japan's power generation over the next 5 years between 2017 and 2022. The forecasts are particularly pessimistic concerning prospects for nonsolar renewables, especially hydropower.

41. Concerning the problems with Japan's FIT, see Ninomiya Y. "Outlook for Renewable Energy Market," The Institute of Energy Economics 426th Forum on Research Work, July 25 2017: http://eneken.ieej.or.jp/data/7575.pdf.

42. IEA, *Renewables 2017: Analysis and Forecasts to 2022*, International Energy Agency, October 2017, p. 48, Figure 2.14.

43. IEA, *Renewables 2017: Analysis and Forecasts to 2022*, International Energy Agency, October 2017, p. 48.

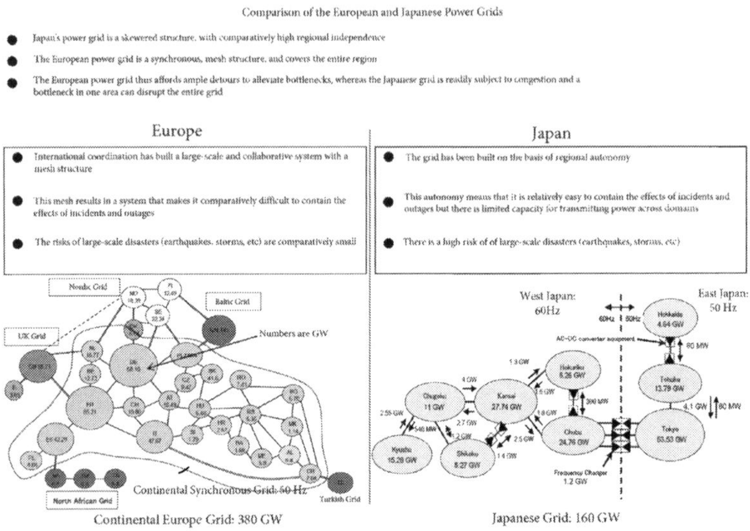

FIGURE 9.9 Europe and Japan's power grids. *From ANRE, 2018.*[45]

One major question is whether these forecasts are actually based on comprehensive assessments or are overlooking important evidence. One example would be the policy integration that is already shaping renewable energy and related elements important to Japan's strategy of mitigation and adaptation. This chapter contends that the forecasts are indeed overlooking a great deal of evidence. For one thing, Japan's Ministry of Land, Infrastructure, Transport, and Tourism (MLIT) has lots of dams used for flood control and is planning to convert some of them to power generation. The MLIT and its advisors believe that Japan's hydro output can be doubled or tripled from its current 9% of power. At the same time, the MLIT aims to bolster the dam network's role in flood control and stable water supply, linking the network advanced radar systems and supercomputers. This project, called the "Dam Revival Vision," was decided by MLIT on Jun. 26, 2017 and is funded in 2018 fiscal requests. It is a clear example of Japan's adaptation-led maximization of synergies between mitigation and adaptation.[44]

One illustration of how and why Japan focuses on adaptation is seen in Fig. 9.9, which compares the European and Japanese power grids. The

44. The details concerning the "Dam Revival Vision" are available (in Japanese) at the MLIT website: http://www.mlit.go.jp/report/press/mizukokudo05_hh_000029.html.

45. Author's translation of p. 10 of ANRE (in Japanese) "Measures for Relieving Constraints on the Power Grid," Agency for Natural Resources and Energy Ministry of Economy, Trade and Industry, January 24, 2018: http://www.meti.go.jp/committee/sougouenergy/denryoku_gas/saiseikanou_jisedai/pdf/002_02_00.pdf.

European Union (EU) system has twice the power flows of the Japanese (Germany and France together equal Japan), but Japan is isolated and still balkanized internally. The same is true of Japan's gas networks.

Japan's power grid structure is balkanized (in areal and frequency terms) owing to path dependence. The country's power technology has German and American roots; this structural difference became embedded during the Pacific War (1941–5). Balkanized power domains and regional separation of frequencies continued in the postwar period. The problem cannot be solved simply by building more transmission links. This is because Japan's demographic, disaster, and other risks undermine the economics of simply building a large, networked synchronous grid as in Europe or even the less integrated North American network.

At the same time, Japan's roads, waterworks, and other infrastructures of the built environment generally exceed what it needs in the face of aging and depopulation. Lower population densities and per-capita incomes erode the fiscal capacity to afford the per-hectare burden of maintaining sprawling infrastructure networks. Add to that a context of rising resource costs as well as increasing costs for health care, pensions, food, and the ravages of climate damage.

Worse, all of these factors exacerbate one another. A changing climate accelerates the wear and tear on the built environment, adds to the medical burdens of aging, and increases the cost of agriculture.

To cope with these challenges simultaneously, the Japanese emphasize integrated planning for the community, including local heat and power networks. The expansion and integration of these projects are accelerated by the coordination of energy policy, spatial planning and disaster planning, information and communication technology (ICT)—Internet of things investment, harvesting of heat from sewers, and bolstering dam networks. Japanese policy makers seek to maximize the productivity of extant assets, cope with the daunting crisis of scarce human resources, raise energy self-sufficiency (from a perilously low 8%), monitor accelerating disaster risks, shrink the spatial and carbon footprint of communities, and enhance the country's export opportunities. The multiplicity of imperatives and interlinked goals facilitates broad, pragmatic collaboration among a range of stakeholders in government (national and local), business, and civil society.

The next section turns to consider the policy background of Japan's smart communities. The argument being developed here is that conventional analyses are overlooking the impact of policy integration on macro-, meso-, and microlevel energy generation, transmission, and distribution.

JAPAN'S SMART COMMUNITIES AND EVOLVING POLICY INTEGRATION

The scale of Japan's smart community policies and projects remains little known or understood, even among experts on smart cities. The flagship "Japan Smart Community Alliance" (JSCA), a consortium of 258 firms (as of Nov. 2, 2017), is remarkably poor at communicating, even in Japanese. The JSCA was established on Apr. 6, 2010 and declares (in Japanese and English) that its role is to promote smart communities in Japan and globally. It defines the smart community as "a community where various next-generation technologies and advanced social systems are effectively integrated and utilized, including the efficient use of energy, utilization of heat and unused energy sources, improvement of local transportation systems and transformation of the everyday lives of citizens."[46] This definition is virtually the same in English and Japanese, and has not changed to reflect the powerful role of disaster resilience after 3−11.

Nor has the JSCA Japanese-language listing of domestic microgrid projects[47] been updated to include the Higashi Matsushima City Smart Disaster Prevention Eco Town[48] and other projects.[49] The site includes no comprehensive list of smart communities. Its most recent Japanese-language publication on smart community projects dates to May 2015 and includes only a brief mention and truncated listing of the disaster-resilient projects that were adopted after 3−11.[50] This failure to provide up-to-date information is especially puzzling because Sekisui House, a member firm of the JSCA, was one of the prime contractors on the Higashi Matsushima City Smart Disaster Prevention Eco Town.

46. The JSCA website is available (in Japanese) at the following internet website: https://www.smart-japan.org/index.html.

47. The JSCA list of microgrid projects is available (in Japanese) at the following website: https://www.smart-japan.org/reference/l3/Vcms3_00000134.html.

48. The Japanese press and other sources run ample material on the Higashi Matsushima City Smart Disaster Prevention Eco-Town. See, for example, Kaneko, K. (in Japanese), "A Japanese First! Solar and Battery Storage via a Microgrid: Higashi Matsushima City Takes on the Challenge of Building Disaster-Resilient Community," *Nikkei BP*, August 5, 2016: http://techon.nikkeibp.co.jp/atcl/feature/15/415282/080200009/?ST=msb.

49. These more recent projects include Panahome's microgrid-centered "Smart City Shioashiya" in Hyogo Prefecture. On the project and its particulars, see (in Japanese) [57]Panahome Implements a Project that Links Homes With a Power System in Hyogo's Ashiya," *Nihon Keizai Shimbun*, September 22, 2017: https://www.nikkei.com/article/DGXLASHD21H2J_R20C17A9LKA000/.

50. See (in Japanese) pp. 14−15 in JSCA, "Smart Communities: Projects by Japanese Firms," Japan Smart Community Alliance, May 2015: https://www.smart-japan.org/vcms_lf/library/JSCA_PR-magazine_web_single.pdf.

As of Feb. 2018, the JSCA's most recent English-language material is a Jun. 24, 2015 document, "Smart Community: Japan's Experience."[51] This material is out of date and emphasizes the four flagship smart community projects of Kitakyushu City, Yokohama City, Toyota City, and the Keihanna cluster in Kyoto Prefecture. Like its Japanese version, described earlier, the translation neglects to include Higashi Matsushima in an incomplete list of post 3–11 projects that stress disaster resilience. The entire document is also replete with inexcusably poor English translations, such as: "Smart Community being addressed in Japan has the concept involving smart grid. Whereas smart grid refers to the state being smarter by information and communication technology ICT for electric power system, Smart Community is the effort of changing social system of a defined area into smarter state with technologies not only for electric power system but also for a variety of public infrastructure including heat supply, water and sewerage, transportation and communications."[52]

These examples reflect shoddy governance in coordinating Japan's domestic and international communication of what its smart communities are and how 3–11 has changed them. The JSCA and its member firms, along with the New Energy and Industrial Technology Development Organization (NEDO) and METI, which serve as secretariat and support organizations for the JSCA, have clearly allowed the crucial role of public relations to fall through the cracks. It is ironic that Japan excels at good governance and policy integration for building resilient smart communities but is poor at communicating the facts.

More comprehensive studies exist. For example, on May 27, 2014, Ernst and Young Institute Japan (EY) released a Japanese-language survey summarizing Japan's over 200 smart-city projects. The EY report was entitled (in Japanese) *Smart Cities: A Study of Issues Involved in Going From Testing to the Deployment Stage*. It was also subtitled *Towards Constructing Smart Cities That Produce a Citizen-Centered "Precious Circle."*[53] The "precious circle" refers to achieving a balance of incentives among the three major stakeholders (citizens, corporations, and governments) such that citizen-friendly and sustainable urban development is achieved.

The EY study did not examine the political economy background to Japan's initiatives, so let us take a short detour on that important aspect with

---

51. The JSCA English-language pamphlet "Smart Community: Japan's Experience" can be accessed at: https://www.smart-japan.org/english/vcms_lf/Resources/JSCApamphlet_eng_web.pdf.

52. See p. 2 and pp. 10–12 in JSCA "Smart Community: Japan's Experience," Japan Smart Community Alliance, June 24, 2015: https://www.smart-japan.org/english/reference/index.html.

53. See Kijou N. and Rure K. (in Japanese) Smart Cities: A Study of Issues Involved in Going from Testing to the Deployment Stage: Towards Constructing Smart Cities that Produce a Citizen-Centred 'Precious Circle', May 27, 2014: http://eyi.eyjapan.jp/knowledge/industrial-research/pdf/2014-05-27.pdf.

Massachusetts Institute of Technology expert on Japanese energy politics, Richard Samuels. Like the EY study, Samuels notes in his 2013 book *3.11: Disaster and Change in Japan* that the METI coordinated the four flagship projects with power, gas, auto, steel and other industrial heavyweights.[54] Samuels adds the important insight that before Japan's natural and nuclear disasters of 3–11, Japanese policy makers aimed to test-bed low-carbon models in a few cities, eyeing export opportunities. METI was aware of innovative developments overseas and wanted to grow new industries. But it had to work within a centralized and nuclear-focused paradigm in the country's power economy. Japan's powerful monopoly utilities were averse to the rise of competitors through deployment of the smart grid and renewable power.

Yet 3–11 delivered a powerful shock to the "nuclear village." The nuclear meltdowns and the rolling brownouts starkly demonstrated the vulnerability of centralized power. They also made conservation and local resilience not mere virtues but urgent necessities. The disasters also expanded non-METI central agencies' and local governments' incentives to carve themselves a role in the accelerating fusion of ICT, distributed energy, efficiency, and other aspects of the smart-city model.

Decidedly apolitical, the EY report did not discuss the tension between Japan's conventional utilities and the producers and consumers (or "prosumers") of renewable power in the business and household sector.[55] Even so, the EY study remains valuable as a snapshot of the state of Japan's smart-city programs and debates.

The EY report pointed out that smart cities had already been an object of specialist attention for several years. Most prominent in Japan, for example, the METI started its flagship "next-generation energy and social systems test-bed" smart-city approach in the four sites of Yokohama City, Toyota City, Keihanna (in Kyoto Prefecture), and Kitakyushu City in 2010. These projects and their particulars are displayed in Fig. 9.10. As can be seen in the figure, energy efficiency and carbon reduction were common threads uniting the projects.

EY's 2014 report was also especially well-timed. Among other developments during the same year, Jun. 2 saw the global giant Apple join a long

54. See Samuels, R. 3.11: Disaster and Change in Japan, Ithaca and London: Cornell University Press, 2013, p. 1.
55. Contrast the willingness of many municipal and cooperative utilities to grow new business models that foster the prosumer. See Paul Hockenos, "Local, Decentralized, Innovative: Why Germany's Municipal Utilities are Right for the Energiewende," Energy Transition: The German Energiewende, September 28, 2013: http://energytransition.de/2013/09/local-decentralized-innovative-why-germanys-municipal-utilities-are-right-for-the-energiewende/.

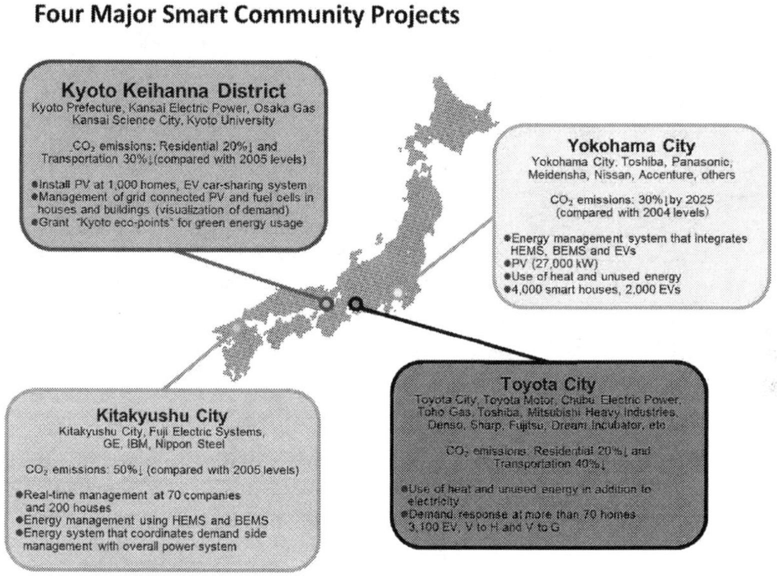

FIGURE 9.10 Japan's flagship smart community projects. *BEMS, building energy management systems; EV,* electric vehicles; *HEMS,* home *energy* management systems; *PV,* photovoltaic; *V to H,* Vehicle to House; *V to G* Vehicle to Grid *V. From NEDO, 2013.*[56]

list of firms including Toyota Home[57] by entering the "smart home" market.[58] The global background includes thousands of smart-city projects, assessed as worth at least USD 650 billion in 2014 and projected to be USD 1.2 trillion in 2022.[59] At over USD 40 billion, Korea's Songdo smart-city project alone is

56. NEDO (New Energy asnd Industrial Technology Development Organization), "Smart Community Projects," May 22, 2013, p. 11: https://www.slideshare.net/IsraelExport/yoko-nedo-smart-community-projects.
57. Toyota Home is just one element of Toyota Corporation's smart city-related initiatives: http://www.toyotahome.co.jp/smarthouse/.
58. See Katherine Tweed, "Apple Launches Home Kit for the Connected Home," Green Tech Media, June 2, 2014: http://www.greentechmedia.com/articles/read/apple-launches-home-kit-for-connected-home.
59. The figure is from the May 2014 MarketsandMarkets report "Smart Cities Market," which estimates 14% annual growth in the sector but also hints at the potential for "exponential demand." The report also identifies "IBM, Alcatel-Lucent, Accenture, ABB, Cisco, Cubic, Honeywell, Intel, Siemens and Oracle" as the "major players" in the global smart-city market. See the report's press release at: http://www.marketsandmarkets.com/PressReleases/smart-cities.asp.

one of the costliest private-sector real estate developments ever undertaken.[60] In 2014, Japan's core smart-city projects in Yokohama, Kitakyushu, Keihanna, and Toyota City were slated for accelerated deployment domestically and overseas. Based on research by Nikkei BP Cleantech Institute, in 2010 Japanese experts estimated that the smart-city market would reach at least a cumulative JPY 5000 trillion in value between 2011 and 2030, when such smart infrastructure as water and housing are included with smart grids, renewable energy, battery storage, and other core elements of the smart market.[61] That 2010 estimate of a cumulative JPY 5000 trillion smart-city market was roughly 10 times the size of the Japanese economy. It is thus understandable that Japanese policy makers began devoting considerable fiscal, administrative, and other resources to the effort.

Cases of Some Smart Green and Healthy Cities in Japan

The EY report's "precious circle" refers to achieving a balance of incentives among the three major stakeholders (citizens, corporations, and governments) such that citizen-friendly and sustainable urban development is achieved.

The EY report points out that smart cities have been an object of specialist attention for several years. Most prominent in Japan, for example, the METI began to lead initiatives on smart cities (or smart communities) years before the 3−11 disaster. An early expression of the METI efforts is seen in Fig. 9.11. The figure outlines a smart community predicated on electrification and efficiency. The concept clearly seeks to reduce reliance on imported fuels by expanding both nuclear and renewables, in the context of a smart community marked by coordinated governance (rendered awkwardly in English as "whole-city task-solution system").

The conception of the smart community changed over the years. As seen in Fig. 9.12, by 2014 the METI's Agency for Natural Resources and Energy was picturing the smart community as focused on energy efficiency, via home energy management, mansion energy management, community energy management, and other systems.

Fig. 9.13 shows how the projects were differentiated by region and application. One category divided metropolitan areas and regional cities to allow for the evolution of approaches suitable to their often-diverging needs. A

60. On Songdo, see Rick Huijbregts, "Songdo Case Study (v 1)," Presentation to Harvard Graduate School of Design, Executive Education," January 29, 2014: http://www.rickhuijbregts.com/wp-content/uploads/2013/03/Songdo-v1.0.0.pdf and Rita Lobo "Could Songdo be the world's smartest city?" World Finance, January 21, 2014: http://www.worldfinance.com/inward-investment/could-songdo-be-the-worlds-smartest-city.

61. See (in Japanese) "The Global Smart City Market Will Be 3100 trillion," *Nihon Keizai Shimbun*, September 9, 2010: https://www.nikkei.com/article/DGXNASFK2401D_U0A920C1000000/.

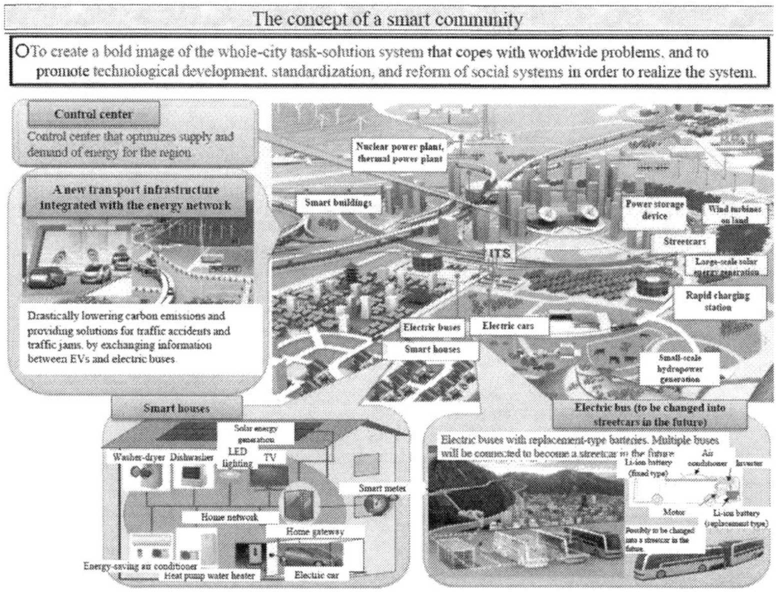

FIGURE 9.11 The Ministry of Economy, Trade, and Industry smart community before 3-11. *EVs*, electric vehicles; *ITS, Integrated transport System*; *LED*, light-emitting diode. *From ANRE, 2014.*[62]

second division was large industrial complexes and regional industrial complexes. As can be seen from the figure, energy systems are the common thread linking all cases.

In fact, the design and implementation of these projects has a decade-long history. In 2008, Japan's Promotion Council for Low Carbon Cities was instituted, with the Cabinet's Office's Regional Revitalization Bureau as its secretariat. The low-carbon program undertook a competition for "Eco-Model

62. See ANRE "ANRE's Initiatives for Establishing Smart Communities," Policy Planning Division, Energy Conservation and Renewable Energy Department, Agency for Natural Resources and Energy, February 2014: http://www.meti.go.jp/english/policy/energy_environment/smart_community/pdf/201402smartcomunity.pdf.
63. See ANRE "ANRE's Initiatives for Establishing Smart Communities," Policy Planning Division, Energy Conservation and Renewable Energy Department, Agency for Natural Resources and Energy, February 2014: http://www.meti.go.jp/english/policy/energy_environment/smart_community/pdf/201402smartcomunity.pdf.
64. See ANRE "ANRE's Initiatives for Establishing Smart Communities," Policy Planning Division, Energy Conservation and Renewable Energy Department, Agency for Natural Resources and Energy, February 2014: http://www.meti.go.jp/english/policy/energy_environment/smart_community/pdf/201402smartcomunity.pdf.

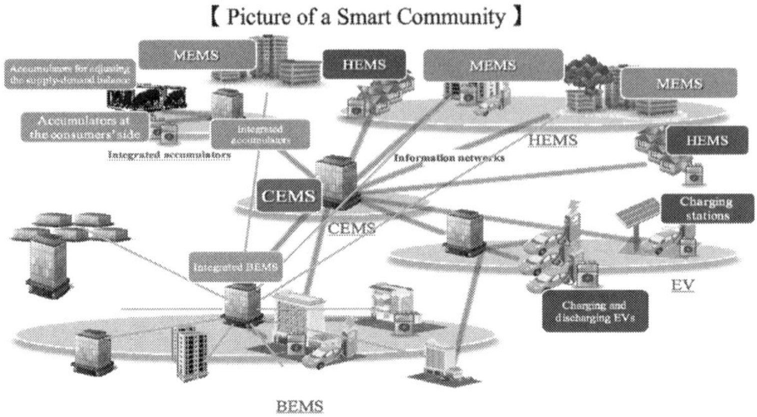

FIGURE 9.12 Agency for Natural Resources and Energy conceptualization of the smart community in 2014. *BEMS, building energy management systems*; *CEMS*, community energy management; *EV*, electric vehicles; *HEMS*, home energy management; *MEMS*, mansion energy management. *From ANRE, 2014.*[63]

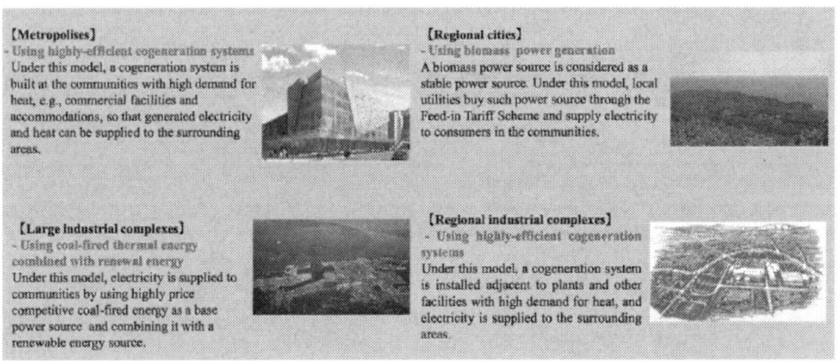

FIGURE 9.13 Agency for Natural Resources and Energy NREregional differentiation of the smart community in 2014. *From ANRE, 2014.*[64]

Cities" between Apr. 11 and May 21, 2008. It received 82 submissions and selected six in July of that year.[65]

By the end of 2008, the number of Eco-Model Cities selected had risen to 13. Later, in 2011, a new category of "Future Cities" was declared. This new designation was not greatly different from the Eco-Model Cities although it included a focus on aging in addition to environmental challenges. The new

65. Edahiro, J. "Conceptual Basis of the Movement to Create and Propagate 'Eco-Model Cities'" Japan for Sustainability, March 9, 2009: https://www.japanfs.org/en/news/archives/news_id028824.html.

category allowed for overlap, such that Kitakyushu was both an Eco-Model as well as a Future City. The most important aspect of the new category was that it included several cities in the Tohoku region that had been hit by the 3–11 disaster.[66]

Subsequent initiatives in 2012 and 2013 selected a broader variety of cases, adding the designation "Future Cities." The 2012 and 2013 selections came with a special focus on the Tohoku region that had been devastated by the 3–11 natural and nuclear disasters.[67]

Defining the Smart City

The EY authors argue that several issues require attention in rolling out the flagship smart cities, not to mention myriad other projects under way. First, they lament the lack of a precise definition of the smart city itself. Second, they express concern that the expanding scope of the initiatives as well as the broad range of stakeholders involved cross over multiple jurisdictions. They believe that such complexity makes the smart city difficult for administrators, voters, and others to grasp, let alone plan for and implement. The published documentation (in Japanese) on smart cities indicated that there was poor understanding of the issues involved in operationalizing the projects. Thus, the EY authors sought to define the elements of the smart city as clearly as possible on the basis of Japanese documentation.

The EY found no concise definition of the smart city. We have already seen that the JSCA provide little guidance in this regard. The EY report remarked that there were various definitions even within the narrow cluster of administrative agencies (such as the NEDO) grouped around METI. One might add that the Japanese smart community paradigm also features a multiplicity of terms including "smart towns," "smart villages," and "eco-towns."

The EY authors show that METI itself defined the "smart community" as a next-generation social system in which households, buildings, transport systems, and other elements are linked by information technology networks and achieve an effective deployment and use of energy across the region. Similarly, the Natural Resources and Energy Agency in its 2011 Energy White Paper described the smart city as one in which electricity supply and demand are monitored with technology, and energy management is done at a regional level with distributed generation or energy systems.[68] The lead agency for test-

66. See JFS, "'FutureCity' Initiative," Japan for Sustainability, no date: https://www.japanfs.org/en/projects/future_city/.

67. See JFS, "'FutureCity' Initiative," Japan for Sustainability, no date: https://www.japanfs.org/en/projects/future_city/.

68. The Agency's glossary for the 2011 Natural Energy White Paper (in Japanese) is available here: http://www.re-policy.jp/jrepp/JSR2011/JSR2011_glossary.pdf.

bedding the smart cities, the METI's New Energy Promotion Commission[69] was even more ambiguous. It defines the smart city as one in which ICT and other technologies are deployed in new ways and residents' quality of life is enhanced while the environmental burden is reduced and the city is able to grow sustainably. These rough definitions are not only tentative but also open to multiple interpretations.

The EY report argues that the various definitions as well as the approaches of the Japanese government and private sector actors offer a rough summary of what makes a smart city "smart." The authors themselves refine the definition by suggesting that "smart" is underpinned by the following factors: (1) increasing the convenience of urban dwellers; (2) pursuing sustainability through a focus on the environment, security, safety, resilience, and other factors; and (3) increasing the efficiency of services and other business.

The EY effort is a useful contribution to achieving a more coherent understanding of what constitutes "smart." What the EY authors overlook, however, was that whereas Japanese smart city definitions were indeed vague, they are nevertheless clearly focused on energy. The projects also increasingly stress disaster resilience. These key features help Japanese smart communities stand out against the thousands of projects then under way or in the planning stages globally. Many smart projects outside Japan (and a few in Japan as well) have at best a tangential relationship to reducing material flows of energy and water, not to mention bolstering their critical infrastructure against the ravages of climate change.[70] The world's reputedly "smartest smart city" in Santander, Spain is focused on energy and water and deeply engaged with its citizens in managing flows.[71] However, it is unclear whether it is Spanish or EU industrial policy. Also, critics have noted, for example, that the Glasgow Future Cities project is overly devoted to their surveillance.[72] By contrast, the Japanese are more inclusive and less intrusive, and they diffuse the projects nationwide as full-fledged industrial policy.

69. The NEPC's material on the smart community is here: http://www.nepc.or.jp/smartcommunity/.
70. On the characteristics of smart cities in the EU, see in particular Chapter 2 of the "Mapping Smart Cities in the EU" study by Urenio, March 7, 2014: http://www.urenio.org/2014/03/07/mapping-smart-cities-eu/.
71. See Newcombe T. "Santander: The Smartest Smart City," *Governing*, May 2014: http://www.governing.com/topics/urban/gov-santander-spain-smart-city.html.
72. See Sally Davies, "Glasgow aims to be the first 'smart city,'" *Financial Times*, June 3, 2014: http://www.ft.com/intl/cms/s/0/d119ac06-e57e-11e3-a7f5-00144feabdc0.html#axzz33y6k5pbG.

Chapter 9

Smart Green Healthy Communities: Cases of Science Parks and Microcities

Woodrow W. Clark, II [a]

Research Professor in Economics, Pepperdine University Graziadio School of Business (PGSB), West Los Angeles, CA Campus, Los Angeles, California

Chapter Outline

a. Woodrow W. Clark II, Ph.D.

Climate Preservation In Urban Communities Case Studies. https://doi.org/10.1016/B978-0-12-815920-0.00009-5

INTRODUCTION

Practical green development is being implemented and needs to be presented. While the focus is on smart, green, and healthy technologies that do not harm the environment, the applications must also use qualitative economics in an interactional scientific perspective. For the green development purposes herein with these theoretical considerations, there are cases that are far from the typical analyses seen in conventional business economic publications. From these theoretical conceptions, empirical data provide clear examples of this perspective to understand business activities, actions, events of people and groups.

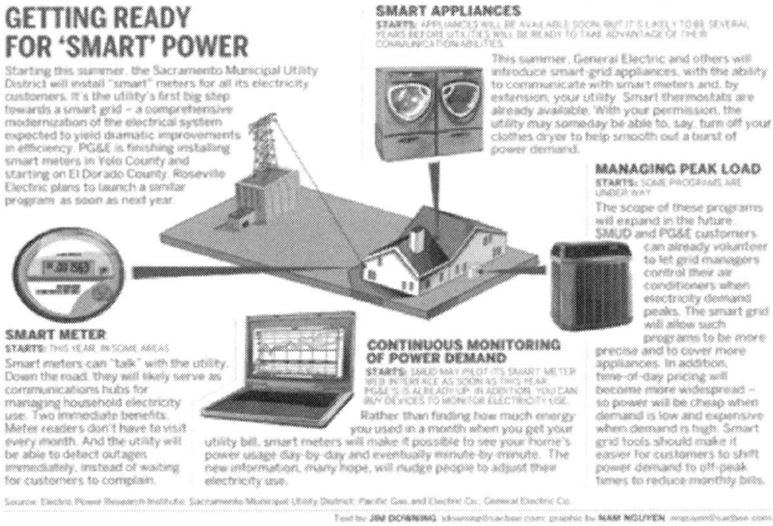

The need is for cross-disciplinary actions with energy in homes and buildings as shown above. As noted in the chapters and appendices, there is a need to look at communities for being smart, green, and healthy to understand how to finance these communities from the community members, interacting not only locally but also regionally and with the national government. The economics in all of these is important as Bonato and Orsini noted in a recent paper (2017) on Urban Circular Economy which "is about our everyday life, it concerns our role as citizens and consumers and it will probably shape our future."

The first case involves the commercialization of a fuel cell technology that could exceed the environmental and political demands. The key, however, for a fuel cell is that it uses only small amounts of water that is converted into hydrogen, *not* natural gas or other fossil fuels that many companies are using. As noted below, there is a "zinc—air fuel cell" (ZAFC) that gets it power form the earth, which is converted into energy. As noted in papers, ZAFC is a technology that generates electricity due to the reaction between oxygen and zinc pallets in a liquid alkaline electrolyte (Cooper and Clark, 1996). What is most important is that the ZAFC can generate power that is environmentally benign due to producing recyclable zinc oxide without gas emission (Clark et al., 2002).

More on that technology is given below. The fuel cell technology was created at a research laboratory during the 1990s at Lawrence Livermore National Laboratory (LLNL) in Northern California and was commercialized into a business application in 2005 for stationary power applications. Commercialization for mobile purposes would occur next, allowing vehicles to exceed legislation for zero-emission regulations in California and other nation states.

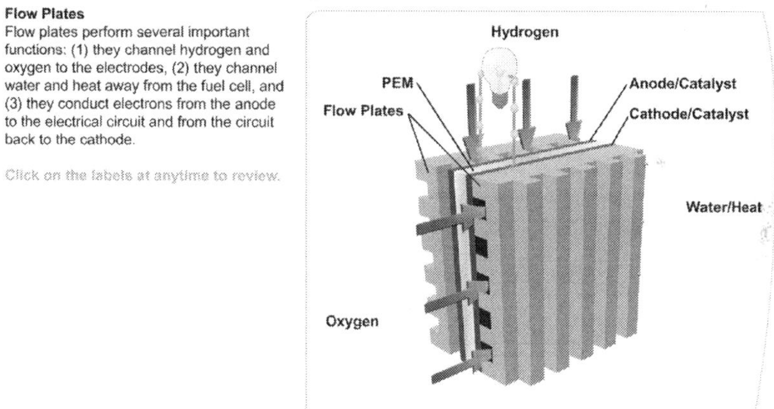

Today in California, electric vehicles (EVs) are back led by the Tesla, and now all car companies around the world are producing and selling EVs. The Tesla and its founder are also moving into other green development areas.

Such a dramatic change in the green industrial revolution (GIR) from the book in 2014 (Mandarin) and 2015 (English) is curbing emissions from power and fuel generation which verifies the reflections of Schumpeter's theories and perspective for "creative destruction." Green fuel cells provide that opportunity as compared with conventional battery technologies which have dominated energy storage with few significant advances or changes for over 150 years. Moreover, fuel cells provide the commercial opportunity to mitigate environment and climate degradation. Well-documented evidence from a variety of economists indicates that research and development is the heart of the capitalist engine through, "organized inventive efforts undertaken by university-trained scientists and engineers, working in special facilities, tied to particular business firms, and focused on advancing their product and process technologies" (Nelson, 1990, p. 199).

With an overview examination not only of the creativity process and technological invention itself but also of the commercial and business applications, one can see how qualitative economics describes, understands, and explains business actions and events. Perhaps the place to start with this analysis is back to the very core of economic capitalism theory with Adam Smith. There is no need to repeat the earlier critique of his perspective of economics. Instead, Smith in *The Wealth of Nations* (1937 edition) did highlight one point of interest for an interactional perspective within qualitative economics when he placed the act of exchange at the very center of economic life or as a universal characteristic of human beings in that "a certain propensity in human nature ...; the propensity to truck, barter, and exchange one thing for another" (Smith, 1937, p. 13).

The act of exchange between human beings is the central theme in the Lifeworld traditions. Qualitative economics approaches the exchange act which is a core concept of neoclassical economics very differently. First is to define exchange from a qualitative economics perspective as the exchange interaction between two or more human beings. Following Blumer and Schutz among others, exchange is not just monetary but physical and symbolic actions and gestures, as discussed in relation to the concept of intentionality and action. A number of countries around the world have become smart, green, and healthy such as Freiburg, Germany (Clark, 2009), as show here:

Some people have green roofs while others are installing solar panels too. Many are doing both. A key solution to stopping climate change is the use of mass transit with buses, walking, or trains as shown here also in Freiburg. The use of cars has been one of the key problems with both pollution and use of fossil fuels.

While now at the end of the second decade in the 21st century there are EVs, the recharging of them often come from fossil fuels. Hence the need for solar and hydrogen fuel cell cars that use only green energy.

Here are some electric cars at responsible prices from Asia, the EU, and the United States.

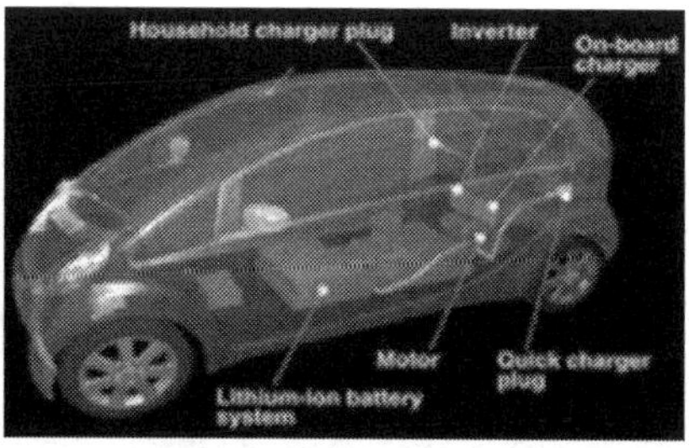

Now (2016) there is an all-solar-powered cars from Hanergy Company in China (Clark driving it).

Then there is the need for bikes to ride that are also protected by special places to ride them as here in Freiburg.

Aalborg, Denmark: Downtown Aalborg City with bike path.

Bike areas for daily bike owners in town when they are working in downtown Aalborg City.

Given the discussion on exchange, the basic issue is about subsets of interactions. Exchange, as in the buying and selling of goods and services, is the foundation of modern economics. Yet exchange is a part of all human interactionism. When a person interacts with another for any reason, that person is part of an act, event, or series of activities to exchange something. Linguists have long seen this with their analyses of discourse and verbal communications. Again, such studies are components and parts of interactionism, which forms the core of qualitative economics.

Interactionism (as needed below) goes into more descriptive (qualitative) details that than move into the exchange of business activities. Interaction, as have seen, is far more complex in that it presumes the ability of the actors to think and reflect upon their actions. In Mead and Blumer, the interactionist perspective involves the actors literally talking to themselves. Or as Blumer calls it, the "generalized other" provides the act of exchange with a far more comprehensive and significant definition. This very basic act is what distinguishes human beings from other forms of life. People interact and, as shown below, this is best understood through the sciencc of linguists.

Anyone familiar with business activities, events, buying, and selling on whatever level knows, for example, that business is not done or consummated at one point in time. It develops over time and is the culmination of multiple business interactions among many actors after considerable reflection and thought. The picture shows community discussions and City Council voted for the green development sustainability plan was enacted (2012):

Case: Solar panels on Beverly Hills City government buildings (gray flat areas).

For some companies, and in more and more countries, business needs transparency also so that it concluded after extensive legal scrutiny as well. Often these interactions (especially the financial and legal aspects) are due to the need for basic definition of words, concepts, and legal terms. Exchange from the interactionism perspective, therefore, is an engagement between actors over time. Two or more people have established relationships. They have engaged each other for some purpose.

It is this notion of engagement that underlies all business activities when understood as exchanges within the interaction perspective of qualitative economics. It is also the understanding of these engagements through *symbolic* and *nonsymbolic* activities that everyday business life is best understood. While some scholars might describe these interactions through quantitative numbers as case studies, a far more descriptive, accurate, and analytical understanding of business interactions are through the subtle and observable situations that surround these business actions with their meanings and interpretations explained. Therefore, the use of varied methodologies including quantitative ones is the appropriate and best approach to describe and understand everyday business life and the phenomenon that influences it.

SCIENTIFIC INTERACTIONISM PROCESS AND PERSPECTIVE

Three cases are presented below. The first case concerns the use of energy deregulation in market economics in the manipulation of a specific infrastructure sector that resulted in the California energy crisis (2000–03). Since Clark had direct involvement, information, and knowledge in this sector during the energy crisis, the data and scientific case made from a scientific interactionism perspective are substantial and well documented.

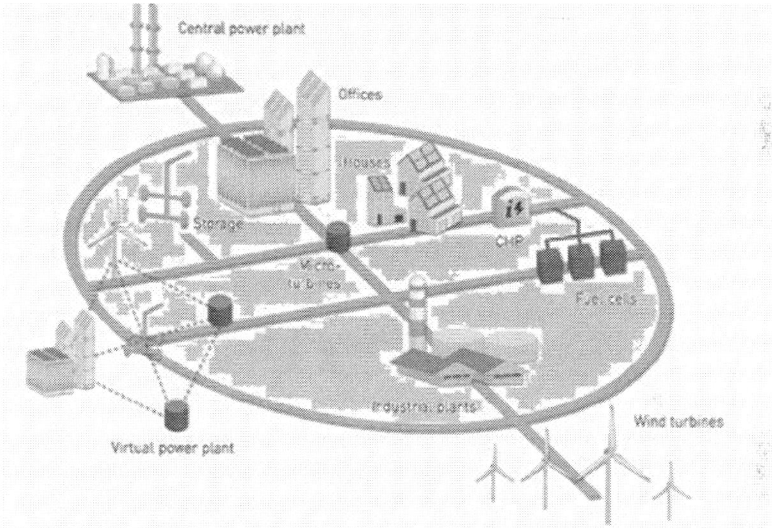

Future: Operation of system will be shared between central and distributed generators. Control of distributed generators could be aggregated to form microgrids or 'virtual' power plants to facilitate their integration both in the physical system and in the market.

In fact, in a related coauthored volume titled *Agile Energy Systems*: *Global Lessons Learned From the California Energy Crisis* with Professor Ted Bradshaw (Elsevier, 2004), Clark looked into public policy and governmental perspectives with both a qualitative economic formalism and quantitative dataset that has on-site distributed and central gird power systems (agile energy).

The second case concerns the commercialization of a new emerging technology into the market. Fuel cells have become increasingly more commercial since the late 1990s. Since technology transfer and business creation are criteria in economic development and job creation, the focus upon one particular fuel cell technology is described and analyzed. Again one of the authors has had direct and continuing involvement in fuel cell commercialization for over 15 years. Moreover, the current "clean air," energy crises, and environmental concerns make the fuel cell a very contemporary qualitative economic model to follow.

The third is the creation of a new business venture, which is critical for both individuals involved and society itself. Entrepreneurship, while dominant in America, has grown worldwide for many reasons. Capturing the essential elements of entrepreneurial firms is key to better understanding how markets and economies grow and expand. Again one of the authors, an entrepreneur himself, had been managing director for an entrepreneurial center in California (Center for New Venture Alliance) and founded his own company over 25 years ago. Moreover, Clark has taught the topic and created global programs for over 15 years.

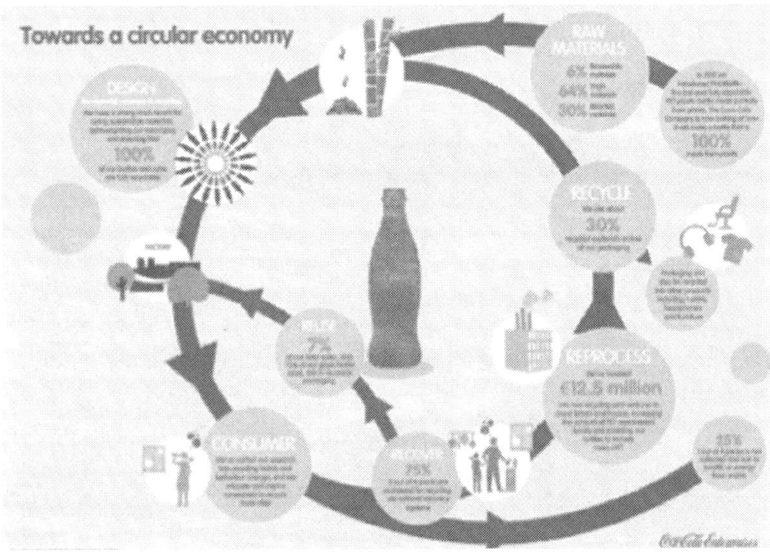

Each of the cases concerns the application of qualitative economics and the related areas of corporate governance and accounting themselves. And can be seen linked together in the "circular economics" model that had spread through the EU and now China.

From the perspective of corporations, the cases involve how numerous American companies went bankrupt during the early part of the 21st century. Corporate cases from the Internet "dot.com" era are *not* considered. The "dot.bombs" present another series of related discussions. Instead, the focus is upon the problems within corporate America that were dedicated and then adjudicated as a result of the Sarbanes–Oxley law. And the results impact government in terms of regulations, supervision, and legal actions.

Some discussion on methodologies and interactionism are presented below and urban communities and cities are in the Appendices. Throughout this chapter the scientific formalist has been advocated for understanding Life-world (or the subjectivist economic paradigm) in everyday business life through the use of "rules" derived from linguistic theory. The fundamental argument herein is that economics can be scientific when it acts like a science in terms of both theory(s) and method(s).

Economists should have the same aspirations. Instead, economists have only a narrow definition of economics which cones from Adam Smith's *Wealth of Nation* in the "Western European" neoclassical tradition using theoretical and methodological theory that leads to ethnocentric definitions, understandings, and results of businesses and their economic systems. The basic arguments have been represented elsewhere. Now consider some discussion among economists. Thus when *The Economist* editorial declares that economics failed (Economist, 1997a, p. 11), as noted earlier, it only means in not conveying strong enough that the market economy "is a marvel" (Economist, 1997a, p. 11).

Citing the failure of command and control economics, the position is that the "market" should be the only consideration in any economic system. Thus, the biggest economic policy mistake of the past 50 years, in rich and poor countries alike, has been and still is to expect too much of government (Economist, 1997a, p. 11). Hence, the blame for this policy in large part is attributed to liberal economists such as Paul Samuelson. Or in another issue, *The Economist* asks rhetorically, we are all capitalists now, are we not? These days the victory of market over state is quite taken for granted (Economist, 1997c, p. 17). And what is surprising is the nearly unanimous support for the idea that government has in fact been in retreat (Economist, 1997c, p. 17).

CASE #1: "MARKET FORCES" IN CALIFORNIA THAT CAUSED ENERGY CRISIS

Situational Analysis

What becomes troubling in these analyses, aside from their influence on the policy makers and the general public, is that they reflect the objectivist philosophical tradition that promotes one paradigmatic view of the world. This particular viewpoint, while even internally contradictory and inconsistent (e.g., Western economies such as France and Germany have a considerable amount of government involvement in their "markets") relies upon statistical methods and probably theory to justify their philosophical (and politicized) position. It is exactly the distinction that Reinert was conveying in his analyses of Anglo-Saxon neoclassical versus German–American Renaissance historical roots in economics (Reinert, 1996, 1997, 1998).

Quantitative models and simulations have not been very successful. Economics is a field dominated by statistical manipulations of numbers. It is not a field that uses mathematical models or more subjectivistic traditions. As McNeill and Freiberger (1993) note in economics, "the mathematical and linguistic realms stand quite apart" (McNeill and Freiberger, 1993, p. 96). Since economics is seen as concerning money, and money is viewed as numerical, then mathematics brings powerful tools to the field of economics.

Yet the precision of math leads to overly crisp estimates and idealized models that seem to describe a society of robots. "Hence, economics also employs verbal concepts like recession. Language handles real-life questions better and treats details more subtly, but it also narrows the scope of models and shortens chains of reasoning" (McNeill and Freiberger 1993, p. 96). The analogy to language is a central theme of Fast and Clark (2012), which explores linguistic theories as applied to describing, understanding, and predicting business and economic everyday life.

Energy is a good example of how neoclassical economics works but does not apply or work in this sector and perhaps in others. The basic parameters are that economic theory presumes a balance or equilibrium between price and quantity. In energy, as the California Energy Crisis reached its peak, two American Nobel laureates along with 10 well-known economists issued a manifesto in the spring of 2001. In general the manifesto argues and has been often repeated throughout 2001, that the energy sector, under deregulation, must be allowed to function on its own. That market power has its own natural way of correcting itself due to the demand and supply of energy. In other words, the argument goes that the market system will balance itself.

Not long after the manifesto, several individual scholars modified their stance and noted that the California market was flawed and "dysfunctional" because it was not a true deregulated market. This was in spite of several of them having helped craft it and vigorously supporting the California form of energy deregulated markets. Space does not allow a full review of the printed opinions in the press or professional journals from this group or others, but similar ongoing discussion around the same perspectives exist well into 2002.

A number of economists have argued (see various issues in *The Economist* in February–March, 2002a) that the collapse of Enron proves how neoclassic economics works. As the argument goes, a company cannot perform in the market and thus fails. Other companies come in and take its place. The free market economy moves on. The problem with that argument is that it ignores what Enron did in the first place—set up and influence the deregulation market, then manipulate it, and used or profited from investors so that in the end, the collapse of the company was based on the value of its shareholders and creditors who lost their investment.

The neoclassical approach to economics, which has become dominant today not just in the energy sector, rests entirely upon the basis of exchange of goods and services. Reinert in a series of articles dismantles the neoclassical (Anglo-Saxon) view of economics by contrasting it with the German tradition whereby manufacturing or the creating of products is the key to economic analyses and everyday life. We concur with that perspective and will present below the subjectivist paradigm, which provides an entirely different approach to economics through the use of interactionism in everyday life. Therefore, economics is *about* everyday life and is *in* everyday life.

The Enron Case: Private Instead of Public Monopolies

The time period is fall 2002, 1 year after 9/11 attacks on the United States by terrorists. All the facts in the Enron bankruptcy will not be known for many years.

Due to extensive legal and accounting litigation, and if even then, a picture has unfolded about Enron that symbolizes the problems with deregulation and privatization. While *The Economist* (various articles in 2002a,b,c) and *Fortune* (January 16, 2002) focus only on the accounting errors of Andersen, the basic fact remains that Enron executives instructed and managed Andersen's activities. In other words, the collapse of Enron is not only an accounting firm gone wrong, but also the core issue of unfettered free market economics in the energy sector.

To be perfectly clear, Enron is (was) a merchant energy company that represented the mythic of neoclassical economics at its best: a "market force (firm)" unencumbered in its business activities due to its political influence, strong impact in forming the deregulated market, manipulating the prices (due to supply), and finally directing the spurious accounting of its cash funds.

As a point of departure for the Enron case, consider an outline prepared by the staff for the Governor of California (Staff, 2002) as they prepare to address the impact of the Enron bankruptcy on the state. For a good updated overview of the Enron case, consult McGraw Hill PowerWeb (2002). As *The Economist* noted a year into the energy crisis:

> *The end was not unexpected, but it was still spectacular. On December 2nd Enron, once America's seventh-biggest company, filed for Chapter 11 bankruptcy. Only days earlier, its bonds had been downgraded to junk and Dynegy, a smaller energy-trading rival, had pulled out of a planned takeover. Enron is the largest company ever to go bankrupt. Disentangling the resultant mess (and lawsuits) will keep legions of lawyers employed for years to come.*
>
> (Economist, 2001)

The monetary impact in California in terms of long contracts was relatively low. For example, about $84 million in contracts was obligated by the University of California system alone. Nevertheless, the prices paid for the supply of energy to the state were controlled by "market" companies or "merchant firms" such as Enron so that about $9 billion in additional costs for power were paid by the state to ensure power in the winter of 2000–01.

In 2000 and the first half of 2001, Enron, among other market force suppliers, reaped enormous profits from California's deregulated energy market. Governor Davis stated in a Letter to the FERC Chairman, "Since the summer of 2000, I have on numerous occasions demanded that the FERC investigate allegations of market manipulation by power generators and marketers, including Enron. In my testimony before the FERC on November 9, 2000, I urged the FERC to investigate and remedy California's dysfunctional electricity market. Last June I called upon the FERC to investigate and refund the detailed overpayment by California consumers and businesses" (Davis and Gray, 2002).

As the Chart 1 below noted, the prices paid for power over this short period of time were by 100s of percents higher than normal or historical prices during this period.

FERC did finally take action in the late spring of 2001 as Governor Davis acknowledged in a press release and noted the actions taken by the State and its citizens to curb the immediate energy crisis:

> *I appreciate the action taken by you and your fellow FERC commissioners to finally reinstate price caps last June. Your action, combined with California's successful conservation efforts, our construction of 11 new power plants, and long-term contracts, caused energy prices to decline.*
>
> Governor Davis Press Release Jan 31, 2002

Subsequent reports and studies have verified the State of California's analysis of how a far more devastating energy crisis was averted over the

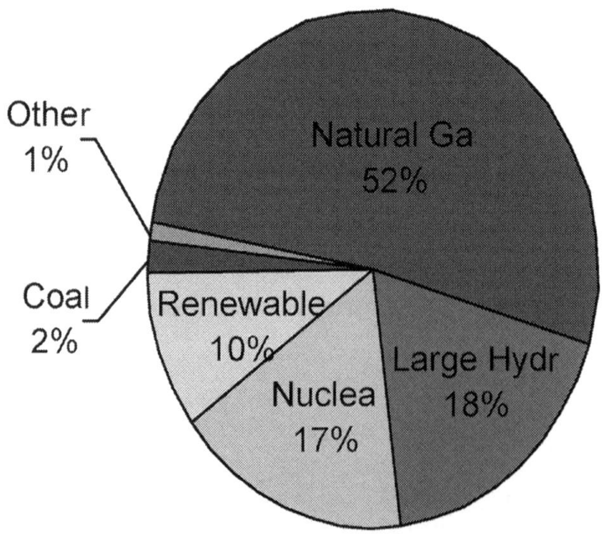

CHART 1 Share of current instate power generation fuel mix. *Source: Rand Institute, February 2001.*

summer of 2001 with these and other policies (California Energy Commission, 2002). While acknowledging the role of more rain in the winter of 01 (hence supplying more water for hydroelectric power), the fact was that the summer of 2001 was one of the 25 hottest in California history. The California government programs and citizens conserved over 12% throughout the summer of 2001. And the market suppliers were forced to keep prices reasonable through price caps at the State (Public Utility) and Federal (FERC) levels. Long-term contracts further assured a constant supply of power, rather than relying upon market manipulation in the spot market.

Yet public records (including emails, see below) confirm that on April 17, 2001, for example, then Enron CEO Kenneth Lay made eight recommendations to Vice President Cheney regarding federal energy policy. One of those recommendations was continued opposition to price caps. A memo containing those recommendations was released the week on March 20, 2002. On April 18, 2001—the very next day—Vice President Cheney told the *Los Angeles Times* that the White House emphatically opposed price caps.

When the White House released the final report of its energy task force, seven of Enron's eight recommendations had been fully adopted by the report. But this was not unusual. Further documentation reveals that Enron staff and senior officers as well as members of its Board had tried to influence the White House during the Clinton administration and systematically pursued its self-interests in States such as California who were considering deregulation. As the chart given below reveals, the interactionism among Enron staff with emails documents which is their "influence" to change and even falsify data.

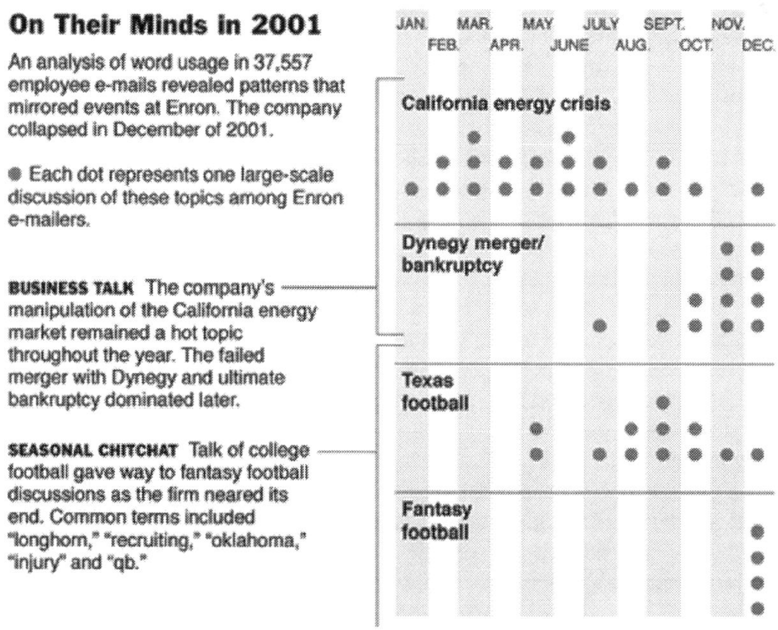

The *New York Times* (Kolata, May 22, 2005) called this "E-Mail surveillance" that documents staff behavior and interactions to influence the energy "market." As Kuttner (March 25, 2002) noted "Perhaps the most damaging effect of all is the eclipse of an opposition politics. Embrace of deregulation by 'pro-business' Democrats is more than a mistaken philosophical conversion."

The White House's energy policy, as enforced by the Federal Energy Regulatory Commission (FERC), particularly its opposition to price caps, helped to sustain the outrageous prices of the wholesale energy market. "This prolonged California's energy crisis and continued the massive transfer of wealth from our state to out-of-state energy companies such as Enron" (Staff, 2002). It is unclear how much direct influence Enron had in the White House.

> *What is clear, however, is that the White House's energy policy was nearly a mirror image of Enron's energy policy. It is also clear that that policy was detrimental to California and the wholesale energy market throughout the West. If Enron was able to use its market power and public influence to unfairly manipulate the California energy market, then FERC must take action.*
>
> (Staff, 2002)

Here is a checklist (Kuttner, 2002) of the icons that collapsed with the fall of Enron:

A pension double standard. Enron's retirement plan was heavily invested in its own stock. Executives cashed out over a billion dollars, while ordinary employees were locked in. Janice Farmer, who retired with \$700,000 in Enron, told a Senate hearing last week that she is left with a \$63 monthly Social Security check.

Bogus accounting. Since the Great Depression, the one form of regulation that even Wall Street has supported is the regulation of stock trades and corporate accounting… (when) once asked an ultra-Chicago economist if there was any regulatory agency that he endorsed. "The SEC," he said instantly, explaining that capitalism itself depends on honest information. Enron's entire game was to make its business plan so complex that neither investors nor regulators nor even its own auditors could penetrate it. While its core energy business made money (at the expense of consumers), it had speculative off-the-books subsidiaries. These borrowed heavily to make risky investments and eventually took the whole company down.

The business press. Enron's breathless cheerleaders included not only its own insiders and stock touts but also a business press that pronounced it the epitome of the new economy. It surely was that, epitomizing all the smoke and mirrors. In the wake of its collapse, Enron's former sycophants have turned on it, with *Forbes* and *Fortune* running scathing denunciations. *BusinessWeek* asked giddy Enron boosters what they thought now. (Gary Hamel, chairman of Strategos, before the collapse: "Enron isn't in the business of eking the last penny out of a dying business but of continuously creating radical new business concepts with huge upside." After: "Do I feel like an idiot? No, but if I misread this company in some way, I was one of a hell of a lot of people.") Well, yes. The one holdout among the business press is, of course, *The Wall Street Journal,* still trying to blame the debacle on regulators.

Deregulation generally. Enron's collapse impeaches the conceit that a market economy can be efficiently self-policing. Enron fleeced consumers by manipulating prices of electricity and gas; it fleeced investors and its own employees. Tycoons do this because they can. Enron should signal a whole new era of reregulation—of everything from electricity to pensions to accounting standards. And it is another warning that Social Security pensioners cannot trust Wall Street.

Nevertheless, The Economist (2001) and economists in general continue the same theme today as it did when the Enron bankruptcy began:

> *The company's opaque accounting makes it hard even now to understand why it got into trouble, and whether the cause was bad luck or worse (see our special report, http://www.economist.com/printedition/displaystory.cfm?story_id=896844) The close links between ENRON's chairman, Kenneth Lay, and George Bush will keep the affair in the political limelight. And Enron's staff, whose retirement fund was, at the company's urging, mostly invested in ENRON shares that they, unlike e the company's bosses, were then unable to sell, deserve public sympathy. But if America's capital markets are to stay the cynosure of the world, some quick lessons need to be drawn.*
>
> (Economist, 2001)

The bottom line to the neoclassical economic community who totally support market power content so that deregulation or privatization will be sorted in order to Enron and other such companies. Some will survive and others will disappear. Business as usual is the economic creed. Or as *The Economist* concludes one article "In the drama of capitalism, bankruptcy plays an essential part—until the next boom" (Economist, 2001). However, in the Enron case, the culprit is the accounting firm that ventured from its traditional and honorable function in the audits to the consulting role now seen in all accountancy firms. As *The Economist* wrongly puts it:

> *The most important concern auditing. ENRON has restated its profits for the past five years, chopping $600m off its earlier numbers. The company's auditor was Andersen, now a target of many lawsuits. Last year ENRON paid Andersen a fat audit fee of $25m; it also paid the firm $27m for consulting services.*
>
> Economist, 2001

The Economist goes on to cite other wrong doing by Andersen. That was December 2001 just after the Enron bankruptcy. Since then a number of firms have confessed to questionable accounting practices; and a number of accounting firms have owned up to their use of "liberal accounting methods." Certainly, *The Economist* is correct in saying that "After ENRON, the SEC should do what its former chairman, Arthur Levitt, has long urged: ban accounting firms from doing consulting work for their audit clients. Accounting rules also need updating" (Economist, 2001). Nevertheless, what is lost in this spurious debate is the core issue of deregulation itself. "Last are regulatory lessons" according to *The Economist*:

> *There is a risk of turning any bankruptcy into an excuse for massive new regulation. Some have argued that energy is too important to be left to markets of the sort that ENRON pioneered; or that, since it was engaged in financial speculation, ENRON should have been regulated like a bank. Neither conclusion is justified. Energy deregulation has brought huge benefits in lower prices and more secure supplies: energy trading will continue to grow regardless of ENRON's collapse. Nor would it be wise to subject all companies with financial arms to stifling bank regulation. ENRON's energy exchange was, however, explicitly exempted from oversight by financial regulators: that should be changed.*
>
> (Economist, 2001)

There has yet to be any evidence supporting these claims from *The Economist* or economist proponents of deregulation anywhere in the world. In the end, the best lessons of all will come from the mere fact of Enron's bankruptcy. *The Economist* argues that:

> *The first is the regulation of auditors. For years the profession has insisted that self-regulation and peer review are the right way to maintain standards. Yet ENRON has shown that this is no longer enough. Second is the urgent need to eliminate conflicts of interest in accounting firms. Andersen collected audit fees of $25m from ENRON, its second-biggest client, last year, but it earned even more for consulting and other work. Lastly come America's accounting standards. GAAP standards used to be thought the most rigorous in the world. Yet under British standards (sic), ENRON would not have been able to overstate its profits by so much. And, once again, although ENRON may have been egregious, it is not a lone offender. The ENRON scandal shows that America can no longer take the pre-eminence of its accounting for granted. That is a far bigger concern than any number of congressional investigations.*
>
> (Economist, 2002c)

The Economist (2002c), *Fortune* (January 16, 2002), other international business journals and most economists worldwide come to same conclusion: Enron was a case of bad accounting (or at least mixing accounting and consulting) rather than a mistake to deregulate public good markets such as energy. As Clark and Dimirag document (2006) the culprit here is the mythology of neoclassic economics that fosters and encourages "market forces" to manipulate and falsify numbers to benefit some people.

CASE #2: THE ECONOMICS AND COMMERCIALIZATION OF AN ADVANCED STORAGE TECHNOLOGY: THE FUEL CELL IS AN EXAMPLE OF THE CIRCULAR ECONOMY

This case focuses upon the commercialization of a new advanced technology from a large research laboratory in the United States. The concern is with the fundamental business notion of how innovation is brought into the marketplace. Hence, the empirical study herein violates many of the commonly accepted economic perspectives in business economics. The interactionism perspective is far more robust for understanding economic phenomenon and is radically different than the subjectivism paradigm. And it meets a need for describing, understanding, explaining, and perhaps predicting events.

This case concerns an advanced innovative technology such as ZAFCs for the automotive industry in the long term, but in the short term (within 4–5 years) for stationary power. In the energy and environmental industries, for a variety of reasons, the data were collected over a 2-year period of time. Clark et al. (Cooper and Clark, 1996) papers as he was a participant–observer with the commercialization of the ZAFC.

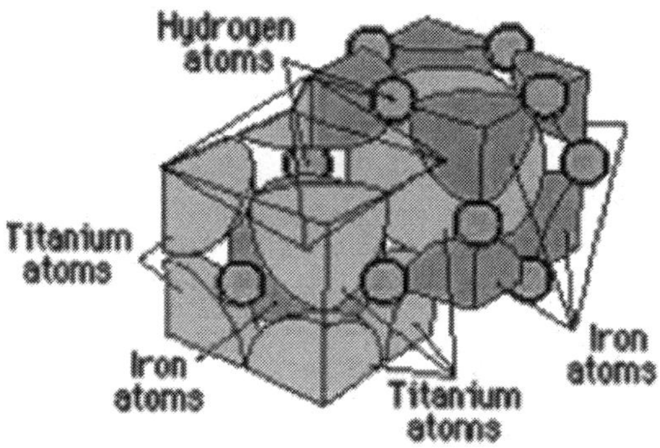

And even more significantly, Lowe notes that "Environmental damage from driving plagues the farthest reaches of the globe—polluting the air in cities, squandering valuable land, and even altering the earth's climate" (Lowe, 1994). While advanced technology has been touted as the solution to this universal human characteristic of "mobility," there are a number of barriers to the commercialization (see Clark and Paulocci, 1997; as well as Nelson, 1990; Teece et al., 1994, 1996 among others).

The case goes even beyond the automotive industry because such new technologies today must also meet the environmental needs and demands of future generations. For that reason alone, this case is significant since the ZAFC and other fuel cell technologies also have applications in the power/

energy industries. Therefore, the commercialization of a new technology from the interactionist perspective provides insights into both the business and technological development of the ZAFC. As such, the case study will violate much, if not all, of the conventional business economic theories as well as challenge most of the current methodologies employed today.

The first violation of conventional business economics is the hypothesis or assumption that environmental regulations provide a positive, if not economic incentive to business development. In the late 1990s, Porter and van der Linde began to make their argument when they noted that environmental laws in Holland that encouraged the growth and reuse of waste for plants stimulated new business growth. Similarly, Clark and Paulocci (1997) argue that new technologies for mass transit vehicles can be environmentally benign as well as commercially viable.

Another violation of the conventional business economic discussion and the functionalist paradigm is the use of multimethods, as explained in more detail later. The conventional focus on uni-methods or markets violates the very nature of everyday business life, which is composed of multi, complex, and nonlinear activities of business interaction among actors. Therefore, the researcher must employ a variety of methods to understand the phenomena under study, in relation to the ontological and epistemological basis of the study as the understanding of business Lifeworld as a social construction of interaction.

Finally, a scientific explanation and even "prediction" of business activities related to the commercialization of the any innovation under study will challenge most attempts by economists to understand business itself. In short, a methodological interactionism perspective is to observe actors and participate in the everyday business events surrounding the ZAFC innovation to capture its application in everyday business usage. If scholars hope to contribute to business economics, they must be engaged by it. Authors as well as researchers, family, and students must be part of the interactionism activities.

The empirical results of the ZAFC case study demonstrate how interactionism with the use of the circular economic model perspective works in everyday life. This chapter applies the interactionism perspective to the commercialization of a technological innovation. Understanding the phenomena of business development in terms of situations and events allows, perhaps, a clear concrete evidential path to be drawn for generalizable research, development, and technology commercialization. The study provides a road map in the form of a business economic model to which other technological innovations and programs can follow with impunity, success, and evaluative documentation. However, the challenge to other scholars is to select the interactionist perspective for the study of business economics.

Fuel Cell Technology: From a National Research Laboratory

"Technology should be regarded a consisting of technique, organization and knowledge." (Lund, 2008). Consider now the knowledge basis for a new advanced technology. ZAFC is an interesting case study, developed at a national laboratory into an environmentally sound technology (EST) to be commercialized. Firms want to commercialize the technology to meet and even exceed the zero-emission requirements (ZEV) for vehicles in California, as well as other American states and a growing number of foreign countries. The ZAFC is a fuel cell invented, patented, and now commercialized for mobile (e.g., vehicles) and stationary power sources (e.g., utilities). ZAFC and the zinc recovery unit (ZRU) are the case study material for this chapter. ZAFC technology is capable of powering a standard car for 250–300 miles between recovery stops, which take 10 min from the ZRU.

The laboratory solution to ZEV regulations was in energy storage devices, such as a flywheel or ZAFC. Therefore, a team created the ZAFC to make use of the high demonstrated energy density capability of zinc–air (>540 kJ/kg [>150 Wh/kg] at up to 100 W/kg). The high energy density in part is due to the high operating voltage of the cell (1.2 V) and the use of atmospheric oxygen as a reactant. Not only is energy density insufficient, but the available batteries require long recharge times (4–8 h) to assure maximum life and acceptable low life cycle cost. While rapid battery recharge is possible, in most cases it is not desirable; to recharge a bus battery in 15 min requires design of the battery for rejection of waste heat associated with 0.5 MW.

The refuelable fuel cell with drops of water has no expensive components other than the air electrode (quoted large volume cost, $\$120/m^2$ [Alupower, Ltd.]). The high surface power ($5\ kW/m^2$) gives the fuel cell a low producer's cost proportional to power: approximately $50/kW, half of which is attributable to the air electrode. Zinc recovery equipment is also inherently inexpensive. Zinc is electrodeposited on cathodic surfaces of a series of bipolar-connected stainless steel sheets, while the opposite surface of each sheet is catalyzed for oxygen evolution. Other components such as a 10 gpm slurry pump and metering devices are also low cost.

Technology as Element in the Creation of New Business Venture

Today, EV propulsion is severely limited by the low energy density of available batteries, of which lead/acid is the only mass produced technology of acceptable cost. Battery exchange requires spare sets of batteries (which more than doubles battery investment), special vehicle designs, and expensive off-loading equipment. All EVs powered by secondary batteries cannot be simply restored to full energy.

Understanding that the ZAFC as a "fuel cell" that converts the zinc into fuel cell rather than a battery is an extremely important scientific and business concept. *The Economist* (October 1997) declared that the "fuel cell" had finally arrived ready for the marketplace. That article further linked both the vehicular and power utility applications of fuel cells to energy generation and storage. Subsequent research indicates that communities can be powered from renewable systems when advanced storage devices are introduced as part of the energy grid (Isherwood et al., 1998). By the end of the 1990s, the US government began to seriously recognize the importance of new advanced technologies that were environmentally begin, completed foreign policy and created new industries.

For 3 years earlier within the laboratory and only after an intensive discussion, the laboratory staff concluded that they had a fuel cell instead of a battery. In fact, many of the publications about the technology initially labeled it as a battery. The basic distinction is that a fuel cell requires refueling (in this case of the zinc) of the cells in order that the chemical reaction (in this case with air) and another substance (electrolyte) will process a chemical reaction generating electricity. For batteries, while the process is similar, the typical battery must be recharged, usually with some sort of external unit. The recharging, unlike the refueling process, which takes 8 to 10 min, takes 4 to 6 h depending upon the extent of the battery discharge.

After the very successful demonstration of ZAFC on a standard-size bus from Santa Barbara, California, the commercial viability (defined as "successful laboratory and demonstration on vehicles of ZAFC for performance, reliability, safety, and cost-benefit") of the ZAFC into a demonstrable prototype would require additional funding for 18–24 months. There were no US federal government resources. Only internal laboratory funds were used. Other economic models had to be found. And into the 21st century, there was *no* market for fuel cells using zinc or other sources.

The issue of how to commercialize ZAFC for vehicles is an economic and business problem normally outside the realm of a research laboratory. Hence, the case is an example of an organization creating a new social construction or reality and operating within its perimeters. Therefore a new business economic model had to be developed and demonstrated for the ZAFC technology within a national laboratory. The interactionism perspective is appropriate, since it agrees for a dynamic and creative approach to a business problem (technology commercialization). Nonetheless, by 2002, the company created in the United States with an Australian partner firm had collapsed into bankruptcy. The individual actors would blame each other and even mention nation or cultural barriers to the new venture. But the reality was clearly something else. As discussed below, the interactionism perspective was able to provide a perspective and approach to understanding the problems for an advanced technology commercialization that provide concrete, positive, and productive direction for other such ventures.

Since the turn of the 21st century, much has happened in the commercialization of fuel cells. Aside from national and state centers being formed to promote the research and development of fuel cells, new companies and established firms have been rapidly commercializing a variety of fuel cells. References and websites are listed herein, but two points need to be made. First is that the commercialization of fuel cells such as all other new technologies has historically involved and been promoted by the public sector. The interaction among scientists and business people with the government as a source of grants, incentives, and tax credits has been critical. These relationships have even turned into formal organizations and associations for both networking and strategic planning. It is worth mentioning two in California that focus on the mobile or vehicle application of fuel cells (California Fuel Cell Partnership) and the power plant uses (California Stationary Fuel Cell Collaborative). Both organizations have been models for similar ones in other states, nations, and regions.

The second point is even more significant: the role of government in setting standards, protocols, or regulations that set public sector signals to the private sector that it must create new products that are environmentally friendly or at least nonpolluting. New alliances and forms are created, known as "civic markets" (Clark and Lund, 2001; Clark and Bradshaw, 2004) that bring together both the public and private sectors to address, define, and solve societal problems. The auto industry is certainly the largest and most impacted industrial sector.

However, there are other industries that pollute the environment such as energy, water, transportation in general, and waste sectors. The public message is that related health and societal impacts on communities and regions can be measured. These externalities must be reversed through government and business actions to commercialize new technologies. The key for fuel cells that needed hydrogen is in the use of very small amounts of water or electorized solar, wind, and other renewable energy sources. To generate power is the fundamental climate change solution and even the economic issue. Given the increasing concerns and public policies for reducing smog, emissions, and even power use from coal, natural gas, and other fossil fuels became the key issue with hydrogen fuel cell—powered vehicles.

Given these concerns, California moved into hydrogen fuel fell vehicles (HFCV) slowly with matching funds for refueling states. However, at the end of the first decade in the 21st century, the California Energy Commission changed its policy from any fuel source (which then were primarily natural gas) to renewable energy sources for the hydrogen. As the chart below shows,

California was predicting how "the market" would grow for HFCV stations due to the increase of cars being licensed in the State by 2015:

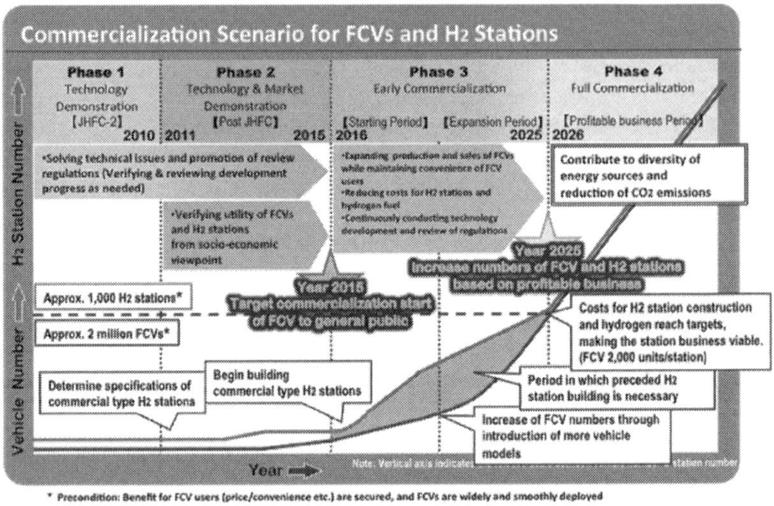

While this then meant a big change in both the HFCV business but also the refueling stations, the fact was within a few years later, the Tesla came out in Northern California and changed the entire "market." The Tesla totally changed the "demand" for cars and even encouraged other international automakers to now produce all-electric cars.

CASE #3: THE ECONOMICS OF ENTREPRENEURSHIP, SMES, AND NEW VENTURES

A variety of economic theories and models today identify the deployment of innovations in the firm. The conventional liberal approach developed by Michael Porter at Harvard University uses a structural analysis of the firm to understand its competitive advantages and plot strategies against other firms (see Porter, 1980, 1990). To accomplish the analysis of the firm, Porter created what he calls the "five-forces" framework: entry barriers, substitutes, buyers' and suppliers' bargaining power, and intraindustrial rivalry. Each force needs to be adequately understood in order that the firm becomes competitive.

The pro-government ideological perspective is reflected in the "push economic model" (Clark et al., 1993, Clark 1996). "The Technology Pipeline" demonstrates how new ideas and research move from the concept stage logically into the marketplace as "stars" or successful industries of tomorrow. All the researcher or inventor needs to do is follow the logical process outlined by business schools: business plan, venture capital, and industrial alliances.

Basically, this approach argues that R&D needs to be pushed out from universities and laboratories into the private sector (Clark 1996).

In basic economic terms, the federal government approach is supply side "technology push" which for business applications is unsatisfactory. Economists will talk about the demand side "technology pull" approach, which is "market" dominated. This implies that the customer knows what it wants, better than the laboratory scientists. A new model of "technology interaction" whereby the research scientist within the laboratory must be in constant interaction with the private sector to create new technologies for a 3–5 year time frame.

On the other hand, the neoclassical economic model argues that the private sector itself (often described as the "market") must see the demand and therefore needs for new technologies come first from the market or industry. In other words, the private sector must "pull" technologies that it defines and sees as important from university and laboratory R&D. The market or industry, therefore, so goes the argument, must define where public R&D funds should be expended. The market defines the most significant research areas for publicly funded R&D.

Both the liberal or neoclassical economic models ignore the give-and-take or interactionism in conducting everyday business activities. In short, businesses are interactive, iterative, and not static or even a linear process. Studies of R&D characterize the demands of industry as being short term and often too narrow to satisfy markets and shareholder pressures. Whereas the universities are long-term oriented in their research programs, in part due to the time needed for graduate students to finish their dissertations (note the time frame is about 5–8 years, often the time needed for students to conduct research, write the thesis, and publish it). Laboratories have a 2–5 year time frame, which places them closer to the commercialization time frame of industry, which is within months of the funding, and certainly not more than a year or two, before R&D must see products in the marketplace. These time frames are important because industry or the market rarely funds R&D that takes over 2–3 years before it is commercialized. Competition and certainly shareholders demand a far quicker return on the R&D investment than the more long-term laboratory and university R&D cycle.

Entrepreneurs enter the R&D scenario as teams or groups who provide the funding for demonstration and final commercialization of a new technology. Schumpeter (1934) viewed the entrepreneur as an individual who is the creator of new business within a society.

While entrepreneurial ventures create new businesses within a community, the new venture or entrepreneurial enterprise is rarely an individual business activity (Clark and Sorensen, 2001), but much more of a team and collaborative effort. Much of the literature attempts to refine Schumpeter's individualistic definition (Freeman, 1994).

Saxenian (1994) tend to see business development much more as community and regional phenomenon. As Barth puts it, "to see the rest of the community as composed of actors who also make choices and pursue strategies ... (and) analyze routinized institutionalized community life in terms of the choices that are available and the values that are ascribed ..." (1964, p. 7). It is the interactionism among the actors that is the concern of this study as in terms of how to analyze any emerging technology becoming a new venture.

In the 1990s and into the 21th Century scholars like Henton portrayed the need to analyze "clusters" of activities on a regional basis primarily in Northern California and Silicon Valley. Researchers at the Milken Institute in Los Angeles take a more Southern California regional perspective as well as state-by-state analyses using a cluster theory approach. While these studies are useful in some ways for understanding the economic development and job creation potential for technologies, they often fail to account for the networks and interactions among the intellectual or knowledge leaders in both the region and globally.

The development of a new company requires many actors constantly communicating, exchanging knowledge, and debating issues as they start the new venture. This far more international and global form of interacting is due in part to the Internet, but also primarily to long time personal relationships from graduate school, international meetings, and conferences. Usually the new company will consist of a technical person, a marketing or business person, and a financial person or backer. As Freeman noted in 1989 and again in 1994, there is a distinction between the "business and technical innovator" and rarely should be the same person.

The "American entrepreneurial spirit" fosters innovation through research and development. The creation and growth of many new firms is directly linked to new discoveries that are financed when they are more "mature" by venture capitalists for entrepreneurs who attempt to commercialize the technologies into product(s) and must get over the "Valley of Death" (lack of funding and cash flow). The organizing and managing of the resulting new company(s) generally only benefit directly from either seed or venture capitalist form of high-risk finance, experienced management, access to funds and markets, and a network of business contacts (Clark et al., 1998).

The outcome is even more significant: new industries and new capital are recycled again into more companies. Rarely, however, are the funds placed into the second "D" (demonstration) and certainly not in the R&D itself. The role of government (and perhaps getting equity for the investment) must continue to interact with new applied technologies especially in the areas of energy and environmental concern (Clark et al., 1997).

METHODOLOGICAL CONSIDERATIONS IN THEORY BUILDING

The scientific interactionism perspective in business economics is an excellent example of qualitative methods and action research. Clark (one of the authors) was a participant-observer in these cases studies. He was the manager of strategic planning at the national laboratory (fuel cells); the senior policy advisor to Governor Davis (California energy crisis); and both a founder of a new company and professor of entrepreneurship. Hence Clark was able to have access to key decision makers, data, and technical staff for all the cases. The participant-observer role also allowed Clark to participate in meetings and interview key staff members. The results of this constant interaction were then implemented into action plans for the further business development.

Qualitative methods are very common within anthropology, linguistics, and some areas of sociology and political science. Some of these fields utilize methods from the conventional functionalist paradigm, such as survey, questionnaires, and statistical methods. Instead, as outlined earlier, participant-observation is one of the preferred research methods along with interviews and action research as other primary modes of data gathering. Where necessary, we provide documents and reports such as emails, court records, and legal documents (Clark et al., 2008). Qualitative methods for gathering data are a significant part of the interactionism perspective.

In anthropology, the product for most research is normally a written ethnography of a culture, group, or class of people. Participant-observation, interviews, and sometimes action research are the common research methods. In more recent years, many anthropologists employ local people, use survey techniques, and take videotape as part of their research. Nevertheless, the basic theories in the field of anthropology follow the functionalist paradigm in a number of ways: (1) provide a static rather than dynamic view of the subjects, (2) find data that fit into some preconceived notions of the people in groups, cultures, subcultures, etc., and (3) fail to understand the interaction between people in everyday life.

Yet, the qualitative methods of anthropologist are useful in understanding the dynamics of interaction within any human organization. When applied to businesses, the outcomes can be dramatic, revealing, and insightful. In the late 1990s, Suchman did just that with her groundbreaking work at Xerox when she applied qualitative methods to understand the interface as business engineers would call then as being between copy machines and humans. Since then, her anthropologically trained team has explored other avenues of research about machine and human interaction. By the mid 1990s, researcher at Apple Computer had completed also an edited volume with articles on the human – computer interactions or interface through the use of qualitative methods while constructing new theoretical perspectives.

An anthropologist conducting qualitative research is directly relevant to this study. In the course of conducting research, everyday discussion and dialogue are often the prime sources for data. Linguistics, a related field to anthropology depends heavily upon qualitative methods since the researchers will solicit words and sentences from native speakers of the language. A native speaker would be asked for sentence examples and then for interpretations and meanings. Alternative sentences would be suggested and critiqued by other native speakers. Often such sessions are in groups or in a classroom setting whereby the participant can hear, ask, and diagnosis the words and sentences.

Now, lets consider the additional information in which there are the need to exercise of clarifying words, phases, and meanings which were initiated by senior staff members, not for linguistic research reasons, but to understand the exact meanings of ideas, definitions of words and numbers, and interactions among technical experts. In other words, while technical people all understood one another on one level, they did not share specific definitions and meaning on another level. Therefore, several work sessions were adjourned where intensive discussion over terms and meanings were clarified and sorted out. In other words, the scientists themselves participated in the research to assist their own technical needs.

Such meetings did serve to clarify ideas and terms and became a consistent and frequent interactive data collection method. Blumer's practical philosophy saw how scientists needed to make indications to themselves within the context of their social world so that they could communicate among themselves as well as to the outside world of other scientists and more particularly to the business community. The need for common definitions and understandings allow for more complex levels of technical and business activities. In short, deriving common meanings and hence understandings creates the social context in which to conduct technical research and by extension commercial business.

The concern of this chapter is to understand the application of the interactionism perspective to a new technology being commercialized into a business venture. The case documents one federally funded technology, the ZAFC along with the environmental concerns for its use and by-products from an industrial ecology framework (Cooper and Clark, 1996). A significant part of action research methods within an interactionism perspective is to argue for a sustainable development or environmental considerations in the creation of any new technology and related business ventures.

By the mid of 1990s, researchers discussed industrial ecology as part of an overall sustainable development consideration. They note that "Researchers in other place began to see and discuss industrial ecology as part of an overall sustainable development consideration." In other words, the global environment is interlinked with business and industrial activities. Everything that human beings do, impacts the environment.

For Graedel and Allenby, the globe is a closed system operating under as if it is biological "life cycle." In short, human beings interact with themselves and their environment. When anyone develops a new business, they must take that perspective into account as well. The ZAFC does just that, since the use of zinc (once mined) is converted into electricity, becomes a slurry in the fuel cell process, and then is reconstituted into pelletes to be fed back into the fuel cell (Cooper et al., 1996; Cooper and Clark, 1996).

PRACTICAL APPLICATIONS IN ENTREPRENEURSHIP AND BUSINESS DEVELOPMENT

The Organization and the Actors

The US government is the primary funding agency for the American laboratory in this case study, which has a contract with a major American university system for the operation of the laboratory. The US government funds over 735 R&D laboratories through various departments and agencies. NASA, for example, has 12 laboratories. The US Defense Department through its various military branches has over 20 laboratories. The US Department of Energy funds more than a dozen laboratories for over US $16 billion annually. Many of the US federal laboratories are world recognized experts in their fields.

The US federal government spends about US $43 billion in R&D annually and the private sector another $30 billion annually (Clark and Jensen, 1997). The Congressional and administrative drive to balance the federal budget in the late 1990s meant, however, a decrease in R&D funding by about 3%–7% annually over the next 5 years. It is this uncertainty in federal funding that is motivating the laboratory researchers and administration to seek "alternative" sources of R&D funding. Clearly one source is the commercialization process from the private sector.

By the end of the 1990s, however, a balanced US federal budget began to show surplus revenues from which both political parties decided to allocate more R&D funds for basic scientific work. Therefore, while small increases in R&D began to appear across federal agencies and departments (from 3% to 6% with the National Science Foundation receiving the largest increase of 6%), they were targeted toward basic research rather than applied and typically had no increase and little impact on energy and environmental R&D budgets (Clark and Jensen, 1997). A "double gap" resulted: one gap between the need for increased funds for environmental and alternative energy research; and the other gap between basic and applied research often inhibiting solid basic research from having demonstration funding for prototyping and field tests.

The LLNL has five "supporting academic divisions" such as engineering, physics, chemistry, and six cores "program divisions," of which environmental and energy are two. The laboratory is operated on a "matrix" model whereby specialists from different disciplines will work on projects within programs for a specific period of time and then move on to another project. For example, any research project may have a mixture of engineers, chemists, mathematicians, and physicists in it. The matrix structure is very unlike the typical university where people and research projects are organized by academic disciplines and rarely inter- or multidisciplinary.

There were over the last three decades, despite changes in both operations and ownership of LLNL, about 8300 employees at LLNL (of whom 1200 are

PhD level scientists and another 2000 are engineers at the collage and graduate levels) for an annual budget of over US $1 billion in FY98. The facilities cover one square mile with over US $5 billion in equipment and buildings representing one of the world's most renown research and development organizations. The laboratory is one among the largest (in terms of annual revenues) research laboratories in the United States and probably the world.

In late 1995, two divisions collaborated with others seeking research projects under the overall title of "industrial ecology." While the ZAFC technology (Clark 1996; Cooper et al., 1996) had been created under the Chemistry and Materials Division, it clearly fits into a broader multidisciplinary effort within the laboratory for industrial ecology-oriented technologies. Fuel cell research and development are administratively within Energy program division since it supplied the research funds to chemistry.

Almost all federal laboratory technologies available for commercial application have been historically taken "off-the-shelf" and "pushed" (some call it marketed) into applied research areas or commercial use. The current laboratory technologies are described in books, magazines, or newsletters, such as the biannual collection of technologies from laboratory. NASA, for example, publishes through a private contractor, "Tech Briefs," which is a monthly magazine featuring new NASA technologies. In most cases, a one-page description outlines each technology available for transfer to businesses, listing the laboratory contact person with telephone, fax, and email addresses.

The actors in the ZAFC example were five people within the laboratory: the lead scientist and inventor of the technology (electrochemist); leader of transportation programs (civil engineer); his marketing assistant (social science undergraduate degree); the lead technology manager in related technologies (mechanical engineer); and the author as participant-observer (qualitative economist) who assist in business planning and strategic alliances. The interaction within this group was frequent and sometimes intense. Several situations arouse in which the group had to or was forced to define some scientific concepts terms because of technical confusion among the various disciplines. Intense all-day meetings, briefing sessions, and casual meetings had to be held frequently as the research changed the definition of the ZAFC and the effort to seek commercialization funds changed business strategies. All were interactive processes demanding constant dialogue and discussion.

ORGANIZATIONS AS THE SOCIAL CONSTRUCTION OF INTERACTIONISM

The purpose of this section is to describe the interactionism within the circular economic perspective as applied to fuel cell technology developed at the laboratory into a business reality for a community, not just the business owners as noted below:

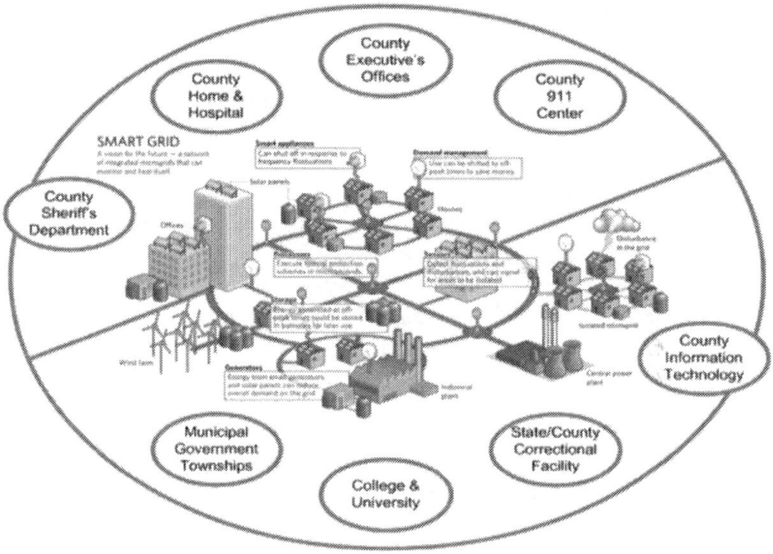

The problem that confronted the LLNL was the need to obtain additional funding for the final research and development of fuel cells. This was a significant issue since as the team proceeded to "market" the technology; it became clear that fuel cells needed some more funding to be "commercially viable." After some internal discussion and debate, it was clear that these "demonstration" (e.g., prototyping, designing, and field testing) funds would only come from sources outside the laboratory itself. The fuel cell technology was not funded initially through traditional laboratory sources from the US federal government and was ignored (if not outright dismissed) by the most obvious American industrial resources: the American Big Three automakers. The laboratory team was forced to find alternative funding routes to commercialization via demonstration of the technology first.

Not until a series of events occurred in late 1997 and throughout 1998, did Detroit announce over $500 million in R&D for fuel cells: First an announcement by Mercedes-Benz from Germany that it formed a partnership with Ballard Corp., a new fuel cell company in Vancouver, Canada. Then came *The Economist* article of October 1997, touting fuel cells as a viable technology for vehicle and power storage. Toyota next demonstrated a fuel cell–powered vehicle at the December 1997 auto show. Soon thereafter, Mercedes announces that it will have 100,000 fuel cell–powered vehicles for sale in the year 2000. The Big Three American firms were losing ground, and fast.

Putting the financial problem for demonstration of fuel cells into an interactionist circular economic perspective meant that new funding sources needed to be sought and the most likely sources would come from Europe or the Far East. Why? The American automakers' definition of economic and business reality is that of the status quo. Change in environmental laws, for example, that impact automobiles are viewed as an economic threat to Detroit. On the other hand, foreign automakers realized that environmental and safety laws in California, and by extension the United States in general, set a trend, which would foretell future American and international market demands.

Thus, the California laws would force new technologies into the marketplace. Mercedes-Benz (MB) established a 40 member engineering R&D group (to track and work with the latest R&D) in Northern California in the early 1997. Volkswagen (VW) followed with a smaller group in 1998. Two decades later VW found to send out false data and information about "diesel" power being good for the environment. Now in 2017, VW had to settle from a law suite in the United States (mostly in California) for over $4.2 billion. Back in the 1990s, Ford Motor Company learned a lot and finally announced the relocation of it Lincoln-Mercury (LM) headquarters from Detroit to Irvine, California, close to the University of California, Irvine campus, which has a major fuel center located on-site. This is the first time a major US automaker has moved its headquarters outside the immediate Detroit region.

There are a number of simple examples of how regulations have compelled companies to introduce new technologies into products. Porter and van der Linde (1995) examine how this government regulatory process worked successfully in Holland with environmental laws and new approaches to technologically improve products (tulips in this case) to be more globally competitive. Consider also the auto industry. The introduction of seat belts and air bags, were all vigorously fought against by the American automakers as being impractical and too expensive. The car companies fought political and legal battles against these laws. They are doing so now with zero-emission vehicles.

In short, the American auto industry's view was based on what the "market" wanted. John DeLorean who led General Motors (GM) to new "markets" in the 1970 with cars marketed in California (due to its "Freeways") featuring beautiful women and handsome men, caused the car manufacturing companies to change. However, DeLorean wanted to be the General Manager of GM in the late 1970s, and the senior executives along with the GM Board rejected him (Wright, 1979). DeLorean quit GM and a few years later (early 1980s) went to start a new car company in Northern Ireland where he created a new car that first used "gull wings" as Tesla picked up then 3 decades to use for its cars. As seen below, this was three decades before Elon Musk and his Tesla with its gull wings (Clark Mass Media Company, CM^2C, 2018): https://www.linkedin.com/in/woodrow-w-clark-ii-b6962214.

CM^2C is now starting to develop a dramatic film or series about DeLorean titled "Detroit: The Sequel" featuring the Wright book and also William Haddad about the DMC car company (Haddad, 1985) which the dramatic film or series would featured too.

In late early 1990s, the American auto industry announced that it too would have EVs by the year 2001. However, the big three automakers will not have zero-emission vehicles as required by California law (and that of New York, Maine, Massachusetts, and Vermont or 10% of the US market). Instead, the new EVs will emit 30% fewer lung disease causing hydrocarbons and 70% fewer smog emitting nitrogen oxides. As Hvelplund and Lund (2008, p. 17) note in energy and environmental industries, "public regulation analysis are often influenced by the interests and thinking of the established energy supply companies."

International automakers, on the other hand, have shown a considerable creative and innovative form of social construction when they examine the reality of future vehicles. By the end of 1998, Toyota all-electric and hybrid car (with the "regenerative braking" technology that they licensed from LLNL) vehicle and Mercedes-Benz announced sale of 100,000 all-EVs in the year 2000. It was clear that the American auto industry was losing technological ground and market share. While the foreign companies may not actually have developed and perfected the new advanced technologies, they are very much in the forefront of implementing new approaches to solving or exceeding environmental and safety regulations. Toyoto after the first decade of the 21st century became the number one car company in the world.

Some non-American automakers, especially VW, for example, before the tragedy of their "diesel data," use safety as marketing tool in the United States with advertising not only about how well their car is built but also about side air bags that are installed for further safety. The strategy from Volvo's social reality works in the United States where they have solid sales because the consumers indeed embrace Volvo's definitions of automotive reality. In short, the American auto consumers have "bought" Volvo's definition of that reality, since Volvo clearly interacts and understands the needs of its customers.

Many foreign automakers take a global "long-term" interactive perspective where they examine government policies with industrial changes to "create" and understand and know what customers wants from their products. This proactive (having R&D offices located geographically close to where the new research is being conducted) means of conducting business is a social construction based on constant and intense interaction.

There are other examples of this within the auto industry and other industries. Suffice to say at this point that the commercialization of new technologies is more a matter of an organization creating its own social construction of reality and then being able to match that with other organizations. The key is to find those "other organizations." As discussed below, in a shrinking world or one that is being globalized rapidly, the task is not that difficult and appears to be of increasing necessity for companies to grow and prosper.

The interactionism perspective posits that if the fuel cell could not be pushed out of the laboratory, other financial sources had to be found to "pull" it into final development for commercial applications through a variety of interactions among actors in a broad array of organizations. The interactionism perspective is the key approach in this case and probably can be applied in most other cases of technical commercialization and business development. Additional technological cases will further exemplify the model.

INTERACTIONISM IN BUSINESS, TECHNOLOGIES, AND ACADEMICS AS CIRCULAR ECONOMICS

Case: Science Parks

In the United States, there are only two science parks, while in Europe, hundreds of science parks presently exist. The background for this rapid growth in the spread of research parks is the research parks' high success rate with regard to the number of company establishments and the companies' ability to survive. The research parks have had significant importance not least as instruments for regional business development. The key is that science parks are different from incubators. The science parks are not only connected to universities and research organizations from which they get technologies and funding to turn new ideas into businesses. Hence the interactionism between students and faculty then along with product development, costs, and marketing. Consider only a few and how they have included the various elements identified above to make them viable. In each case, the science parks are considered successful in terms of the metrics (interactionism, circular qualitative economics plus others) identified earlier.

NOVI A/S

Aalborg, Denmark

From a translation of the original (1987) planning document for NOVI titled: The science park concept and the initiative of North Jutland.

The Science Park of North Jutland (Nordjyllands Videnpark), NOVI, is a research park. By research park is understood, according to "the International Association of Science Parks," a property possessive initiative which has formal and reel relations to a higher educational facility or research institution such as Aalborg University located just outside of Aalborg, Northern Jutland, Denmark, is called NOVI which in Danish means "Science." NOVI was created to promote the establishment and development of science-based companies that have management, which is actively engaged in transfer of technology and skills related to business to the clientele of the science park.

Here is what NOVI looks today (left picture) from when it started in the early 1990s with only two buildings.

Research parks have long been an unknown concept in Scandinavia, but within the last 4–5 years a series of initiatives have seen the light of day—most with tremendous prosperous results. In recent years, Denmark has entered with the research park Symbion in Copenhagen and Århus Forskerpark.

The research park initiative of North Jutland at Aalborg University (AAU) outside the City of Aalborg, which was initiated in January of 1986 by rector Sven Caspersen, AAU, has had a strong backing from the business environment of North Jutland and from authorities, who see the park as a vehicle for promoting the redirection of the regional business structure and creating industrial growth in the region. Clark knows it well as he was affiliated with AAU as a Fulbright Fellow for 6 months in early 1994. Then he came back once a year to lecture until in 1999, he was asked to be a visiting professor at AAU and also with NOVI.

Within half a year under the umbrella of "The self-owned institution for the establishment of The Science Park of Northern Jutland," the North Jutland Research park has undergone intensive planning, with active participation from a broad circle of company and authority representatives, researchers, etc. The consultant firm Sverdrup-Jensen & Partners A/S, whom has also formulated the total concept for NOVI, has coordinated the planning work.

Today, NOVI opened in 1989 and has "attained a unique competence in the area of knowledge-based development projects. A total of 40 companies employing app. 450 people occupy the NOVI 1 and 2 buildings and the adjacent business park called NOVI Park. A multimedia centre for both academic and commercial activities was inaugurated in 1999. The multimedia centre will be housed in new buildings, NOVI 3 and 4." From the 1999 Annual Accounts, NOVI consists of three major elements.

The NOVI Science Park

The 9600-m^2 NOVI Science Park houses entrepreneurs, development projects, and technological advisor companies as well as the Aalborg University Centre of Personal Communication. Companies typically want to transfer part of their development activities to NOVI to have peace and be quiet for their research and to benefit from the expertise and services offered by NOVI.

The 55,000-m^2 NOVI Park is situated close to the NOVI Science Park. The area is reserved for the establishment of knowledge-based and high-tech companies. So far, three companies, namely Maxon Cellular Systems (Denmark) A/S, Force Institute, and L.M. Ericsson have started activities in this area. Maxon has stated that the environment of NOVI, Aalborg University, with the Centre of Personal Communication, and the North Jutland telecommunications industry as a whole, was the reason for Maxonís establishment in NOVI Park. The NOVI Science Park has grown and expanded from its core buildings next to AAU into other areas of Aalborg City.

Technology Transfer

NOVI mediates knowledge from Aalborg University and other research institutions in Northern Jutland to industrial companies in the area, ensuring that the latest research and knowledge is incorporated in the development of new projects. Conversely, industrial research requirements are mediated to the academic institutions.

NOVI has been appointed "Innovative Environment" by the Danish Ministry of Business and Industry, allocating an amount of up to 750,000 Danish kroner per project for feasibility studies and related activities.

Venture Capital Company

NOVI is the only science park in Denmark with an angel and venture capital system that is earmarked for investment in commercially attractive development projects. Seed investments are made in start-up companies utilizing the Danish university environment and the special environment of the Danish science parks. Seed investments are regarded as high risk and require thorough prior investigation. The investment can be based on a product or business idea of an individual (inventor and/or entrepreneur), an existing company, or research establishment. Highest priority is given to ideas that have novelty value, are unique, with a high knowledge content and promising commercial potential.

NOVI has up till now invested about 54 million Danish kroner in 35 companies.

NOVI aims to withdraw as an investor after a limited period, usually of 5 to 8 years. Upon withdrawal, NOVI disposes of the shares to the originator of the idea or finds a new investor, preferably an existing company that can contribute to further success of the development company.

The finance area is what makes NOVI unique among most science parks. In its own words from the Annual Accounts 1998 (1999), the Innovation Environment funds from the Danish Agency for Trade and Industry mean that about DKK 20 m are available over the coming 3 years for the identification of propitious ideas conceived in the research environments in Northern Jutland and subsequently for the investment of publicly financed preproject capital in the most propitious ones.

The combination of NOVI formerly having been appointed a state-approved seed venture company subject to a 50% loss guarantee on investments, and NOVI in a new role as an innovation environment offers a unique opportunity to create a profitable business based on the commercial exploitation of research results.

Thus to assure a sufficient number of commercial ideas for NOVI venture activities contracts have been formed with some of the other science park–based innovation environments. These contracts have led to the establishment of joint companies in the individual innovation environments which aim at continuing the most propitious projects right through to final commercial optimization. This financial approach has lead to an innovative method such that

NOVI can select the most appropriate projects from among the volume of ideas created on the basis of publicly financed preprojects to the tune of DKK 50 m annually. Once fully developed, NOVI will act upon the basis of its three-legged structures to combine its anchoring to the regional know-how centers and regional trade and industry with a purposeful openness toward the need for venture capital experienced by other Danish innovation environments (e.g., groups or organizations).

A third dimension in NOVI strategy is a global aim, both in terms of attracting international businesses to the region and in terms of determining the best possible business opening to realized project ideas. Unlike the situations in other science parks, NOVI will try to have the raising of additional capital covered by the rules governing state-guaranteed seed venture companies, preferably in a modified form, which would allow for the great differences found in the risk profiles of investments in projects that are 3–5 years old and those which NOVI is mainly involved in, i.e., seed investment in virgin project ideas.

The results have been impressive. At the end of 1998, for the Annual Accounts, NOVI could report for the coming year to achieve a stable income, which is expected from the current activity area of Forskerparken NOVI. The completion and putting into operation of NOVI 2 will have a favorable impact on the second half-year of this financial year.

As a consequence of this approval as an innovation environment, NOVI innovation will progress to a considerably higher activity level. Activities within this field are expected to be sustainable, thus providing some cover of NOVI capacity costs. NOVI A/S made a profit in 1999. After two decades, NOVI had built eight full sized buildings for students and faculty to work, plus added four full sized buildings as a hospital attached to NOVI as well.

Consider a case of a new community being rebuilt due to weather issues that destroyed it. And the need for academic, business, and science parks in it.

Climate Change and the Future of Coastal Cities: A Texas Microcity Case Study

Naved Jafry[a, 1], Garson Silvers[2]

[1]*La Jolla, California, United States;* [2]*US Green Building Council, Beverly Hills, California, United States*

INTRODUCTION

Our team has been working in the coastal bend of Texas for over four decades now. The City of Port Lavaca situated in the middle of the coastal bend of Texas 275 coast line is our subject with a current population of around 12,000 citizens. This chapter is a case study of being prepared to deal with the effects on a modern coastal rural city in North America. This small Texas town situated on the Texas Gulf Coast region has always been plagued with hurricanes, in fact there are places in this region where whole towns have been wiped off the face of the earth in modern times. Indianola was one such town that was wiped off the face of the earth twice once in 1875 and a second time in 1886 located on the Matagorda Bay in Calhoun County, Texas, United States. This community, once the county seat of Calhoun County, was part of the Victoria, Texas, Metropolitan Statistical Area. The city had a population of 5,000, but on September 15 of that year, a powerful hurricane struck, killing between 150 and 300 and almost destroying the town. Indianola was rebuilt, only to be wiped out again on August 19, 1886, by another intense hurricane.

Leveraging the public federal recovery funds from HUD grants, Texas Civic Code 380, disaster recovery funds, and engaging the private sector, the City of Port Lavaca has expressed the desire to be a hub of new age buildings and living technologies. Learning the lessons from the past, the City of Port Lavaca is gearing up to be a catalyst in saving lives and a way of living by being a part of a green paradigm shift on how we should build environmentally friendly and defensive cities using off the grid and microgrid systems that may be able to withstand the super storms and have a quicker recovery (Fig. 1).

Some of the concerns are to create new educational programs that are "applied" into businesses and even new companies. The city and its team are aware of "science parks" and how the can be created next to the education building areas that also create new businesses.

---

a. Naved Jafry, Ph.D.

FIGURE 1 Proposed Microcity Arial view of the City and port of Port Lavaca.

For example, partnering with Zeons off the grid building systems and microcity protocols, the city plans to encourage buildings to now individually generate their own energy and process their individual sanitation needs. Due to the growing challenges of food deserts the city would start catalyzing the urban agricultural movement and encourage its residents to grow their own food through techniques such as aeroponics, hydroponics, aquaponics, and vertical farming.

THE ENGINEERING AND DESIGN COMPONENT

With a mandate to establish a formidable and beautiful infrastructure, city engineer, planner, and business development leader Jodi Weaver has certain microlevel questions about the idea of decentralization of utilities and wastewater with aerobic system spacing and permitting. Her mind-set is to bring a new job base to grow her city while considering many challenges and bringing in new perspectives. However, Bill Dilibero, the City Manager, also has to look at the big picture on what the city can do to grow its population, increase and maintain the quality and level of services with a tight city budget along with state regulation that could cause permitting challenges (Figs. 2 and 3).

The Business Case

The water front village with a variety of small businesses with live work lofts above the retail and public service facility will employ workers who will live in the lofts above the store fronts, in eco-friendly environmentally defensive multifamily and single family affordable homes made from Zeons Millennial Units in the heart of this modern village. Surrounding the seaside town center of this proposed modern fishing village on the shores of Matagorda Bay on the Gulf of Mexico are thousands of acres of flat rich soil farm land as far as the eye can see. The current local agriculture in the area seems to be composed

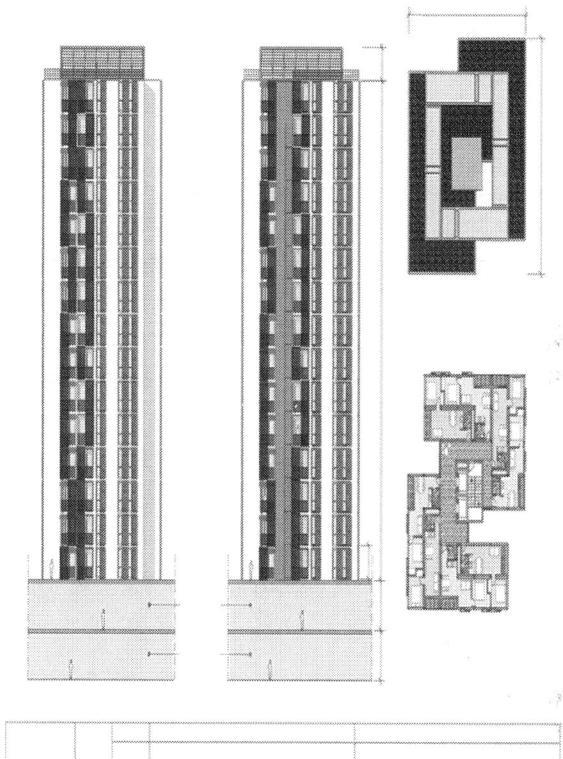

FIGURE 2 One of the 2000 ZMU buildings set to be deployed in the Port Lavaca Microcity. The ZMUs are one such dwelling that are 100% off the grid, hurricane, flood, earthquake, and fire resistant to withstand future catastrophes without compromising on sustainability.

FIGURE 3 One of the super buildings set to be deployed in the Port Lavaca Microcity. These structures adapt to nature while keeping the buildings heating and cooling costs down.

mostly of cattle grazing ranches mixed with crops of cotton, corn, maze, sorghum, soybean, and beans.

So, this small Texas agricultural town like many is losing population of its youth who go off to college and then to metropolitan cities to get employment. The City of Port Lavaca has a port, and port authority mostly a wet terminal for the petrochemical industry providing an employment base, and recently lost an Alcoa Aluminum plant as a major employer. The key challenge here is to grow the economy by attaching both tourist to enjoy the beautiful beaches of north Padre Island, sports fishing, boating, water sports, sailing, and bird watching to track major migratory routes and nesting area for whooping cranes and other colorful exotic water fowl.

Through the engagement of citizens involvement, the city has aggressively pursued individuals such as Danny Richards, a local football star and an army veteran turned pastor who has a vision for his hometown. A seaside village that includes a boardwalk with Ferris wheels, roller coasters, and other amusement park feathers with a Texas flavor for entertainment for tourist, locals of Calhoun and the surrounding counties as well as building new educational institutes including a new junior college, new high school, and include institutes such as the marine biology institute to study and explore local aquatic ecosystems. His vision includes placing the earthship ZMUs underwater like a space station or Atlantis, an underwater institute with windows for observing the sea life.

Exorbitance and the Rise of the Next Generation

Market conditions and buying patterns are changing dramatically as we live in exponential times and an era of disruption. Our top down command control systems are quickly being replaced by a shared and decentralized economy. With knowledge and technology doubling its influence every 1000 days the old gods and guards are being replaced by the new and bold. Just imagine 10 years ago, virtually unknown firms such as Amazon, Netflix, Facebook took out market share, weakened, or undermined well-established and entrenched giants such as Walmart, Blockbusters, and print media. Uber and Air BnB do not have a single rental car or a hotel room, respectively, but yet through the power of a shared economic model rent more cars or rooms than the likes of giants such as Hertz and Hilton hotel combined.

With so much change the new generations have evolved with the new times. Gone are the days where laying roots and foundation while accumulating assets was the right thing to do. The youth are more interested in experiences and making new connections. This in turn has transformed the buying habits on all levels. This new shifts perfectly adapts to the value systems for the millennial as most of the developments within a microcity includes buildings that are off the grid and allow for lower operational costs and red tape. Since the uniqueness and excitement looking for millennials and GenZ move each 4–7 years for occupations, enterprise, or their renditions of small retirements, they would prefer not to be secured by government charges, rent, home loan, and HOA expenses as their salaries are unusual. They need to gather their own particular nourishment, deliver clean vitality, and gather water as they need to live economically off the matrix and be naturally cognizant with no carbon impression. They need to have a choice to just back, guarantee, keep up, manage the cost, and transport their homes. This implies for a run-of-the-mill one room, a conventional size live workspace ought not to surpass USD 25,000 or more than USD 4500 every year as installment.

The worth recommendation relying upon the locale is anybody on any financial plan can now own/travel and live serenely in their own homes situated in numerous areas without the cerebral pains of administration and upkeep. They can likewise pick to have an individual cook, servant, driver, companion, and masseuse in their bundles. It resembles being on a getaway day in and day out/365 days a year. That implies any individual who lives in a microcity can live on a financial plan with convenience, sustenance, utilities, transport, security, cleaning, and relaxation services, with short or long possession in or around a few nations and districts of the world. In general, wherever there will be microcities there will be an event of candid flexibility. My organization's real estate division has now dispatched a few of these urban areas and has been extremely well known with each demographic and pay horde. It would be best if you visit www.zeons.org to view few of the microcity brochures to perceive how it is done in this present reality and acclimate with this idea (Fig. 4).

FIGURE 4 Rule, command, and legal perspective of a Microcity.

The uniqueness of microcities lies in the understanding and execution of productive laws with strict punishments and outcomes for organizations and open officials encouraging debasement. Before we start talking about the significance of laws in a microcity, we should first comprehend where city laws begin. Lawrence F. Keller (2002) contends that, "[m]unicipal sanction can be seen as the plan for a powerful government" (p. 55). "Civil" is another name for city, a term which started from the Roman Empire. The Romans trusted that neighborhood laws could be executed without adversely influencing the Roman Empire; this is the place the possibility of a city was considered. The urban areas represented themselves yet were required to stay inside the legitimate parameters of the Roman Empire. The laws made in a city were called "municipal charters." This established the framework for American law. Keller later states, "Nearby governments in the U.S. are creatures of the state government; contracts resemble laws made in the city; and urban communities can hone some home principle" (p. 57) rules, while as yet having nearby authenticity.

The development of a microcity is the ideal thought for enhancing current states of previous urban communities where the general manageability and prosperity of all gatherings included turn into a need. By analyzing models of created urban communities, a microcity could encourage a prosperous city contract inside or close to the outskirts of a current city. The new city chiefs of microcities could actualize key laws to empower decency, straightforwardness, protection, well-being, and profitability to satisfy the necessities of every one of its occupants. The microcity idea requires comprehensive plans that are sustainable and which draw in financial specialists and firms who aspire to form the city's infrastructure. By and large, microcities are usually modeled like a foreign embassy lands in host countries. They are 100% off the grid, efficient, effective, and exclusive full-fledged city developments in and around existing city areas with their own governing laws (Fig. 5).

Why Do We Need Microcities?

There are a few reasons where a microcity can be an immaculate enchantment shot to address a hefty portion of our cutting-edge challenges, yet the main three zones where it can have the most effect is by turning around outdated government strategies, out of line business rehearses, and actually implementing and catching up to the most recent trends within our different and outsized society.

The Impacts of Flawed Government Approaches

Executing awful fiscal strategies may prompt extraordinary retreats bringing on high unemployment, fall in salary level, and abatement of economic movement. In any case, the subsidence can have long haul outcomes, for

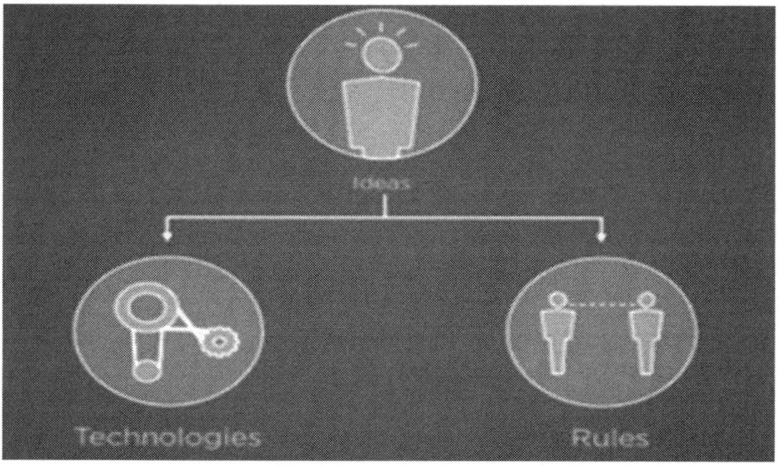

FIGURE 5 Components of a microcity. Standards and developments fitting to the new city's new thoughts are the real segments of an effective microcity. Propelled from Paul Rome's idea of Charter City (2012).

example, job loss and insufficient profit can compel individuals to give up education for their children. It might even restrain credit markets and lower buyer consumption, which can at last end the manufacturer of generally active small enterprises. Also, bigger partnerships may defer or lessen their expenses on innovative work. In each of these cases, a financial retreat can be entirely scarring, in light of the fact that there would be long haul misfortune to individual's economic conditions.

With declined family wages, the level and nature of schooling offered to their kids could likewise be bargained. What's more, there is proof that early adolescence sustenance can influence mental development of a teenager. Also, schoolings in developing countries have demonstrated that great support can prompt better score accomplishment, perusing capacity, mental aptitudes, and livelihoods well ahead in life. Hence, if there is a retreat then families may endure financial upsets and scarcity can hurt youth sustenance.

Financial portability: Bad monetary approach may bring about monetary versatility or the deficiency in that department can offer ascent to unyielding effects of subsidence. For this situation, poorer social orders can deteriorate financial results for their children, for example, absence of nourishment and poor instructive accomplishment. Furthermore, monetary stagnation will influence the pending prospects of families and kids. Additionally, if teens are grown in poverty, they may potentially have low livelihoods, commit crimes, and have low triumph in their impending life.

Business and monetary approach: Executing awful economic strategies will at last strike business advancement and foundation, in light of the fact that new ventures require new clients. Henceforth, if the economy strikes there is a lesser amount of outflow in the general fiscal apparatus. Along these lines, if people anticipate to raise another industry, they may need to defer their missions till interest comes back to standard levels. In like manner, new businesses have need of investors also, leasers. In this condition, because of lower wage level the company will find tough to raise new investors and creditors. On the other hand, as indicated by a most recent study, the credit imperative in the interim currency market had a staggering outcome on the financial plan and small organizations. Plus, the average manufacturing of products and services is just about backed off.

Furthermore, study of credit officers recommends that mechanical advances are significantly solidified. Not just does awful monetary approach make it testing to begin another endeavor, they additionally can deteriorate new businesses that are battling to get by. Therefore, there are numerous businesses that could have been effective, however, and are inept to succeed due to the usage of awful financial policies. Since these microcities can be either worked without any preparation or changed from a current group it can guarantee governments and metropolises can now be run like world-class organizations. This thus will curtail waste and debasement through the force of innovation, straightforwardness, and network.

With the above data in mind Port Lavaca is well positioned to expand and prosper in this new era of constant transformations, and thankfully the leadership has deployed and engaged key players from all areas to rebuilt and rehabilitate their city to the best of its abilities.

SOCIAL IMPACT

The key pillars of the city will be laid on the foundation of community development and economic empowerment. One of the plans that are under consideration is to purchase surrounding farmlands and subdivide these parcels into 10- and 20-acre microorganic vegetable farms that can be sold to homeless veterans with zero money down, low subsidized interest rates using the Texas Land Veteran's program to purchase an off-the-grid ZMU housing solution, a John Deere tractor. Plans also include establishing a veteran's outreach facility and a medical service facility to attract retired and disabled veterans on pensions. The State of Texas Agricultural Commission extension programs is also being engaged to help determine the best cash crops to be grown in the area that may provide income for the homeless vets. This single initiative will help solve multiple social issues related to unemployment, sustainability, and homelessness.

VOCATIONAL AND TECHNICAL EDUCATIONAL INSTITUTE NEXT TO A SCIENCE PARK

With the city's public–private partnership with companies such as Zeons, the city anticipates that approximately 100 new jobs will be created at the Regional Community Facility. To provide exceptional local economic benefits to the Regional Community, additional activities could be performed to further increase the economic and social impact of such partnerships. For example, the company anticipates that even after the city is completely rehabilitated, they will continue to generate revenues from various streams of income on a long-term basis:

1. The sale of units to FEMA and related government entities to replace the three quarters of a million homes lost to domestic disasters in United States.
2. The sale of ZMU product line sales territories.
3. North American ZMU product line sales transaction through these sales territories offices.
4. North American ZMU product line sales to the North American Territory.
5. Installation fees, associated with the sale of North American ZMU product line.
6. Servicing fees, associated with the sale of North American ZMU product line.
7. Financing (through principal and interest payments), associated with the sale of North American ZMU product line.

Hence the need for opening up vocational and technical institutes, to provide a significant amount of the labor force for the final assembly of the ZMU. The city is committed to creating and providing long-term value to the vocational and technical institute, resulting in the output of technically skilled employees overall by leveraging the private sectors experience and investments in production management, quality management, and project management.

The city can attract and retain talent and ensure that the new Community Facility and the locally hired personnel have the ability to fabricate high-quality ZMUs, on time. This labor force will work under the direct supervision of Zeons's experienced team as described above. Training will be delivered by Zeons staff from India in a "transfer of technology" approach and information on production, welding, assembly and test processes and procedures will also be provided to this labor force.

Staff working in the Regional Community Facility will have the opportunity to get trained in the facilities located in India, in preparation for the work. Zeons will utilize the many avenues available to hire the appropriate personnel and provide the necessary support to their technical skills development. Zeons has reviewed and understands and appreciates the importance of the Workforce Initiative new model and will make every effort to ensure

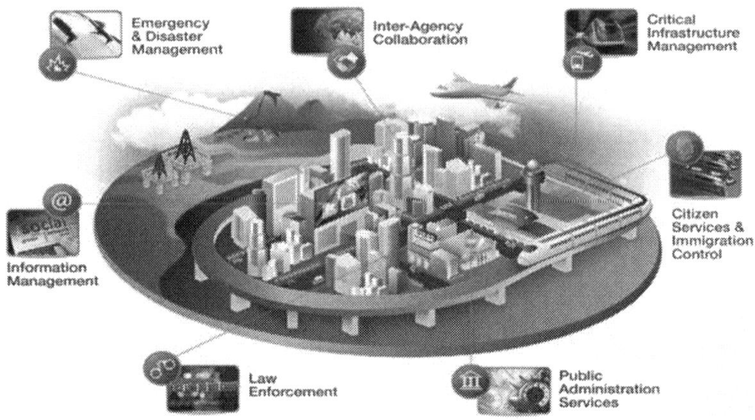

FIGURE 6 Essential components of Port Lavaca Microcity.

that all potential employees, regardless of community or background, have access to the open opportunities at the Regional Community Facility.

BRIEF BACKGROUND AND CONCLUSION

In conclusion, we believe that this case study can show how strong-willed leaderships and common sense partnerships can transform even devastated communities through trade, tax relief, and an availability of trained and skilled labor (Fig. 6).

To address modern day and anticipated future challenges, new and practical laws should be implemented that could facilitate a prosperous city charter located within or near the borders of an existing city. By examining models of developed cities, a microcity should implement only the items that fulfill the needs of its residents and refrain from adding useless laws to the new society.

The microcity concept requires sound policies that are sustainable and attract investors and firms who are willing to build the infrastructure and absorb new residents into the area. Moreover, preventative measures and improving the quality of freedom, liberty along with personal privacy are becoming values for this new living space. The new legislative proposals are geared for effectiveness and through its deployment; the microcity vision will be fulfilled. Ultimately, the creation of the Port Lavaca Microcity is the perfect idea for improving current conditions of preexisting cities where the overall sustainability and well-being of all parties involved becomes a priority.

REFERENCES

Applied Statistics for Public and Nonprofit Organization.

Bee, S., April 2001. Gorging Profits. California Energy Commission, Sacramento, CA.

Blumer, H., 1969. Symbolic Interaction - Perspective and Method. Prentice-Hall, Englewood Cliffs, NJ, p. 1986.

Bonato, D., Orsini, R., Summer 2017. Urban Circular Economy: The New Frontier for European Cities' Sustainable Development.

Brown, R.H., September 1978. Bureaucracy as praxis: towards a political phenomenology of formal organizations. Administrative Science Quaterly.

Buchholz, T.G., 1989. New Ideas from Dead Economists. Penguin Press, New York.

Chomsky, N., 1957. Syntactic Structures. Mouton & Co., The Hague.

Chomsky, N., 1975. Reflections on Language. Pantheon Books, NY.

Chomsky, N., May 8, 2012. Plutonomy and the Precariat: On the History of the US Economy in Decline. TomDispatch.com, pp. 1–5.

Clark II, W.W., Cooke, G., November 2011. Global Energy Innovation. Preager Press.

Clark II, W.W., July, 1996. A Technology Commercialization Model. Journal of Technology Transfer, Washington, DC.

Clark II, W.W., Demirag, I., 2006. US financial regulatory change: the case of the California energy crisis. Journal of Banking Regulation 7 (1/2), 75–93.

Clark II, W.W., Paolucci, Emilio, 1997. Environmental Regulation and Product Development: issues for a new model of innovation. Journal of International Product Development Management.

Clark II, W.W., Fast, M., 2008. Qualitative Economics: Toward a Science of Economics. Coxmoor Press, 2nd Ed from Springer Press, 2018.

Clark II, W.W., Isherwood, W., 2007. Energy Infrastructure for Inner Mongolia Autonomous Region: Five Nation Comparative Case Studies. Asian Development Bank, Manila, PI and PRC National Government, Beijing, PRC.

Clark II, W.W., Isherwood, W., 2009. Report on energy strategies for inner Mongolia autonomous region. Utilities Policy. https://doi.org/10.1016/j.jup.2007.07.003 in Woodrow W. Clark II, and William Isherwood, (Authors and Co-Editors). Special Issue of Utility Policy Journal: China: environmental and energy sustainable development. Winter 2010.

Clark II, W.W., Dan Jensen, J., 1997. Economic Models: The Role of Government in Business Development for the Reconversion of the American Economy.

Clark II, W.W., Li, X., 2003. Social capitalism: transfer of technology for developing nations. International Journal of Technology Transfer. Inderscience, London, UK.

Clark II, W.W., Lund, H., September 2006. Sustainable Development: the economics of energy and environmental production. Journal of Cleaner Production.

Clark II, W.W., and RaeKwong Chung, "Transfer of Publicly Funded R&D Programs in the field of Climate Change for Environmentally Sounds Technologies (ESTs): From Developed to Developing Countries – A summary of six country studies", UN Framework Convention for Climate Change, Report. Bohen, Germany: 1997

Clark II, W.W., Norman Smother, Ricardo Singson, and Lawrence Udell, The Pipeline Model: from Invention to Product Development, Washington, D.C.: Technology Transfer Business Magazine. Spring 1993, pp. 45.

Clark II, W.W., Olav Jull Sorensen, Entrepreneurship: theoretical considerations, International Journal of Technology Management, Inderscience, London, UK, Dec 2001.

Clark Mass Media Company (CM2C), Clark, W.W., PhD. Founder and Executive Producer. https://www.linkedin.com/in/woodrow-w-clark-ii-b6962214.

Clark II, W.W., 2003a. Science parks (1): the theory. International Journal of Technology Transfer and Commercialization 2 (2), 179–206. Inderscience, London, UK.

Clark II, W.W., 2003b. Science parks (2): the practice. International Journal of Technology Transfer and Commercialization 2 (2), 179–206. Inderscience, London, UK.

Clark II, W.W., December 2006. Partnerships in creating agile sustainable development communities. Journal of Cleaner Production 15, 294–302. Elsevier Press.

Clark II, W.W., 2007. Eco-efficient Energy Infrastructure Initiative Paradigm. UNESCAP, Economic Social Council, Asia, Thailand, Bangkok.

Clark II, W.W., 2008. The green hydrogen paradigm shift: energy generation for stations to vehicles. Utility Policy Journal. Elsevier Press.

Clark II, W.W., June 2010. In: Sustainable Communities Design Handbook. Elsevier Press.

Components of a Charter City, 2012. Retrieved March 30, 2012, from: http://www.urenio.org/2009/11/21/video-paul-romer-on-charter-cities/.

Cooper, John F., Clark II, W.W., 1996. Zinc/Air Fuel Cell: an alternative to clean fuels in fleet electric vehicle applications. International Journal of Environmentally Conscious Design and Manufacturing 5 (No.3–4), 49–54.

Dubai. Retrieved March 30, 2012, from: http://ssqq.com/archive/vinlin23dubai.htm.

Economist, January 4, 1992. Examining the Mystery, p. 17.

Economist, July 16, 2009. Modern Economic Theory. Special Issue.

Economist, January 12, 2012. State Capitalism. Special Issue. http://ezproxy.nu.edu/login?url=http://search.proquest.com/docview/1373219091?accountid=25320.

Fan, X., Yakita, A., 2011. Brain drain and technological relationship between skilled and unskilled labor: brain gain or brain loss? Journal of Population Economics 24 (4), 1359–1368.

Fortune, January 16, 2002. The Letter to Ken Lay.

Fast, M., Clark II, W., 2012. Qualitative Economics—A Perspective on Organization and Economic Science. Theoretical Economics Letters 2 (No. 2), 162–174.

Freeman, C., Hagedoorn, J., 1995. Convergence and divergence in the interantonalization of technology. In: Hagedoorn, J. (Ed.), Technical Change and the World Economy: Convergence and Divergence in Technology Strategies. Edward Elgar, Aldershot, UK.

Freeman, C., Perez, C., 1989. Structural Crisis of adjustment, business cycles and investment behavior. In: Dosi, G., Freeman, C., Nelson, R., Silverberg, G., Soete, L. (Eds.), Technological Change and Economic Theory. Pinters Publ., London, pp. 38–66.

Fuller, B., Romer, P., 2010. Cities from scratch. City Journal 20 (4).

Grabher, G. (Ed.), 1993. The Embedded Firm: On the Socioeconomics of Industrial Networks. Routledge, London, UK.

Hasset, W.L., 2011. Reviewing a city charter. National Civic Review 100 (2), 48–57.

Heilbroner, R., Thurow, L., 1994. Economics Explained. Touchstone Press, New York.

Heilbroner, R., 1989. The Making of Economic Society: Revised for the 1990s. Prentice-Hall International, London, UK.

Hill, M.G., 2002. PowerWeb, "Entrepreneurship". http://register.dushkin.com.

Hirschman, E.C., Holbrook, M.B., 1992. Postmodern Consumer Research: The Study of Consumption as Text. Sage, New York, NY.

Holland, J.H., 1992. Adaptation in Natural and Artificial Systems. MIT Press, Cambridge, MA.

HUD-VASH. Journal of Community Psychology, 41(4), 505–514. https://doi.org/10.1002/jcop.21554.

Husserl, E., 1962. Ideas. Macmillan, New York.

Imershein, A.W., 1977. Organizational change as a paradigm shift. In: Benson, J.K. (Ed.), Organizational Analysis - Critique and Innovation. London.

Isherwood, W., Ray Smith, J., Aceves, S., Berry, G., Clark with Ronald Johnson II, Woodrow W., Deben, Das, Douglas, Goering, Richard, Seifert, 2000. Remote power systems with advanced storage technologies for Alaskan Village. University of Calif., Lawrence Livermore National Laboratory, pp. 1005–1020. UCRL-ID-129289: January 1998. Published in Energy Policy, Fall v. 24.

Jehenson, R., 1978. A phenomenological approach to the study of the formal organization. In: Psathas, G. (Ed.), Phenomenological Sociology - Issues and Applications. John Wiley & Sons, New York.

Jerry, J., June 2010. Transformational relationship of renewable energies and the smart grid. In: Clark II, W.W. (Ed.), Sustainable Communities Design Handbook. Elsevier Press, pp. 217–232.

Jones, V., 2008. The Green Collar Economy: How One Solution Can Fix Our Two Biggest Problems. HarperOne, New York, NY.

Kahn, J., January 7, 2002. One plus one makes what? Fortune.

Keller, L.F., 2002. Municipal charters. National Civic Review 91 (1), 55–62.

Kremel, F.M., 1958. Police, prosecutors, and judges. The Annals of the American Academy of Political and Social Science 320, 42–52.

Kuttner, R., January 1–14, 2001. The ENRON economy. The American Prospect 13 (1).

Li, X., Clark II, W.W., 2009. Crises, opportunities and alternatives globalization and the next economy: a theoretical and critical review. Chapter #4. In: Xing, L., Winther, G. (Eds.), Globalization and Transnational Capitalism. Aalborg University Press, Denmark.

Lund, H., 2008. Special Issue Sustainable Energy and Transportation, Utilities Policy.

Lund, H., Ostergaard, P.A., 2009. Sustainable towns: the case of Frederikshavn – 100% renewable energy. In: Clark II, W.W. (Ed.), Sustainable Communities. Springer Press, pp. 155–168.

Lundvall, B.-A., 1993. Explaining interfirm cooperation and innovation: limits of the transaction-cost approach. In: Grabher, G. (Ed.), The Embedded Firm: On the Socioeconomics of Industrial Networks. Routledge, London, UK, pp. 52–64.

Mead, G.H., 1934. Mind, Self, & Society - from the Standpoint of a Social Behaviorist, vol. 1962. The University of Chicago Press, Chicago.

Meier, K.J., Brudeny, J.L., Bohte, J., 2015. Applied Statistics for Public and Nonprofit Adminstration, ninth ed. Wadsworth Cengage Learning, Belmont, CA.

Morris, A., 2008. Taking leadership in the green economy: a conversation with Van Jones. Harvard Journal of African American Public Policy 14, 73–78.

Mowery, D.C., Oxley, J.E., 1995. Inward technology transfer and competiveness: the role of national innovation systems. Cambridge Journal of Economics 19 (1).

Nagpure, P., 2008. Homebased Healthcare: Issues and Challenges. Retrieved from: http://dspace.mit.edu/handle/1721.1/45232.

Nelson, R.R., Winter, S.G., 1982. An Evolutionary Theory of Economic Change. Harvard University Press, Cambridge, MA.

Neuwirth, R., 2011. Stealth of nations: the global rise of the informal economy. Publishers Weekly 258 (34), 58–59.

Perkins, L., 1996. John, Senior Physicist, Lawrence Livermore National Laboratory. University of California, "What is Physics and Why is it a "Science"?", Lecture presented at Graduate Physics Seminar. University of California, Berkeley.

Quah, J., 2008. Curbing corruption in India: an impossible dream? Asian Journal of Political Science 16 (3), 240–259.

Reinert, E., Daastol, A.M., 1997. Exploring the genesis of economic innovations: the religious gestalt-switch and the duty to invent as preconditions for economic growth. The European Journal of Law and Economics 4 (3/4), 1–58.

Reinert, E., 1994. Symptoms and causes of poverty: underdevelopment in a schumpeterian system. Forum for Development Studies (1–2), 73–109.

Reinert, E., 1995. Competitiveness and its predecessors – a 500 year cross-national perspective. Structure Change and Economic Dynamics 6, 23–42.

Report, R., Bernstein, M., January 2001. California Energy Market Report. Winter 2000-2001, unplublished paper presented to Staff in California State Legislature. Rand Corporation, Sacramento, CA. Santa Monica, CA.

Chomsky, N., 1980. Rules and Representations. Columbia Un. Press, NY.

Ring, P.S., Bigley, G.A., D'Aunno, T., Khanna, T., 2005. Perspectives on how governments matter. Academy of Management Review 30 (2), 308–320.

Romer, P., July 2009. Paul Romer's Radical Idea: Charter Cities [Video File]. Retrieved from: http://www.ted.com/talks/paul_romer.html.

Romer, P., 2010a. For Richer, for Poorer. Retrieved from: http://www.prospectmagazine.co.uk/2010/01/for-richer-for-poorer.

Romer, P., 2010. What parts of globalization matter for catch-up growth? The American Economic Review: Papers and Proceedings 100 (2), 94–98.

Romer, P., March 2011. Paul Romer: The World's First Charter City? [Video File]. Retrieved from: http://www.ted.com/talks/paul_romer_the_world_s_first_charter_city.html.

Schumpeter, J., 1942. Capitalism, Socialism and Democracy. Harper & Brothers, New York, NY.

Schutz, A., 1972. The Phenomenology of the Social World. Heinemann Educational Books, London.

Schutz, A., 1978. Concepts and theory formation in the social sciences. In: Emmet, D., Macintyre, A. (Eds.), Sociological Theory and Philosophical Analysis. The Macmillan Company, New York.

Searles, C., 2011. Beyond health care reform: Immigrants and the future of medicine. Ethnic and Racial Studies 35 (1), 135–149.

Shadow Cities by Robert Neuwirth, 2012. Retrieved March 30, 2012, from: http://www.21stcenturychallenges.org/focus/robert-neuwirth/.

Siegal, A., 1983. Plain English. Vital Speeches of the Day 49 (8), 249–252.

State of California, Press Release, January 31, 2002. Governor Davis Calls for Federal Investigation of ENRON's Role on California Energy Crisis.

Steffen, A., 2012. How to save the global economy: build green cities. Foreign Policy 191, 1–3.

Urban Population Living in Slums, 2012. Retrieved March 30, 2012, from: http://www.economist.com/node/15766578.

Weber, M., 1964. Social and Economic Organization. The Free Press, New York. http://www.zeons.org/.

Conclusion: Transforming Economic Theory Into Practical Applications

As noted, the definition of the zinc-air (green) technology as a fuel cell, rather than as a battery was a matter of constant interaction between the representations made on the surface structure organizations and actors who must create a common set of definitions in the deep structure. The same was true of "regenerative braking" that was perfected in research laboratories and then became the key to "hybrid cars" led by Toyota at the beginning of the 21st century.

Once the economic and legal terms were agreed to, the expansion into other related technologies for a zero-emission vehicle become clear and far less complex to implement. The interaction rules allowed the fuel cell to be commercialized since there was common definition of ideas, concepts, and terms. Thereafter, agreement, on the need and application for the technology, could be readily made and advanced through proposals, contracts, and seeking other consortia members.

The results of interactionism between and among the people and organizations over fuel cells lead directly to other cooperative technological commercialization. It is only a matter of time. There will be other technologies that the consortia will purse as well (such as in other fuels, materials, and intelligent systems). While economists debate "market demand" and "market supply" issues, the facts are that technological innovation and commercialization are very much outside these simplistic theories. Business, in general, when seen as a dynamic and ever changing collection of people (defined as a firm or company) will change due to the interaction of its actors.

Change is the norm in business. The question is how do people and businesses know when and where the change will come? Clark sees qualitative economics as one area where the future can be seen, not then, but today. Defining what numbers are all about helps to create and enact plans that are needed for a business (not just through an MBA) but due to the interactionism results between people, groups and communities which was a key in creating Google and many other tech, internet and wifi companies today, the key is that the neoclassical model is only a theory and rarely occurs to predict the economic future.

The process of business innovation through technological commercialization, using the transformations between surface and deep levels of representations, is interactive. It is not simply a pull or push economic approach. The use of interactionism perspective provides for a better understanding of everyday business phenomenon. Furthermore, much can be learned about conducting research and implementing business practices. This is one key reason that China has now been creating science parks over the last decade. Most science parks in China are also connected to universities. There are over 30 science parks now in China with more coming. Given the "Green Development" and "Silk Road" through civic government policy, China will be reaching out to other countries via interactionism to the world connecting all these ideas. The picture shows below one of the science parks in China among many (Furlong, March 24, 2016). Here is the Shenzhen High-Tech Industrial Park where more than half of China's manufacturing activities are conducted in industrial parks and export processing zones (Furlong, March 24, 2016).

Image Credit: Wikimedia Commons/Brücke-Osteuropa From Hannah Furlong.

The discussion starts with how to understand the constitution of what can be called the "firm." A more detailed discussion of "interactionism" is provided in the development of a "new economics" later in this volume. A few issues are worth presenting in the context of the development of the energy "market" and restructuring or creating firms in that sector. One issue that the Enron case has clearly demonstrated is how important "knowing" people can be. That is, close friendships can be useful for political influence. In academic terms, this has been called "networks" (Håkansson, 1993, 1994).

In other chapters, the firm is discussed and how it should be understood as the actions and knowledge of the individual actor and of the actors' collective actions over time. These actions and interactions have some consequences, i.e., the organizing of activities and knowledge. In this organizing or intersubjective moving, pictures of what the firm and the environment are will be created among the actors.

The manipulation of markets by firms is rooted in the notion that the firm acts and behaves as seen above in the Enron case in its (actors) own self-interests. In this case, the firm involved and subjected its accounting firm to do its bidding. Individuals communicating in an interactive or face-to-face manner, where the relations consist of concrete meetings between members in the firm, conduct all business.

So to improve an understanding of the firm, it best to start with an understanding of what constitutes the "firm": What is a firm, or rather what is a business organization? The word organization is a noun and also a myth. If you look for an organization, you will not find it. What you will find is that there are events, linked together that transpire within concrete walls, and these sequences, their pathways, and their timing are the forms which are erroneously made into substance whenever anyone talks about an organization (Weick, 1979, p. 88).

When organizations are examined, especially the larger, older, famous ones, they seem solid, they seem permanent, and they seem orderly. This is, after all, what scholars and the public often call organizations. Images of organizations as solid, permanent, orderly entities run through many textbooks. But they only tell half the story. They obscure the other half: the chaos which looms behind the order, surfacing from time to time, such as when computer systems break down, when products are sent to the wrong destination, or when bookings are made for the wrong dates. They also obscure the immense human efforts and energies, which go into keeping organizations solid and orderly.

The word "Organization" is (only) a concept, which is used to describe a phenomenon. It is a conceptualization of what people believe and do and what they orient toward actions. Organization is a concept in the same way as the concepts of family, class in a school, a football team, a union etc. In other words, an organization is a phenomenon that people experience when and where they see more than one person involved in activities over time. Thus, organization becomes a collective arrangement where people try to give the situation and the activities meanings. In line with Blumer (Fast and Clark, 2012), organizations consists of the fitting together of lines of activity—the interlinking of lines of action. Actors mixing, sharing, competing, and cooperating are parts of the interactive process that define groups and organization. And that is why most organizations, by definition, change and move dynamically in space and time.

By fitting together the lines (what can be called the "rule") of action and interaction as logically prior in organization, actors are discouraged from mistakenly regarding organizations as "things" or simply "solid entities" such as a building or structure. Some scholars argue that organizations are "living things" akin to the biological sciences. And refer to organizations as "evolving" with "life cycle" attributes (Graedel and Allenby, 1995).

An organization should be understood through the actors who by their actions and knowledge create the firm in their everyday pursuit of life. The focus and start of understanding organizations is the members or the actors within the organizational life. In this the relation between action and knowledge is the central issue of interaction.

Action expresses the behavior of people, expressed in advance by the actor whose behavior is then based upon an anticipated project. The expression "the act" expresses the result of the ongoing process or the fulfilled action. Action could be hidden or open intervening into the outer world. All projection consists of an anticipation of future behavior with help of "fantasy" (thinking-in-future) or as Blumer puts it, acting in relation to "the generalized other." But it is not the ongoing action process, but the fantasized act, *as it* was fulfilled, that starts all the projection.

The collective actions constitute the firm, but it is understood in relation to interactivism, interpretation, and intersubjectivity. The constitution is formed by the actors, but at the same time, it is a restriction for the actors through the moving picture of reality that they have built and act in relation to. This is central, because the relation between the development of knowledge and the actions is the foundation for the process of interaction. All this is about how to understand everyday of life processes in the actors' construction of the firm.

People understand themselves retrospectively and act accordingly, but additionally, they are thinking-in-future: What are the projects they are thinking upon? In which way do they try to realize them? And how do the projects change through the process of action and interaction? Cases in point are buildings where people live, work, and visit. The home needs to be integrated with smart, green and healthy technologies, systems and interconnection with others in the neighbor, community, and even regions or states. For example, here is what needs to be done in every place where people interact from Clark (2004 and now 2017):

Note the green technologies and smart systems such as solar roof, efficient energy use even with green trees and grass.

People construct their organizational reality through actions in everyday life and they build paradigms to orient themselves to their own reality. Individual actors have to relate themselves to this discussion if it is the empirical reality and not the theoretical "reality" in which one is interested. In other words, understanding of the social construction of people's organizational life and activities is the context of their everyday business life within the firm.

The basic issue for most actors in firms is how to make the company survive and grow during any particular point in time. They must be free to move in the marketplace, but also be secretive enough to protect its privacy. More importantly, firms must have concern for others and their environment. The public good must be protected because it is in the best interests of the firm to maximize its profits for shareholders and executives alike. The protection of the public good is essential in various infrastructures and sectors (Clark and Lund, 2001). In recent months, the protection of the transportation sector must be embedded in the government. The energy crisis in California has clearly shown that the energy sector is a public trust. The same could be argued for environment, water, and telecommunications.

REFERENCES

California Energy Commission, February 2002. Report on Summer 01. Sacramento, CA.

Clark II, W.W. (Ed.), 2009. Sustainable Communities. Springer Press.

Clark II, W.W., Bradshaw, T., 2004. Agile Energy Systems: Global Lessons from the California Energy Crisis. Elsevier Press, London, UK.

Clark II, W.W., Paulocci, E., Cooper, J., 2002. Commercial Development of energy – environmentally sound technologies for the auto-industry: the case of fuel cells. Journal of Cleaner Production. Special Issue.

Clark II, W.W., Lund, H., November 2001. Civic markets. International Journal of Global Energy Issues X (X).

Clark II, W.W., Bline, D., Demirag, I., 1998. Financial markets, corporate governance and management of research and development: reflections on US managers' perceptions. Chapter in Book edited by. In: Demirag, I. (Ed.), Comparative Capital Systems. Pinter, London.

Cooper, J.F., Clark II, W.W., 1996. Zinc/Air Fuel Cell: an alternative to clean fuels in fleet electric vehicle applications. International Journal of Environmentally Conscious Design and Manufacturing 5 (3–4), 49–54.

Davis, Gray, G., January 31, 2002. Letter to FERC Chairman Wood.

Economist, August 23, 1997a. The Puzzling Failure of Economics, p. 11.

Economist, E., September 20, 1997c. The Visible Hand, p. 17.

Economist, December 6, 2001. The Amazing Disintegrating Firm.

Economist, January 19, 2002c. The Real Scandal.

Freeman, C., March 2–7, 1994. Lecture Series: "Interfirm Cooperation and Innovation; Technical Change and Unemployment: The Links between Macro-Economic Policy and Innovation Policy; Schumpeterianism: Between History and Economic Theory". University of Aalborg, Denmark.

Furlong, H., March 24, 2016. How the World's most wasteful Country is leading the shift to a circular economy. The Bridge to Better Brands.

Haddad, W., 1985. Hard Driving. My Years with John DeLorean (Now on Amazon).

Hakansson, H., Snehota, I., March 1994. Developing Relationships in Business Networks. University of Uppsala, Sweden.

Hakansson, H., Johanson, J., 1993. The network as a governance structure: interfirm cooperation beyond markets and hierarchies. In: Grabher, G. (Ed.), The Embedded Firm: On the Socio-economics of Industrial Networks. Routledge, London, UK, pp. 35–51.

Kuttner, R., March 25, 2002. The road to ENRON. The American Prospect 13 (6).

McNeill, D., Freiberger, P., 1993. Fuzzy Logic: The Discovery of a Revolutionary Computer Technology – and How it Is Changing Our World. Simon & Schuster, New York.

Nelson, R.R., 1990. Capitalism as an engine of progress. Research Policy. Elsevier Science Publishers B.V., North Holland.

Porter, M.E., 1980. Competitive Strategy: Techniques for Analyzing Industries and Competitors. Free Press, New York.

Porter, M.E., 1990. Competitive Advantage. Free Press, New York.

Reinert, E., 1996. Economics: 'the dismal science' or 'the never-ending frontier of Knowledge?: on technology, energy, and economic welfare. Norwegian Oil Review (7), 18–31.

Reinert, E., 1998. Raw materials in the history of economic policy. In: Cook, G. (Ed.), The Economics and Politics of International Trade. Rutledge, London & NY, pp. 275–300.

Reinert, E., 1997. The Role of the State in Economic Growth. Working Paper #1997.5. University of Oslo, Centre for Development and the Environment, pp. 1–58.

Schumpeter, J., 1934. The Theory of Economic Development. Harvard Un. Press, Cambridge, MA.

Smith, A. An Inquiry into the Nature and Causes of the Wealth of Nations. 1776 and Bantam Classics republished frequently.

Staff, G., Davis, G., 2002. Talking Points: ENRON. Sacramento, CA.

Teece, D.J., Bercovitz, J.E.L., De Figueiredo, J.M., 1994. Firm Capabilities and Managerial Decision-making: A Theory of Innovation Biases. Unpublished. University of California, Berkeley, USA. Haas School of Business.

Weick, K.E., 1979. Cognitive processes in organizations. In: Research in Organizational Behavior, vol. 1.

Wright, P., 1979. On a Clear Day You Can See General Motors (on Amazon now).

FURTHER READING

Economist, August 23, 1997b. Europe: Remodeling Scandinavia, pp. 37–39.

Economist, February–March, 2002a. RE: ENRON.

Economist, March 4, 2002b. Social Science Comes of Age, pp. 42–43.

Chapter 10

Overview of Global Urban Cities

Woodrow W. Clark, II

Research Professor in Economics, Pepperdine University Graziadio School of Business (PGSB), West Los Angeles, CA Campus, Los Angeles, California

Chapter Outline

INTRODUCTION

The book has gone through the ideologies, theories, and applications especially in the area of economics for planning in implementing programs that stop and reverse climate problems. Preservation of the climate is important to everyone; thus, interactionism is important at all levels and across all communities. It is critical after the early chapters in the book to give some specific cases here: California and Japan. Then, in the appendices, more information and cases are given as well.

Climate Preservation In Urban Communities Case Studies. https://doi.org/10.1016/B978-0-12-815920-0.00010-1

The discussion in the book has primarily focused on agile energy systems that also need to be integrated (smart), as illustrated by the Siemens company.

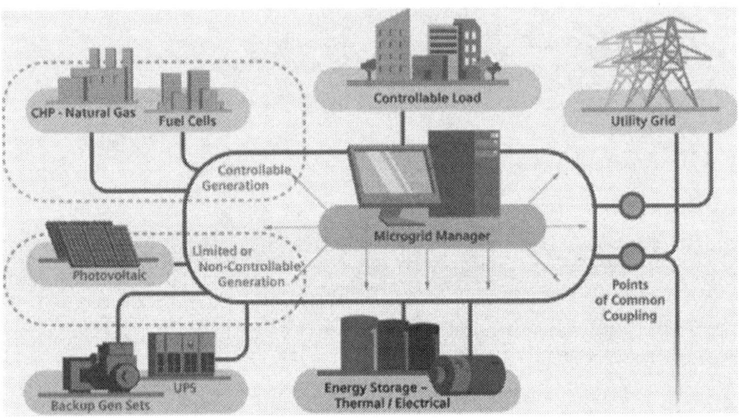

The key here is that each system (generating energy, being efficient, and conserving energy) needs to be economically viable as well as able not to pollute.

Aside from cities and communities, there has arisen a need for "nation-states" also to be smart, green, and healthy. Because of the results (foreseen by many of us), California Governor Jerry Brown decided before the United States pulled out of the United Nations (UN) G20 Summit to have California be a "Nation-State" member. Why not? California is the fifth largest economy in the world, and for the most part it is trying to be a smart, green, and healthy "nation."

There are issues and problems. Governor Brown's term as leader of the state was not as smart and green as he promised. For example, fossil fuel energy was not curtailed in the state and fracking was not capped. Governor Brown's term will end in 2018. The questions are, who will be the next California governor and what will that person be doing to have the nation-state follow (or even lead) the UN G19 to enact the UN Paris 2015 Accord?

Meanwhile, Governor Brown gets into the press through some actions that are not green, smart, and healthy. A case in point follows:

BREAKING NEWS

California Governor Seeks $2.5Billion for Clean Cars Push

Anne C. Mulkern, Energy and Environment (E&E) News reporter.

Published: Monday, January 29, 2018.

California Gov. Jerry Brown (D) wants the state to spend $2.5 billion to accelerate adoption of clean cars.

The basic issue is what does "clean cars" mean? If the past actions from Governor Brown are correct then "clean" means the use of natural gas. And even energy for electric vehicles (EV) and hydrogen fuel-cell vehicles (HFCV) coming from fossil fuels like natural gas or recharging energy for EV from coal.

In short, the world needs to stay tuned to what is next in California.

THE NATION-STATE OF CALIFORNIA

The California energy crisis, and similar energy crises globally, make it clear that extreme or polarizing public policies (regulate versus deregulate and public ownership versus privatization/liberalization) are not the answers. Behind these policies are economic theories that provide misleading policy approaches that miss the actual realistic needs of communities and environmental concerns of society itself. Infrastructure systems have the critical role. They must be a combination of both public and private concern, ownership, finance, and operations.

Such an approach is known as creating a "civic market" (Clark and Lund, 2001) or "civic capitalism" (Clark and Li, 2004) with the need to have "agile systems" that thus encourage and meet public standards, rules, and companies for civic market goals such as renewable energy portfolio standards, economic accounting for society that includes health, welfare, and education, as well as economic development and job creation.

Today (2018), China is pursuing "green development" around the world through the "Silk Road." The goal is to control, stop, and reverse climate change, as *China Daily* notes.

CHINA DAILY

EUROPEAN WEEKLY

Finding a voice in humor

GREENER CHINA

The continuing demand for energy fuel at the pump for vehicles and for power in the home and businesses for heating and cooling has placed the United States in its most vulnerable geopolitical and now economically vulnerable position since before World War II. The US Department of Energy (USDOE) charts energy use in the United States as 45% for fuel transportation and almost 43% for stationary power in buildings. The remaining 12% is for other uses. In the transportation and stationary sectors alone, there is over 53% loss of efficiency from the primary sources of energy generation, which are predominantly from coal, oil, and gas. Renewable energy power generation was only 7% of the total (including hydroelectric power from dams) by fall 2008.

Nonetheless, "energy independence" must be enacted as a national energy policy to stop global warming and mitigate climate change. When it is, energy independence will not mean drilling for more gas and oil in the United States or off its shores, or using "clean technology" and "clean energy" such as natural gas and coal or nuclear power within America's own borders. The "next global economy" is just that: a dramatic economic paradigm change from the neoclassical economics of energy (fossil fuels and nuclear power) to renewable energy (green technologies) that is becoming increasingly commercially viable and cost-effective as consumers install solar and wind energy systems in their homes, offices, and farms. The chart demonstrates the traditional second Industrial Revolution central power grid paradigm, which depends on fossil fuels and is common globally today.

It is critical for public policy makers and the general public to understand the distinction between words such as "energy independence" and especially "green" versus "clean technology" and their definition (Clark and Fast, 2008). What has become even more significant in leaving the solution to energy production under the economic theory is that there is no need for an energy plan and markets can set costs and prices for energy. However, there was little concern regarding the impact on climate change that "deregulation" (Europeans called it "privatization" or "liberalization") or "market forces" would cause, as California and other nation-states discovered in early 2000.

Historically, created public energy monopolies such as utilities controlled power generation from one location or centralized supply source, known as a "central power grid." When these utilities were "deregulated" in California and 19 other states, private companies took over and created "private monopolies" that bought power-generated resources and controlled energy supplies. The results were disastrous because the same central grid economic paradigm prevailed, but now as private monopolies. Government oversight, let alone regulations, was almost entirely gone.

The traditional (second Industrial Revolution, during the 20th century) power system is illustrated in a diagram from the USDOE.

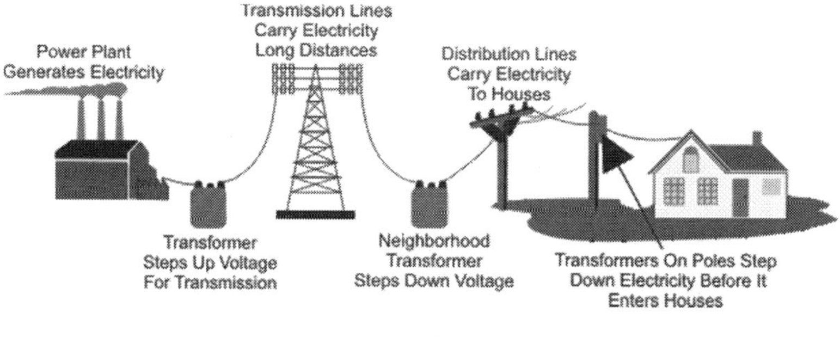

Agile energy systems (the green Industrial Revolution, during the 21st century) provide a energy and economic infrastructure different from that of centralized grid systems. Furthermore, agile energy makes economic sense in the long run compared with the associated costs of modernizing and replacing transmission and distribution networks of the existing system, let alone the issues of security and climate problems. For example, in May 2008, the USDOE (Energy Information Administration [EIA]) projected that electricity demand in the United States would grow by 39% between 2005 and 2030. The USDOE also projected that at least 20% of the increased demand could be met using wind resources. The USDOE estimated that total capital cost to meet the 20% wind scenario would be nearly $197 billion. However, electric utilities would save at least $155 billion by not having to purchase as much coal and natural gas for conventional power plants. That would bring the net cost of the 20% wind-power scenario to about $42 billion (EIA, ibid).

This statistic does not take into account the added savings for not having to construct costly transmission lines. However, as in northern Europe, when wind turbines are installed in towns and villages or remote areas, there is no need for massive transmission lines along with their high costs. For example, Denmark plans to have 50% of its energy come from wind power by 2020. By 2009, it already nearly achieved that goal with close to 38% wind power generation from a combination of centralized wind power farms (including some in the ocean) along with local onsite wind power sites (see the subsequent chart). This is an example of an "agile energy system" that has solar, wind farms, and local wind turbines generating power as well as storage devices such as batteries and fuel cells.

THE PROBLEM CAN BE THE SOLUTION

The "next global economy" today is the generation of energy for vehicle and stationary use through the renewable power generation of sustainable resources found locally, such as sun, wind, geothermal, oceans, and rivers. There are no drilling costs and few construction, operation, and maintenance costs. In fact, the costs for wind turbine power today are lower around the world than for natural gas after it has been drilled, piped, and processed for consumption. As noted earlier, far more significantly and rarely calculated in the costs for fossil fuels is the transportation, transmission, or pipeline flow, which can match even the initial discovery costs or more for these fuels. Although there are transmission costs and the intermittency of renewable energy must be accounted for (especially solar and wind), these are far smaller than actual fossil fuels.

Herein lies the advantage of renewable energy for agile systems (onsite power generation). Solar power as well as geothermal, wind, and biomass (recycled and waste materials) can be generated and used in local communities. There are no costs for securing wind or sun power. Furthermore, with local renewable energy generation, there is no need to be transmitted or piped from a central power plant to distant consumer locations and communities. Some utilities such as Southern California Edison have taken the lead in seeking local, onsite power generation from solar systems places on storage facilities and other large roofs where the power generated can be sold to local consumers.

Unlike traditional power generation, localized agile energy systems such as solar photovoltaics (PV) can be situated near load centers (see Transmission Chart from USDOE, EIA). Generating at the point of use avoids usual grid-based costs such as supply costs, transmission and distribution costs, and the cost of energy loss. Transmission and distribution costs can be two to three times the cost of bulk power. Simply using agile energy systems can result in estimated realized savings of $30–50 per MWh simply from the avoided transmission costs (Beach and McGuire).

The result of local and central power generation is the creation of "agile energy systems" (Economist, 2004; Clark and Bradshaw, 2004) that are sustainable and environmentally compatible with the local community. The Los Angeles (LA) Community College District is one of the most progressive cases. The map shows the locations of the nine college campuses. All of the campuses depend on a central grid for power. However, by the end of 2009, all of them were "off the grid."

THE 21ST-CENTURY "GREEN ENERGY ECONOMIC" PARADIGM

Agile Energy Systems

Agile sustainable energy systems and communities are flexible infrastructure systems that are able to change quickly. An agile system can adapt owing to change, where innovation is welcomed rather than opposed. Agile energy systems are resourceful and are adept at developing ways to avoid or solve conflicts while deconstructing socioeconomic barriers that slow effective solutions to problems. See the chart of an "Example of an Agile Smart Energy System" for the individual homes and customers of all kinds (Clark and Eisenberg, 2008). (Note: "Smart" is used because wireless communication and the Internet are employed, which also create higher energy costs but are offset with onsite power generation from renewable energy sources.)

Agile systems are dynamic and progressive. Agile systems are not limited by political pressure or constraints imposed by lobbyists and special interests. Instead, such agile systems foster and promote diversity along with dynamic growth from which there is change using knowledge, intellectual capital, financing mechanisms, and advanced technologies such as wind power, geothermal, solar thermal, photovoltaic, combined heat and power, fuel cells, and hydrogen options, together with conservation and load management.

All of these and other integrated or hybrid technologies are discussed in light of being components in agile energy systems. There is no one standard or uniform set of circumstances or technologies that fits all communities and regions. Geothermal is not found everywhere. Nor are sunshine and wind, but the combination of these renewable resources along with new technologies for storage such as fuel cells and flywheels provide for hybrid technologies that create firm base loads and dependable power. Such systems can replace or complement conventional fossil fuel system or be ready in the transition to fully energy-independent onsite and distributed energy systems.

Agile energy systems are a combination of local or regional energy systems (including onsite, distributed, and self-generation) central grid systems that will be used in the future, primarily for redundancy and backup purposes. These agile systems are not particular technologies or market mechanisms, but rather a paradigm shift that can be described as a new civic market orientation. Agile sustainable energy systems have the following features:

- **Integrated diversity.** Using diversified or hybrid integrated renewable energy sources make agile systems less vulnerable to disruption and more reliable, especially because they are less reliant on distant suppliers that depend on declining fossil fuels. This diversity also builds in needed redundancy.

- **Economic balance.** Agile systems emphasize the best use of energy, not just the amount of supply and demand. The balance involves promoting conservation, encouraging shifting of energy use to non-peak times, and reducing consumption.
- **Interdependence and interconnection.** Agile systems find ways to avoid bottlenecks in delivery and integrate energy use, which are traditionally separated. For example, cogeneration integrates electricity and heat systems. Hydrogen from renewable energy sources pair technologies: e.g., wind-produced electricity can be stored in hydrogen fuel cells for stationary and transportation fuel. Today, local campuses and other similar systems can be interconnected with wireless and related technologies. This is being done in Japan, as the image illustrates, in which personal as well as community needs are being met.

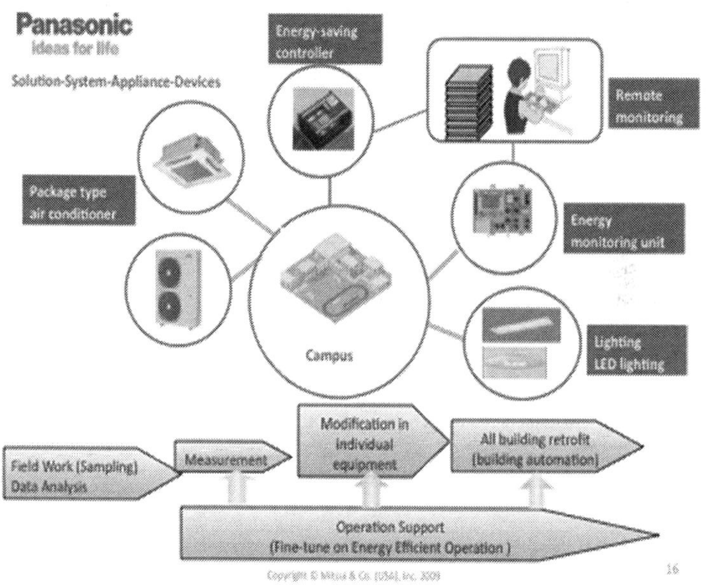

- **Spatially appropriate.** Agile energy systems are smaller and located onsite for buildings but also for vehicle power. They should locate close to where energy is needed and be coupled with renewable resources. Building specific or onsite generation is critical. Neighborhood scale systems or distributed generation are ideal because they factor in environmental costs but can be linked to the grid to act as storage for times of intermittent use as well as to supply the central grid with renewable energy generation.

- **Community, regional, and nation-state.** An agile system links the community to regional, national, and global societal levels. The agile system is the infrastructure that transmits energy and also supports other infrastructures such as transportation, telecom, water, and waste.
- **Social and public good.** Agile systems have the good of society and the public as their primary goal. This is called the "civic core" or the need for the energy market to be government-driven for the local community and citizens.

In short, agile sustainable energy systems make up the basic infrastructures for the new energy economic paradigm of the 21st century, which makes every community energy independent and carbon (particulates and pollution) free. Stopping climate change and global warming must start at the local community level. Agile systems for every home and for work and leisure activities will soon be the norm.

CALIFORNIA CASE STUDIES

Bay View: Hunters Point in San Francisco

In mid-2001, but starting many years earlier, community activists in Bay View/Hunters Point, a southeastern part of the city and county of San Francisco (SF) approached the California State government for assistance in closing two power plants in their poor, working-class, mostly ethnic minority neighborhood, which they claimed were polluting and causing severe illness in children and adults. Evidence abounded regarding the high health risks from living in the area; the city and county had already agreed to "phase out" the power plants, each of which was over 35 years old, but the community wanted reliable clean energy replacements that would serve the SF city and county but also the local community in terms of creating jobs and providing far more viable and livable community as well. Numerous local and state agencies have jurisdiction over the power plants and their replacement. South SF City area toward the SFO Airport, Hunters Point had been one of the poorest regions and areas in SF. That is not true today; the illustration shows it in 2018.

Known as Bay View and Hunters Point, this area of the City of SF was where cargo and shipping were historically the economic base for northern California. Now production is technology via dot-com in SV (SV) and it has become the basis for both the city and others in the Bay Area.

The issue was contentious on many levels. Moreover, the California energy crisis, which started the year previously (2000), added to the drama in SF. However, when the issue of clean power for SF was elevated to one of "environmental justice" the state had the mandate to act: not punitive or controlling action (as some activists demanded), but leadership one centered on "civic markets" In other words, state government under the governor's office would step in and get the public good proponents (activists, community leaders, SF officials, and state/local resources) together with private sector market forces (the California Independent System Operator [CAISO] who buy and sell power and are responsible for reliability; the Pacific Gas and Electric Company (PG&E), local business owners and leaders, and other private sector interests) in a series of meeting to address and solve the issues. Shown is a map of the City and County of SF, which are the same but with two different legal governing groups. SF is viewed southwest from Marin County.

Next is a map of the SF Bay Area with Marin County to the north, Oakland to the east, and SV south of both areas and just north of San Jose City. Stanford University is in the heart of Silicon Valley (SV) and acts almost as a science park for new companies, especially in the technology world. SV has other universities in the area as well as major research laboratories. Across the Bay Area, just north of Oakland, is the University of California, Berkeley (UCB), where Clark earned his PhD. When he was there in the 1970s through 1990s, he started the Clark Communication Company and taught at UCB. Then, in 1994 (to 1999), he was the first manager of energy technology transfer at the Lawrence Livermore National Laboratory, about 60 miles east of Berkeley.

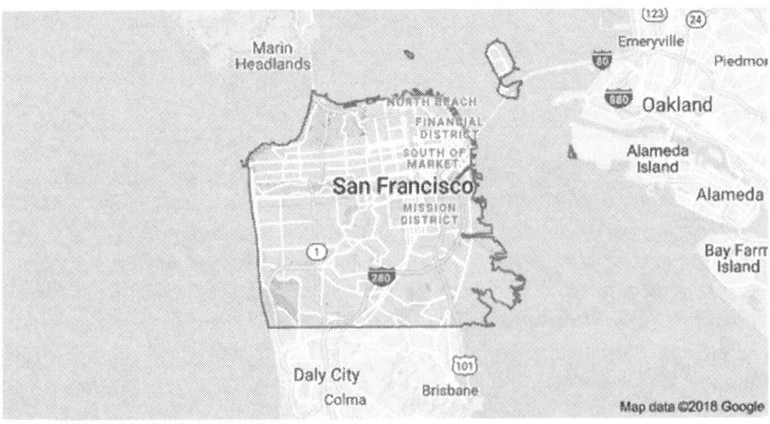

Consider what SF went through to be a sustainable city. Here is a case from detailed discussions about what occurred between spring 2002 and winter 2003, illustrating what civic markets can do to accomplish mutually shared goals for the public and private sectors: clean reliable energy that provides jobs and businesses for the local community. This approach illustrates how agile power systems can generate clean and renewable distributed energy for a major city. Moreover, it defines what sustainable development is in economic terms for the local community. Finally, civic markets that create agile power systems are financial, as SF clearly illustrates, and tailored to meet the future demands of the region. In 2018, the city and almost all areas in it have become green, smart, and healthy.

MUNICIPAL INVOLVEMENT BY THE CITY AND COUNTY OF SAN FRANCISCO

As the SF Public Utilities Commission (PUC) (SFPUC) Energy Plan outlined (Nov. 2002), the City and County of SF are at a cross-roads on energy policy because it faces the harsh realities of how, by, and for whom electricity will be produced, delivered, and used for the 20 to 30 years. Concern regarding environmental justice at the local level has focused concern on the city's economic and environmental health, so the community in south SF got together, as noted in the chart.

Dramatic environmental changes are driving this effort as much as are the drives for more local control and lower costs. An important initiative in that direction is the "Human Health and Environmental Protections for New Electric Generation" ordinance passed unanimously in May 2001 by the SF Board of Supervisors. The ordinance directed the SFPUC and the Department of the Environment to develop plans to implement all practical transmission, conservation, efficiency, and renewable alternatives to fossil fuel generation in the city.

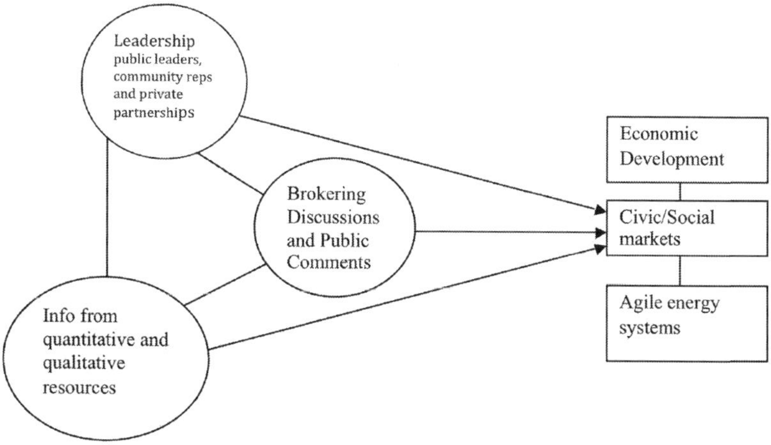

The ordinance was created in response to community concerns over the construction of a large (540-MW) new power plant at the Mirant facility in the southeast sector of the city. The communities in these neighborhoods have already been experiencing the unfair burden of the price of modern society's dependence on electricity. Mirant's facility already has one 207-MW plant (Unit 3) and three 52-MW "peakers" for use during times of peak demand. The only other power plant in SF is nearby at Hunters Point, where PG&E operates one 163-MW plant (Unit 4) and one peaker. These plants are old, inefficient, prone to failure, and extremely polluting. People living in these neighborhoods experience an exceptionally high incidence of respiratory ailments, especially among children and the elderly. (SFPUC, 2002).

Central to the debate over the proposed plant, aside from its size, is whether the city will close the Hunters Point power plant. In 1998, Mayor Brown and PG&E had agreed that the plant would be shut down as soon as replacement power became available. Since deregulation took place in 1996, the CAISO imposed this restriction owing to its responsibility to ensure the state of reliable energy. However, the final decision on the plant has yet to be made because of precise data from PG&E.

By spring 2003, reworking and refinement of the data for analysis continued. At the same time, the SFPUC built renewable capacity to replace the older power plants. In short, the shut-down became more of a political statement than an energy issue. There is clear consensus within SF that the plants need to be shut down and replaced with renewable energy power generation.

Because the city is located at the end of the SF peninsula, importing electricity to SF is highly constricted. Only about 60% of the power needed for SF can be provided during periods of peak demand over existing transmission lines, and the Hunters Point and Potrero Hill power plants meet the balance of need. SF uses about 5000 GW hours (or 5 billion KW hours) of electricity per year and reaches a peak demand of about 900 MW in a given year. SF depends on power plants located within the city limits.

The neighborhoods of southeast SF have historically received a disproportionate burden of environmental, health, social, and economic impacts. Southeast SF is predominantly the home of low-income, non-white residents living in public housing and industrial warehouses. Many local residents have developed ailments related to these environmental impacts. The health problems come from a number of sources including air emissions from the Hunters Point and Potrero Hill power plants. "Children in Southeast San Francisco have higher rates of asthma than in other parts of the City. While these communities bear the negative environmental and health costs of these plants, they do not have access to the financial benefits that accrue to the owners of the plants nor to the businesses outside their neighborhoods whose success is so reliant on electricity" (SFPUC, 2002).

Growing scientific and anecdotal evidence points to severe and costly health problems in the community associated with power plants. Both Potrero Unit 3 and Hunters Point Unit 4 were subject to more stringent NOx emission limitations beginning Jan. 1, 2005. However, to meet this requirement and continue operating, both plants may have to shut down in 2004 and be retrofitted. The plants may be able to continue operating for a time using emission reduction credits if the plant owners acquired and "banked" such credits. Without the retrofit, the Potrero plant could be derated from 207 to 47 MW in 2005 to be compliant.

As the SFPUC Energy Plan (2002) indicates, the possible shutdown of Potrero in 2004 presented two problems for SF:

> *First, during the period of time that the plant is out of operation, electric reliability in San Francisco will be highly dependent on the operation of the Hunters Point Power plant. Given its age this plant has a higher rate of forced outages than do newer power plants. If the Potrero Plant is not available because of its retrofit schedule and the Hunters Point Power plant experiences a forced outage during a period of peak demand, then it is likely that blackouts would be experienced throughout San Francisco.*
>
> *Secondly, the retrofit of the Potrero Unit 3 would be costly. Mirant would only make an investment in such a retrofit if it were guaranteed by the CAISO that it could recover those costs through an RMR contract or similar mechanism. This investment would be amortized over time and could result in the operation of Potrero Unit 3 instead of the development of more efficient and reliable sources of generation.*
>
> *PG&E, as the owner of the Hunters Point Unit 4, has indicated a strong desire to avoid having to invest in emission reduction retrofits at this 44-year-old plant. PG&E may be able to operate Hunters Point using emission reduction credits into 2005 and 2006. The number of hours they can operate will be determined by the Bay Area Air Quality Management District.*

As voiced at public hearings, and by using the civic markets process, the most pressing issue is closing old polluting power plants and preventing the construction of any new polluting sources of electricity generation in southeast SF. More renewable sources and the expansion of energy efficiency programs were proposed as alternatives. This approach and strategy within the goals of the community lead to economic development as the most productive, vital, and long-term solution to the issue of environmental justice.

In addition, diesel-fueled generators are used to meet peak demand and provide emergency backup power for critical facilities. There are four 52-MW diesel-fueled combustion turbines, or "peakers," located in the Potrero Hills and Hunters Point area. These power plants are limited in their operation to 10% of the hours in a year by the Bay Area Air Quality Management District. Diesel fuel produces many times more pollutants than does natural gas. Thus, significant improvements in air quality could be achieved if these units were not used to meet peak demand but were reserved only for emergencies.

The consultation and negotiation process took over 10 months and has yet to be finalized with all of the stakeholders, which include regulatory agencies, commissions, and local government (see Appendix A). However, the final results, as reported in this chapter, look promising. Moreover, what started as an environmental justice issue soon encompassed an agenda and helped to shape it for sustainable economic development: that is, helping to expand local businesses and training workers for future jobs. Although economic development and job training can become part of any community plan, the SF community focus on clean energy and environment gave them a particular direction.

The results of the local community energy plan will be to shut down polluting power plants as well as to put clean energy-generating facilities online. More important, the local community will be able to train workers, employ them at this and related facilities, and create and grow new businesses in the energy and environmental sectors. The community is calling this the "94124 Plan" after the local postal zip code. It is strongly supported by local SF community college staff and students. With some leveraged support from the state, the local government, and the private sector, the plan appears to be on its way to success for a "sustainable community."

LOCAL (ON-SITE DISTRIBUTED) CONTROL OVER ENERGY RESOURCES NOW

SF has continued to lead the way to clean energy; several voter initiatives were passed in 2002. Among the most significant was Vote Solar, which yielded the city $100 million for solar systems for its buildings. A number of programs support the effort and major new construction are ramping up daily. The city has had plans to create its own energy power company, which would be the holding company for local energy generation from communities and public and commercial buildings.

Where local control can be most effective is in promoting and developing small or on-site energy generators. Permitting power plants under 50 MW in size is subject to local review under the California Environmental Quality Act. In California, plants over 50 MW are required to be granted permission by the California Energy Commission in Sacramento.

The siting of these smaller power plants requires a local agency to assume the lead in determining the environmental impacts of a proposed power plant and to provide for their mitigation. Hence, it is possible for the city either to own smaller power plants through the Hetch Hetchy Dam (about 70 miles east of the SF Bay Area) Water and Power System or to enter into contracts for power. Very small generators, particularly those that produce renewable energy, are completely within the domain of oversight for the city.

PROMOTE OPPORTUNITIES FOR ECONOMIC DEVELOPMENT

The choices SF makes about its energy future will have a bearing on the local economy. It is important for homes and businesses to have reliable and affordable energy but also for some of these businesses to become part of the local energy supply infrastructure. This goal would consider to what extent different energy technologies are able to keep dollars in the local economy by supporting local community or city manufacturing, production, distribution, and installation services. Local employers would need local workers who are trained at the community colleges and who are available for work and advanced education.

Some advanced technologies are available outside the local economy, such as combustion and combined-cycle turbines, microturbines, fuel cells, and large wind turbines. However, with the exception of wind turbines, because they can be installed, operated, and maintained, the fuel needed to run these other technology systems involves revenues leaving the city. Solar technologies, peak load management systems, and energy-efficient products are all available locally in the SF Bay Area and can reduce spending on fuel.

Communities within SF saw new opportunities for local economic development and growth in an increased demand for solar products and energy efficiency, especially with the passage of local citywide energy propositions. An expanded and growing market in SF would support new business enterprises, encourage training and education programs, and create jobs while reducing air pollution.

The basic strategy for the civic markets process in SF was to create and implement a local "agile energy system" that consisted of generating clean energy, lowering energy demand through conservation and efficiency, and providing a new economic basis for a sustainable community with job training, advanced education, and business creation.

Sustainability Program: Smart, Green, and Healthy City

The Sustainability Report shows that the city of Beverly Hills (BH) has accomplished many tasks and projects that support the city's goals of sustainability. The next steps for the city are to evaluate the success of these programs, projects, and policies and to develop and implement a monitoring program that includes identifying measures and a standardized reporting mechanism. In addition, staff will evaluate interest in developing an implementation plan, including monitoring and reporting mechanisms, to carry out a sustainability plan.

In 2009, the city council adopted the city's sustainability plan ("Plan"). The Plan defined "sustainability" as the city's capacity to provide desired levels of service to all residents and businesses using ecological sensitive practices, employing resources in an efficient and effective manner, and preserving the local environment as both an ecological asset and a benefit to the city. The Plan provides the city with guiding principles and goals that the city can use to develop and implement sustainability programs.

The Plan does not direct the implementation of any specific actions but provides a foundation upon which the city can build a unified sustainability strategy and a list of potential programs. The Plan is not a state requirement, but it represents the aspiration of BH to become more sustainable. In the event of a conflict, the general plan takes precedence over the sustainability city Plan. The map shows the location of BH west of downtown LA and just east of the city of Santa Monica on the Pacific Ocean.

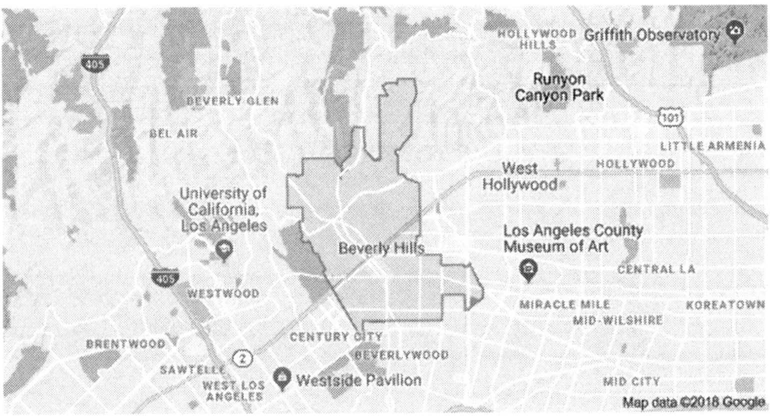

The map of BH shows the University of California, Los Angeles to the east of the city, as well as other landmarks in the area. For example, the city is mostly in a flat valley area with hills to the north that have wealthy homes and areas of LA.

Sustainability Factors

In 2002, the city established the Environmental Sustainability Committee. Clark was asked the City Public Works to join them as a member because he was the governor's renewable energy advisor at the time. This was one of several general plan topic committees that met to maximize public involvement in the general plan update process. The committee identified nine factors that affected sustainability and developed goals, objectives, and policies to support these factors.

Community participation and civic duty is interactionism in practice: to nurture an engaged, educated, and actively participating community and increase the opportunities for community participation by implementing comprehensive education programs, integrating sustainability considerations into city commission decision making, and maintaining transparency in the city's initiatives and programs. BH has parks and play areas. However, until the turn of the 21st century, there were few public activities. One that started at La Cienega Park in the southeast area of BH for movies there and later at several BH public elementary schools. Due to interactionism in Clark's family, working on creating this and other activities for the community.

1. Climate protection and air quality: To set policies and programs meeting the objectives of AB 32 to reduce and encourage the reduction of air emissions in city operations and citywide.
2. Energy: To encourage the use of energy in a clean and efficient manner and the use of renewable energy sources by reducing the use of nonrenewable fuels through efficiency and an increase in the use of renewable energy.
3. Water: To set policies and programs that would reduce water use and expand its water resource portfolio to achieve sustainability. A part of this potential portfolio is to manage storm and wastewater in a beneficial manner.
4. Land use, transportation, and open space: To lessen vehicular dependency and provide a mode of transportation that is environmentally and economically beneficial to the community, improve access to goods and services to the transportation-disadvantaged, set policies and practices requiring best management practices on construction waste, reduce air pollution and energy consumption in new developments, and set policies to preserve and improve parks and open space that provide environmental, social, and economic benefits to the community.

Related to this area is the need in BH for protected bike paths that get people to exercise, out of their cars and hence do not use fossil fuels (unless they have EV or HFCV) to get around the city and its area. LA is flat but has only a few bike paths. Now more and more are being designated as a result of activism in the city through committees and commissions. An example is Clark's wife and son, who are riding their bikes (Feb. 2018).

Note the green areas (trees and grass) around them on the sidewalk along the bike path along one road in BH.

5. Materials and waste: To set programs and policies to reduce the amount of nonrecyclable waste in the landfill and increase the amount of recyclable materials produced from waste, and to reduce the use of nonrecyclable materials in the community. Processing waste is associated with the production of carbon dioxide and methane in the environment. By reducing waste, less carbon dioxide and methane is emitted into the environment.
6. Environmental and public health: To provide access to healthy goods and lifestyles, regulate the use of unhealthy products and activities, and encourage the public to undertake activities to improve health.
7. Sustainable local economy: To support a stable and diversified business community by providing high-level services and programs and adopting policies that would be beneficial to the growth of commerce while positively affecting the environment.
8. Social equity: To address community needs in fair and equitable fields by ensuring public safety, equitable access to basic needs, and the fair treatment of all community members: in a larger sense, to maintain and improve the quality of life for all community members in BH.

INTERACTIONISM: PHILOSOPHY INTO PRACTICE: SOME RESULTS

Since the BH sustainability plan was enacted, a number of achievements occurred including the installment of solar panels on city roofs. Although more needs to be done, there were other achievements. One was in 2015, when the city of BH was awarded with Google's eCity Award, ranking the top eCity in California; the award had been in place for only 3 years in each of the 50 US states. Scoring criteria are based on the strength and robustness of Web, social media, and e-commerce presence as well as high user experience. The award to the city of BH is shown; the Google site is https://www.google.com/economicimpact/ecities.html.

Clark remains a member of the BH City Technology Committee, which focuses on the city being smart as well as green.

See the BH Google News: http://www.canyon-news.com/beverly-hills-receives-2015-google-ecity-award/46603.

Another city council action item came from Mayor Lili Bosse in BH, who started two other programs that are part of smart, green, and healthy communities. These include Walk With the Major, which has taken place each Monday morning at 8:30 a.m. since 2015.

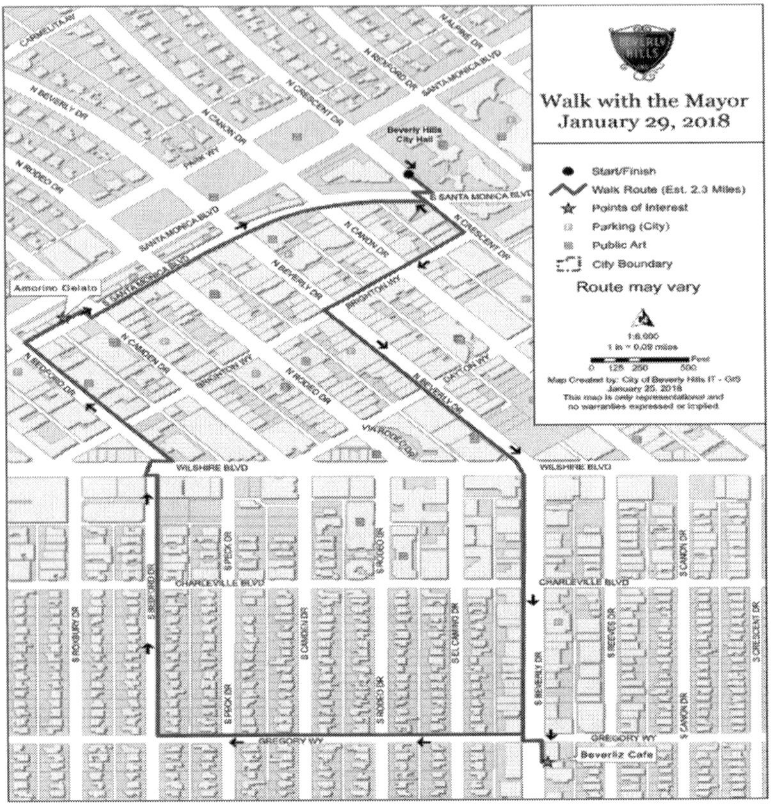

The areas and items of interest each week for the Walks With the Mayor are posted the Friday before each one takes place.

Sustainable Local Economy

BH Mayor Lili Bosse also started the BH Open Later Days (BOLD) program in spring 2016 to promote local commerce and shops in BH by keeping them open over the weekend after midnight.

There is also a need to make tourism buses follow sustainable smart, green, and healthy practices as in the example of BH. Many businesses work these areas, so there could be a big increase in economic returns for the city.

REFERENCES

Clark II, W.W., Bradshaw, T., October 2004. Agile Energy Systems: Global Lessons from the California Energy Crisis. Elsevier Press, London, UK.

Clark Woodrow W. II and Larry Eisenberg, "Agile sustainable communities: On-site renewable energy generation, Utilities Policies Journal 16(4) 2008: pp.262–274.

Clark II, W.W., Fast, M., 2008. Qualitative Economics: Toward a Science of Economics. UK (Coxmoor Press, London.

Clark, W.W., Li, X., 2004. Social Capitalism: transfer of technology for developing nations. International Journal of Technology Transfer and Commercialization, Inderscience 3 (1). London, UK.

Clark II, W.W., Lund, H., December 2001. Civic markets in the California Energy crisis. International Journal of Global Energy Issues. Inderscience 16 (4), 328–344. UK.

USA, Department of Energy (USDOE), Energy Administration Data Bank.

FURTHER READING

Aki, H., Yamamoto, S., Kondoh, J., Murata, A., Ishii, I., Maeda, T., 2004. Analyses on cost reduction and CO_2 mitigation by penetration of fuel cells to residential houses. Transaction of IEE Japan 124-B (3).

Clark II, W.W., 2003. California's Next Economy. Governor's Office Planning & Research.

Clark II, W.W. (Ed.), 2009. Sustainable Communities. Springer Press, New York, NY.

Clark, W.W. II, W. Isherwood. Energy Infrastructure for Inner Mongolia Autonomous Region: Five Nation Comparative Case Studies, Asian Development Bank, Manila, PI and PRC National Government, Beijing, PRC, 2007. To be published in Utility Policy Journal, Spring 2009.

Clemmer, S.K., Donovan, Nogee, note above, Clemmer, S.L., Nogee A., Brower, M. "A Powerful Opportunity: Making Renewable Electricity the Standard," Union of Concerned Scientists, January 1999; and Nogee, A., Clemmer, S., Paulos B., Haddad, B. "Powerful Solutions: 7 Ways to Switch America to Renewable Energy," Union of Concerned Scientists, January 1999.

Clemmer, S.L., Donovan, D., Nogee, A., June 2001. Clean Energy Blueprint: A Smarter National Energy Policy for Today and the Future, Phase I. Union of Concerned Scientists and Tellus Institute.

EDA. Electricity Distributors Association. www.eda-on.ca.

European Union (EU) Parliament Rule (#116), May 2007. Smart and Sustainable Communities. Brussels.

Garvey, M., September 14, 2000. Electric update workshop. In: Presentation at the Association of Bay Area Governments.

Inter-laboratory Working Group. Scenarios for a Clean Energy Future (Oak Ridge, Tenn.: Oak Ridge National Laboratory; and Berkeley, Calif.: Lawrence Berkeley National Laboratory), ORNL/CON-476 and LBNL-44029, November 2000.

Matteson, G., 2008. An international model for nation, state, province sustainable energy. accepted Energy Policy Journal (Elsevier Press).

PG&E, Pacific Gas, Electric, Electricity Rate Schedule A-1 Power Markets Week, October 22, 2007. New Method of Financing Lines for Renewable Receives Approval from California ISO Board, p. 16.

PIU, 2001a. PIU. Working Paper on Generating Technologies: Potentials and Cost Reductions to 2020 by Rob Gross and Jake Chapman. UK Cabinet Office. http://www.cabinet-office.gov.uk/innovation/2002/energy/workingpapers.shtml.

Rifkin, J., 2004. The European dream. The European dream: how Europe's future is quietly eclipsing the American dream. Tarcher/Penquin.

THES. Toronto Hydro Electric System. www.torontohydro.com.

Tsuji, K., Mizuno, M., Saeki, O., Sano, F., Ueno, T., 2001. Reports on the Monitoring of End-use Demand for Residential Houses in the Kansai Science City, vol. 1.

USA, Department of Energy (USDOE), 2008. Energy information Administration (EIA). Chart of Transmission Systems.

Chapter 11

Conclusion: The Global Green Paradigm Shift

Woodrow W. Clark, II[a]

Research Professor in Economics, Pepperdine University Graziadio School of Business (PGSB), West Los Angeles, CA Campus, Los Angeles, California

Chapter Outline

OVERVIEW

The purpose of this chapter is to stimulate discussion and concerns over how the "field of economics" is both studied and used to understand, use, and help communities, cities, and regions become smart, green, and healthy (SGH). The chapters and appendices review some of the history of economics with a strong focus on the neoclassical economics from Adam Smith and how it has not performed, or probably better said, failed to perform today. There is a need to change economics from only being theoretical and involving only numbers into a science.

That focus and perspective is a major paradigm shift for the field of economics, but also in how it can be applied and used in creating, planning, building, and maintaining SGH around the world. The world is round. And now economics needs to be round (e.g., circular) too. What happens to the environment, water, land, and air in one area of the world impacts others too.

a. Woodrow W. Clark II, MA[3] Ph.D.

Climate Preservation In Urban Communities Case Studies. https://doi.org/10.1016/B978-0-12-815920-0.00011-3

We see that pattern more and more in the last two decades. We cannot ignore or do nothing.

Ironically, a new paper from John Ross in *Learning from China* (February 5, 2018) noted that China's economic growth is the fastest by any major nation in over 40 years of reform there. As Ross put it, China "has produced the greatest economic achievement in human history" and "that statement is not meant as an exaggeration or as polite words, it is the objective fact." See the chart below and the website link with all the facts documented.

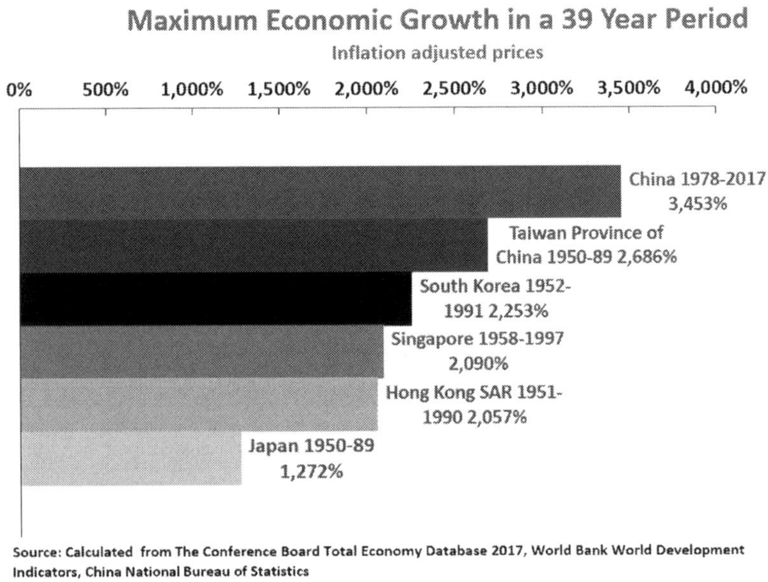

https://www.learningfromchina.net/blog/chinas-is-the-fastest-growth-by-a-major-economy-in-human-history-40-years-of-chinas-reform-and-opening-uphttps://www.learningfromchina.net/blog/chinas-is-the-fastest-growth-by-a-major-economy-in-human-history-40-years-of-chinas-reform-and-opening-up.

And the actions that we take are more than just being resilient. Our actions must protect but also return to our historical pre—fossil fuel days of over the last century or two.

TOWARD ECONOMICS AS A SCIENCE

The last four decades or so of supply-side economics with even the revisionists ranging from Thomas Friedman ("the world is flat") to Jeremy Rifkin ("the world is entropic") both fail to consider the basic problem with economics: it is *not* a science. *The Next Economics* is about how economics can become a science.

Although most economists, including many in this book, make the point that economics can or should include environmental externalities and even concerns on how energy is measured and used, they fail to note the basic problem: economics is *not* a science.

This was brought home to me personally again in June (2012) at a meeting for the UCLA Cross-Disciplinary Scholars in Science and Technology (CSST). I was one of them and was involved in teaching and administrative work. We had a UCLA pre-summer school meeting of over 40 mentors (all scientists from field ranging from physics to chemistry, biology, medicine, and engineering with combinations thereof) to the CSST students. I was the only economist along with one sociologist.

After I was introduced and later in the meeting during a discussion, I made the comment: "And I am here as an economist, because I want to learn to become a scientist." The entire room laughed and some applauded my presence there. Later several people got into conversations with me and hence gave talks in my class on sustainable communities at UCLA over the summer. Their goal and mine were to teach social scientists what science is and for the social scientists to become scientists.

My book, *The Next Economics* (Springer Press, 2014), was one of the key goals for all the chapters. But even more significant, underlying that goal has been the need for economists to understand what science is about and how it works. Every scientist is trained the same way in terms of thinking, creating, and proving hypothesis so that they become repeatable theories for application and use in the future. To say that economics is scientific because Adam Smith followed Sir Isaac Newton misses the point.

Science has a set of processes and procedures to which every scientist follows. The conclusions of every scientific study must be subject to challenge and replication. That is why climate change issues, without a doubt, are now significant and proven theories. The evidence and replicable data are now daily events. But that was only after over two decades of work by the UN Intergovernmental Panel on Climate Change and UN Framework Convention on Climate Change. I worked in both of these groups, starting almost 20 years ago.

The basic conclusions from *The Next Economics* are how society (primarily political leaders and decision-makers) and academics look into the definition and meaning of "science" itself. The review of modern economic theory with its failure to be a science means that the field of economics must find a new paradigm and philosophical roots. For one thing, to address climate change, communities and nations must set plans. They need to state visions, which have measurable objectives as well as frequent updates and revisions. On the business level, the same criteria and needs take place. Whereas a community or government entity is accountable at election time, businesses need to be so on monthly, quarterly, and annually in verified accounting reports. Nonetheless, all organizations have leadership that is held accountable.

Consider physics: "The 'rules' of physics are also of fundamental help in understanding the associated sciences of chemistry, biology, metallurgy, astronomy, etc. For example, before we understood the physics that govern the behavior of electrons in an atom, chemistry was purely a phenomenological science; that is, we knew we could repeat what we had done or seen before, but we didn't know why, nor could we predict what would happen in a new situation. Now, because we understand the physics of the atom, we understand fully why there are around one hundred basic chemical elements and around a few million chemical compounds" (Perkins, 1996, p. 1).

THE ECONOMICS OF FINANCE

On a recent trip to east Africa to talk about climate change for the US State Department, I met a group from the World Health Organization (WHO) in London. Over several meals, this issue and need was discussed. Yet there were little or no data and certainly none of us had been able to get either funding or commitments to do the research and data gathering necessary.

Consider some cases where renewable energy for designated areas has worked or is being considered. The basic issue is economics. How to get renewable energy installed through long-term financing? For example, in 2008, the City of Berkeley, California, created the PACE (Property Assessment for Clean Energy) program for renewable energy systems through additional payments on local tax bills. The program meant that an owner of a building could chose to tax themselves to pay for the renewable energy to be installed. The monthly tax bill would be the way to pay off the funds needed to buy and install the renewable energy system. Should there be a sale or transfer of the building, then the added tax would also be part of the sale or transfer.

Herein lays the problem for the program that in the summer of 2010, Fannie Mae and Freddie Mac ruled against the PACE program for individual homes: if there was a default on the home mortgage or taxes and hence a lien for those funds, the Fannie Mae or Freddie Mac owner would be placed in the second position. The situation for commercial properties was different and depended on a variety of other variables. But PACE was stopped due to that decision by the key economic source for financing homes.

However, this meant that other creative economic mechanisms needed to be found for financing renewable energy and sustainable communities. One example that worked in other areas to fund new but higher cost technologies was through government purchase programs or other nonprofits who could pay over long periods. Some of these programs would establish competitively bid master contracts for vehicles, college campus, and government buildings, infrastructures, and other public needs.

The strategy was to establish a master contract with one company that successfully won the bid. From that one contract, other customers could buy the products. This was done in some cities and states to reduce carbon emission because natural gas cars polluted less than gasoline-fueled vehicles. But as the master contract took effect, it then allowed other local government entities to purchase off the same contract and save large sums of money. The key is bulk purchasing through the use of competitive bids for "large group costs" that cover a community, buildings and even shopping malls, so that the costs are greatly reduce the costs for innovative technologies to mitigate climate change.

Applications of this financing mechanism could be used for other nonprofit organizations such as religious groups, hospitals fire stations, and public schools, among others. Additionally, some renewable energy companies have provided a very interesting economic addition: provide employees, staff, students, and administrators or even subcontracts the ability to buy off the master contract for a lot less money than if they were purchasing the systems on their own. Group and large-scale economic contracts yield lower costs than single-item purchases.

In more conventional economic areas also there are creative approaches to financing renewable energy and technologies to make communities sustainable. One is the mortgage for any building or complex. The other is a lease. In both cases, the cost for a building must be seen as to include renewable energy much as the plumbing, lighting, heating, and cooling systems are all parts of building costs today. Not more than 50–60 years ago, after WWII, mortgages for homes and buildings did not include bathrooms or air conditioning.

These "modern" technologies (21st century) over time were then included in the mortgage or cost for a home or building. The logic seems apparent to do for renewable energy what heat, plumbing and other systems were over the last 5-6 decades. Instead of considering solar panels, for example, as an add-on cost, they could be made part of the mortgage cost for a new or refinanced home or building. The impact on reducing climate change and establishing sustainable buildings and communities would be a large step in reducing reliance on fossil fuels and making communities and regions energy independent and carbon neutral. The new economic paradigm that includes long-term cost reductions can occur and provide solutions to climate change.

HYPOTHESES, PLANS, RULES, STANDARDS, MEASUREMENTS, AND ACCOUNTABLE RESULTS

Economics must become a science.

> *One very important job in physics is to determine the fundamental constants of nature that are needed to understand the "rules" and apply them to predict and calculate things. A good example of such a constant is the speed of light (approximately 186,000 miles a second). Typically experiments are used to measure these physical constants. Theoretical studies can also be used to infer their values but an experimental test is usually ultimately required. A remarkable achievement of physics is that we know the value of virtually all the constants of nature that we are aware must exist for our present understanding of the "rules."*
>
> Perkins (1996, p. 3)

Consider a very recent example of science at work. In early July 2012, it was reported in newspapers that "Physicists are celebrating their Higgs boson 'triumph'" (Brown, 2012a). Over the entire 20th century, scientists have tried to learn more about the standard model and thus answer one important question, "why does matter exist?" (Brown, 2012a, p. 8). For physicists the answer was similar to "landing on the moon or the discovery of DNA" (Brown, 2012a, p. 1). The basic question was posed by British Physicist Peter Higgs who heard the scientific beginning of the answer to his question from over 48 years ago, at the European Organization for Nuclear Research (CERN) in late June 2012 by two independent research teams (CMS and ATLAS). Both teams at the CERN Conference independently reported data from December 2011: "uncovered 'tantalizing hints' of a Higgs boson with the a mass of about 125 billion electron volts" (Brown, 2012a, p. 8).

However, as with all scientific research, this is not the end. The next step all of the physicists stated "would be to figure out whether the particle is indeed the single Higgs boson described by the Standard Model or some exotic variant" (Brown, 2012a, p. 8). Some physicists believe that there are multiple Higgs bosons requiring more research, data, and evidence. Hence, the need for a massive proton collider that needs to be updated in the next 2 years (Brown, 2012b, p. 9). Scientific research never ends.

What does it then mean for economics becoming a science? One basic concern is even with the report on the Higgs boson. Aside from what is Higgs boson in terms of both definition and meaning (smallest atom that could be the basis for all matter). But then, consider some of the discussion reported above: what does "tantalizing" and even "hints" mean in science? Clearly not something that is proven beyond a doubt; thus not subject to further investigation. Just the opposite, as the last quotation indicates, "figure out" the next steps and how that relates to the "standard model" used in physics. In short, science is a never-ending search for truth through theories, data collection that have conclusions, yet with a continuing need to investigate, valid, and double check. Economics needs the same set of rules.

A very simple application of economics is with "everyday life." All forms of groups, families, businesses, and governments need sets of rules. People and families need rules as much as all businesses and governments. Although some people complain about government rules and codes, the fact is that they have the same thing for their families and children. Science follows that same line of thinking. To discover something or investigate an idea, there must be some set of hypotheses that turn into a plan with rules and standards that are measureable. If not, then science tries another set of hypotheses with rules. Even when the results are proven, it is also critical to replicate those hypotheses and test them again, and again, and again.

For example, I teach entrepreneurship in the United States, China, and European Union since the term was new in the early 1990s around the world. One of my first published articles was about entrepreneurship in Silicon Valley and how it worked. And my reason for teaching entrepreneurship was so that people who want to start businesses need to understand that there are sets of rules to follow from other NewCo experiences. I knew from own personal experiences because I had started my first business, a landscaping company with my younger brother (Wayne) in the late 1950s called "Wayne—Wood Nurseries" when we were not even in high school yet, that soon paid our way through college. We sold the company after we both went on to graduate school outside of Connecticut, where we grew up. We were both eager to move "west," gradually as we spent 5—6 years in Illinois before going to Berkeley. Neither of us had loans or had to pay back for our undergraduate and graduate education. But we wanted to continue our education.

The best example of where an area of social sciences or field of study like economics can be seen as a science is linguistics. In the field of economics, as described and outlined in several chapters in this book as well as *Qualitative Economics* (Clark and Fast, 2008), linguistics changed into a science under the leadership of Noam Chomsky with what he calls "The Galilean Style," which comes from the natural sciences and is the process to construct "abstract mathematical models of the universe to which at least the physicists give a higher degree of reality than they accord the ordinary world of sensations" (Chomsky, 1975, p. 28). I had read and followed Chomsky's work while working on my PhD degree in Anthropology at the University of Illinois, Urbana, in the early 1970s. Chomsky became a long-distant friend and mentor.

Later in Chomsky's book *Reflections on Language* (1975) and then in far more detail with *Rules and Representations* (1980) as well as a number of books since then, he states that language (a field of study that needed to become a science) is seen as "A comparable approach (which) is particularly appropriate in the study of an organism whose behavior, we have every reason to believe, is determined by the interaction of numerous internal systems operating under conditions of great variety and complexity." (Chomsky, 1975, 1980, p. 218). Chomsky called all of his work in linguistics transformational grammar that led an intellectual revolution, while at MIT in the early 1960s, that continues to this day.

In short, language moved from being a field of study to a science. Therefore, creative aspect of language is a characteristic species property of humans. Language serves as an instrument for free expression of thought, unbounded in scope, uncontrolled by stimulus conditions though appropriate to situations available for use in whatever contingencies our thought processes can comprehend (Chomsky, 1975, 1980, p. 222).

The challenge for linguists is to approach language and its study as a science. Therefore, what language means to the linguist, grammar (as distinct from speaker-hearer's grammar) is "a scientific theory, correct insofar as it corresponds to the internally represented grammar … The grammar of the language determines the properties of each of the sentences of the language. For each sentence, the grammar determines aspects of its phonetic form, its meaning, and perhaps more" (ibid., 220). And therein became a challenge to linguistics, not as a science, but as needing more than just structure that determines meanings.

The biggest challenger was George Lakoff, whom I had for classes while at the University of Illinois, Urbana, and then again when he came to California for a Summer School program at the University of California, Santa Cruz. After that, Lakoff went to the University of California, Berkeley, in 1972 where he is still today. I chose to leave Urbana and follow Chomsky to MIT. But he urged me to instead go to Berkeley due not only to the work of Lakoff but also due to the "politics of Vietnam" at the time. MIT was not the "best place" to be as a student or faculty member, if you were interested in politics as I was and as was Chomsky. Lakoff was in a more friendly campus in Berkeley.

My classes and studies with Lakoff and other faculty at Berkeley, however, led me in a different direction. Lakoff become a critic of Chomsky on transformation grammar because from Lakoff's perspective it only dealt the rules and structures of language. Lakoff and others at Berkeley (Lakoff and Núñez, 2000) were far more interested in the meanings and definitions of words and even phrases in language. It was with the "deep structure" of language that both the structures for language and their impact had the most significance (Lakoff, 1999, among earlier articles and books). What I got from all this, was seeing how science worked in linguistics, but also another perspective to other social sciences, including anthropology, sociology, and political science.

With these theoretical concepts in place from linguistics, the actual transformation rule-making process can be seen. That is, the business relationship becomes successful or unsuccessful because she/he draws on the defining characteristics in the deep structures (universals or common properties across cultures) of the new business creation interactive process (surface structure) and applies the proper rules. For example, when a business agreement appears to have been very successful (e.g., material wealth, power, or head of large company), there were many transformational rules that got the business actors to that place (surface structure interactions). The transformational rule-making process is often intuitive and based on common sense. It is the surface structure in scientific terms.

However, there are also the deep structures of words, concepts, numbers, and phrases that need to be explored in depth. I did that since the early 1980s in a number of areas but really decided that understanding deep structures and meanings was a core element needed in the social sciences. That led me to take classes from Herbert Blumer, the founder of the Sociology Department at Berkeley. His concerns were more psychological than mine with symbolic interactionism, but the notion of understanding how people think and act is critical and becomes a core interest of mine. My book with Professor Michael

Fast, *Qualitative Economics* (2008), reflects those ideas and concerns in practical ways for economics to be a science.

In the end, I changed directions at Berkeley and even took law classes at Boalt Hall wanting to understand how law gets involved with determining and defining meanings of words, phases, and numbers. Unfortunately, there were not too many law professors then who cared about science. So I decided to take a PhD in an applied area that I had experiences in as well as a social science. Hence, I got an inter-disciplinary degree in anthropology, education, government, economics and linguistics. The topic gained a great deal of interests over 30 years ago and still does today. The basic issue is what causes violence in public schools, then and now.

The scientific understanding of surface and deep structures helps provide an examination of schools and other institutions, which leads to solutions, especially because the problem is repeated and can thus be examined repeated in a scientific manner. What I did, however, after the PhD was go into business, applying some scientific insights that I had gained over the years while in academics and then practice as a teacher. What I had learned with Wayne—Wood Nurseries was now needed to be put into practice. In the academic world, if you receive your PhD from an institution, you *cannot* then teach there but must go elsewhere for 6—7 years before coming back. I had offers to go to other universities, but I liked the San Francisco Bay Area. So I stayed.

What I did was start a mass media company in San Francisco that produced dozens of documentary and educational media. My dissertation topic on "Violence in Public Schools" was a perfect topic for me when I started producing documentaries due to my giving talks on radio and television shows. Moving forward 4 decades, I updated my thesis to be a book that came out in December 2017 from NOVA Press as "Violence in Schools, Colleges and Universities". In the 1980s, Clark Communications was earning over $1 million annually, until the early 1990s when the end of Cold War came and California was hit particularly hard with the need to downsize, convert, and sell-off military bases and restructure the entire American (global economies). The point is that as an entrepreneur, I knew what it meant and how to teach the subject. Was starting a business, let alone, economics a science? No. But I had studied chemistry and physics in high school and at my graduate universities to know what the basics of science were. Then later on, I was asked to go to Lawrence Livermore National Laboratory to train scientists in technology transfer and commercialization. While there, I learned a lot more about science from the scientists there with over 1300 physicists, 1200 chemists, and over 3500 engineers, plus others.

However, it was when I went to Aalborg University (AAU) in Denmark as a Fulbright Fellow in early 1994 that I learned even more about the basics of science as I taught in the Faculty of Natural Sciences while teaching entrepreneurship in the Business Economics Faculty. I was asked to start a program at AAU that involved their science park (NOVI) across the street from the university. This was *not* an incubator and very different from what Americans were doing then (and even now) to commercialize new technologies and ideas. I have written and published many articles on Science Parks from over a decade ago. But the articles are not as well as received in the United States, as they are in the European Union and Asia (Clark, 2003a, 2003b). Science parks have been successful and moved communities into the green industrial revolution.

The application of my three graduate MA degrees with the PhD made me unique then and even now. Today, I refer to my academic studies and publications as "cross-disciplinary studies" with my focus then and now on SGH "sustainable communities." And even more so since I had an entrepreneurial series of businesses. However, even more significant was my working with scientists at LLNL and also at AAU. I became a sort of translator from science to business and then back to science again. My experience and work on science parks in Denmark and then throughout Northern Europe over a decade of research and publications led to a number of other ways in which economics can become a science in practice.

My books with Grant Cooke on "Global Energy Innovation" (2011) and then "The Green Industrial Revolution" (GIR) in Mandarin 2014 (short version) and English 2015 (longer more in depth) reflect the concerns with the environment and also the impact and need for a science of economics to do that. It is one thing for people, communities, and governments at all levels to talk about climate change and the environment, but they all have very different discussions about what to do. The biggest and most concerning barrier for taking any action is always economics. It is too expensive to save our earth, reverse climate change, and stop pollution. That is foolish and frankly just playing politics with the "field of economics"—not the "science of economics."

Let me end with a personal case that proves economics as a science can help resolve and stop environmental (and other) societal issues. Consider energy efficiency and conservation. Today, LED lights are available that take a faction, if any electricity. The issue is that these LED lights cost too much. Any and every new invention always has a slightly economic cost when first brought out and into any market. LED bulbs are now recognized around the world as being a significant "game changer." Companies and consumers all only produce, buy, and use LED bulbs. Furthermore, LED bulbs can be an example of where that neoclassical Adam Smith theoretical economic process (never been enacted) is considered in scientific terms, makes the energy costs minimal, and the scientific process to do so, patentable.

In the attachment below (Appendix A), that scientific economic process is now as a patent pending (February 2012) done by two electrical engineers and myself through a NewCo, called Nularis. The scientific economic process for LED lights is patented that shows how to conserve and safe energy. With this case, the rules and formulas are laid out in detail. These patents set a new standard for economics and application to business for *The Next Economics*. More and similar ideas for the science of economics will be forthcoming and welcome. The time and need have come for economics to become a science. That is key factor as my book on Qualitative Economics (with Michael Fast, PhD) in 2008 is updated to be published by Springer Press as a 2ed for 2018. From there, Prof Fast and I are writing a book on both Qualitative and Quantitative Economics (**Q²E**) for Palgrave Press due out later in 2018. Stay tuned for more.

APPENDIX #1

Patent Invention Disclosure Overview: Applied Dated: February 12, 2012

Economic Efficiency Through Lighting (Nularis Corporation)

Title: Method and design of software and related systems which couple amortized loan terms and payments with predicted and actual energy cost savings.
Inventor(s): Woodrow W. Clark II PhD, Wendell Brown, and Jonathan Fram.

Background of the Invention

Energy efficiency and conservation are important to achieve international goals for reduction of greenhouse gas emissions, fossil fuel usage, grid load strain, costs, and a wide range of other benefits. However, many approaches to energy efficiency and conservation involve significant capital outlays that create financial management risk and provide undetermined return-on-investment rates and payback periods, which often hinder their adoption.

These inventions relate particularly to methods and design of software and systems which run in computing environments (computer hardware, virtual CPU environments, servers, computers, tablets, wireless mobile devices, etc.) that couple and integrate amortized payment terms and amounts with predicted and actual energy cost savings. The inventions are thus novel, innovative, and useful in that they provide a mechanism for financial risk reduction/management and predictable cost outlays (loan repayment terms that are directly linked to energy savings), thus serving as an enabler for the financing of such energy efficiency and conservation projects.

Description of the Inventions

The inventions relate to the methods and design of software and related systems which couple amortized payment amounts with predicted and actual energy cost savings.

Features as Formulas: 1−14 14: The formula for the periodic payment amount A is derived as follows. For an amortization schedule, we can define a function $p(t)$ that represents the principal amount remaining at time t. We can then derive a formula for this function given an unknown payment amount A and $r = 1 + i$.

$$p(0) = P$$
$$p(1) = p(0)r - A = P_r - A$$
$$p(2) = p(1)r - A = Pr^2 - Ar - A$$
$$p(3) = p(2)r - A = Pr^3 - Ar^2 - Ar - A$$
$$p(t) = P_r^t - A\sum_{k=0}^{t-1} r^k$$

Applying the substitution,

$$\sum_{k=0}^{t-1} r^k = 1 + r + r^2 + \ldots + r^{t-1} = \frac{r^t - 1}{r - 1}$$

After substitution and simplification, we get

$$\frac{p(t)}{P} = 1 - \frac{(1+i)^t - 1}{(1+i)^n - 1}$$

The annuity formula is

$$A = P\frac{i(1+i)^n}{(1+i)^n - 1} = \frac{P \times i}{1 - (1+i)^{-n}} = P\left(i + \frac{i}{(1+i)^n - 1}\right)$$

REFERENCES

Brown, E., 2012a. Physicists Are Celebrating Their Higgs Boson 'triumph'. Los Angeles Times, Los Angeles, CA, pp. 1–8.

Brown, E., 2012b. Massive Proton Collider Is a Great Big Hit after All. Los Angeles Times, Los Angeles, CA, p. 9.

Chomsky, N., 1980. Rules and Representations. Columbia Un. Press, NY.

Chomsky, N., 1975. Reflections on Language. Pantheon Books, NY.

伍德罗·克拉克,格兰特 with Anjun Jerry JIN & Ching-Fuh LIN 库克,金安君,林清富 Clark II, W.W., Cooke, G., August 2015. China's Green Development Paradigm (Mandarin). Ashgate and China Electric Power Press.

Clark II, W.W., Cooke, G., 2014. Green Industrial Revolution. Elsevier Press.

Clark II, W.W., Cooke, G., November 2011. Global Energy Innovation. Preager Press.

Clark II, W.W., Fast, M., 2008. Qualitative Economics: Toward a Science of Economics. Coxmoor Press.

Clark II, W.W., 2003a. Science parks (1): the theory. International Journal of Technology Transfer and Commercialization 2 (2), 179–206. Inderscience, London, UK.

Clark II, W.W., 2003b. Science parks (2): the practice. International Journal of Technology Transfer and Commercialization 2 (2), 179–206. Inderscience, London, UK.

Lakoff, G., 1999. Philosophy. In: The Flesh: The Embodied Mind and its Challenge to Western Thought. Basic Books.

Lakoff, G., Núñez, R., 2000. Where Mathematics Comes from: How the Embodied Mind Brings Mathematics into Being. Basic Books.

Patent Invention Disclosure. Inventor(s):Clark II., W.W., Ph.D., Brown, W., Fram, J. Applied dated: February 12, 2012. Method and Design of Software and Related Systems Which Couple Amortized Loan Terms and Payments With Predicted and Actual Energy Cost Savings. .

Perkins, L.J., 1996. Senior Physics, "What Is Physics and Why Is it a 'Science'?". In: Lecture at University of California Physics Seminar. University of California, Berkeley, pp. 1–3.

FURTHER READING

Bonato, D., 2017. Circular Economics. ReMedia Company, Milan, Italy. Remedia. www.consorzioremedia.it.

Blumer, G.H., 1969. Symbolic Interactionism: Perspective and Method. Prentice-Hall, Inc., New Jersey.

Chomsky, N., 1957. Syntactic Structures. Mouton & Co., The Hague.

Clark II, W.W., Dec 2017a. Global Green Smart Sustainable Cities and Communities, second ed. Elsevier Press.

Clark II, W.W., 2017b. Agile Energy Systems: Global On-site and Distributed Energy. Elsevier Press.

Clark II, W.W., Cooke, G., March 2016. Smart Green Cities. Routledge Press.

Clark II, W.W., 2014. Global Sustainable Communities Design Handbook. Elsevier Press.

Clark II, W.W., Fall 2012. The Next Economics: Global Cases in Energy, Environment, and Climate Change. Springer Press.

Mead, G.H., 1932. The Philosophy of the Present. Prometheus Books.

Appendix

Smart Community and City Environmental Responsibility (C^2ER) With Green Computing

Hemantha S.B. Herath[1], Tejaswini C. Herath[2] and Athula Ekanayake[3]
[1] *Department of Accounting, Goodman School of Business, Brock University, St. Catharines, ON, Canada,* [2] *Department of Finance, Operations and Information Systems, Goodman School of Business, Brock University, St. Catharines, ON, Canada,* [3] *Faculty of Management, University of Peradeniya, Peradeniya, Sri Lanka*

INTRODUCTION

One of the biggest challenges facing the environment undoubtedly is global warming caused by carbon emissions. The Energy Information Administration reported that over 90% of CO_2 emissions is primarily due to energy consumption. With the growth of the Internet and globalization, private and public organizations and individuals now own and/or use computers and peripherals extensively. As a result, energy consumption in computing has increased tremendously and will likely continue to increase as worldwide access to the Internet and computing requirements grow. Consequently, it is crucial for many organizations, communities, and cities to commit to decreasing the environmental impact (EI) and power consumption due to computing. The new norm of proactive social responsibility in computing for organizations, communities, and cities is an environmentally sustainable infrastructure design. As part of community and city environmental responsibility (C^2ER), many organizations need to consciously consider operating in a "green," that is, environmentally responsible, manner by utilizing methods of enhancing green computing in cities and communities in addition to adopting and disclosing environmental practices and sustainability principles.

According to the Pew Research Center study titled "Smartphone Ownership and Internet Usage Continues to Climb in Emerging Economies" (Poushter, 2016), Internet access is increasing worldwide but remains higher in advanced economies, with the highest rates of access in countries such as South Korea (94%), Australia (93%), and Canada (90%), and rates of over 80% in the United States, the United Kingdom, Spain, Israel, and Germany

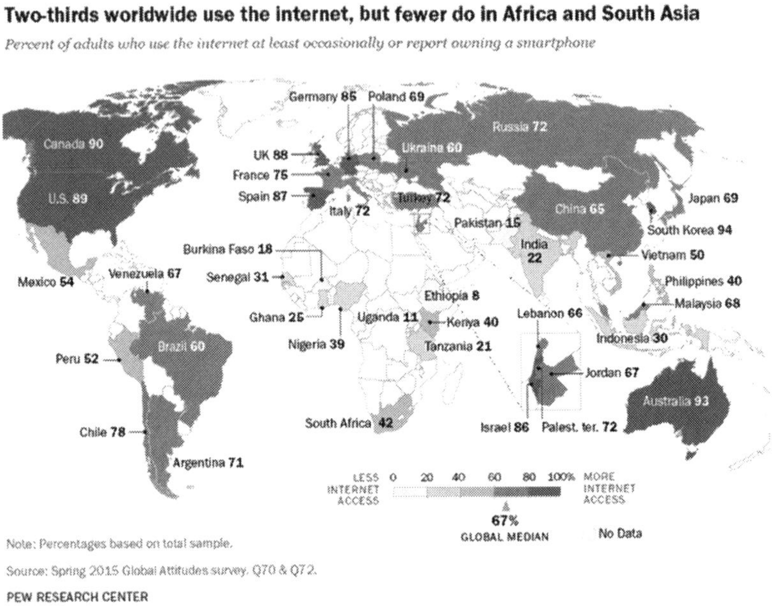

FIGURE 1 Worldwide use of the Internet.

(see Fig. 1). While many large emerging economies have rates of access above 60%, the rates of access in general are lower in developing countries (Poushter, 2016). The lowest rates of access are concentrated in sub-Saharan Africa and parts of Asia, with the lowest rates in Burkina Faso (18%), Pakistan (15%), Uganda (11%), and Ethiopia (8%). The Pew Research Center report also finds a strong correlation between country wealth, measured by per capita gross domestic product on a purchasing power—adjusted basis, and Internet access. Generally, greater access to the web worldwide is found to be associated with younger, more educated, and higher-income individuals. The Pew Research Center study (Poushter, 2016) finds significant gender gaps in half of the countries surveyed, with men having greater access to the Internet than women in many nations.

The number of personal computers (PCs) per 100 people in different regions around the world is compared worldwide and with the Group of seven (G7) countries in Fig. 2. The average number of PCs per 100 people for G7 countries, which include Canada, France, Germany, Italy, Japan, the United Kingdom, and the United States, has increased from 24 in 1997 to 64 in 2005. Of the G7 countries, for 2005, Canada shows the highest number of PCs per 100 people (88), while Italy shows the lowest (37) (World Bank, n.d. World Development Indicators and EconStats™). The worldwide average has grown from 5 PCs per 100 people in 1997 to 13 PCs per 100 people in 2005, which is

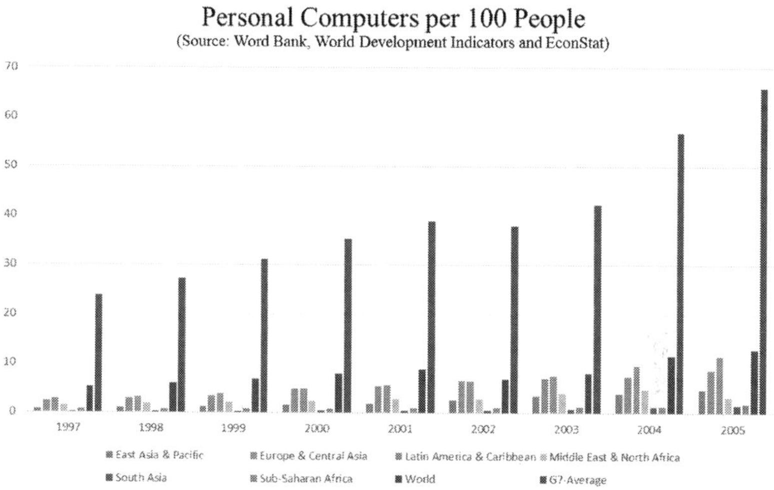

FIGURE 2 Worldwide growth of personal computers compared with G7 countries.

strongly correlated with the worldwide growth in Internet access (see Fig. 1). The PC growth in different regions in the world varies significantly but indicates increasing trends overall. During the period 1997 to 2005, the highest growth is shown in Latin America and the Caribbean, increasing from 3 PCs per 100 people in 1997 to 12 PCs per 100 people in 2005. Europe and Central Asia show the next highest increase, with 2.5 PCs per 100 people in 1997 to 9 PCs per 100 people in 2005. The lowest growth in number of PCs per 100 people is shown in South Asia and sub-Saharan Africa. The increasing trend in PCs per 100 people and access to the Internet indicates that worldwide computing needs at the household level are likely to increase in years to come. Although ideas such as "one global-scale shared computer" to run the entire Internet are discussed based on the thousands of networked machines or computer grids (Carr, 2008), at least in the reasonable future, individual PC ownership is expected to continue to grow, which will increase worldwide computing requirements. This increasing PC ownership/use trend will be even more prevalent in developing countries because while the developing nations have a lower number of PCs per 100 people, when population is factored in, the number of PCs may be much higher in absolute terms. Consequently, as energy consumption in computing increases as part of C^2ER, environmentally sustainable infrastructure design becomes a critical issue.

Numerous factors, economic and societal, can be attributed to the current and future need for C^2ER "green computing." The rising cost of energy consumption due to diminishing finite scarce resources; the emergence of some form of carbon taxes for CO_2 emissions; efforts by consumer rights advocates, policy makers, and influential industry leaders to compel organizations to minimize the negative effects on the environment from their computer use; and the emergence of worldwide regulations to enforce sustainable environmental policies at all levels—national, provincial or state, and municipal—are some of the factors influencing C^2ER. For example, British Columbia, Canada, continues to be a leader in climatic action with its 2008 revenue—neutral carbon tax implementation. It is a comprehensive tax covering approximately 70% of provincial emissions. "Revenue neutral" means that every dollar generated is returned to British Columbians through reductions in personal income tax rates, tax credits, and corporate income tax reductions.

Cap-and-trade programs also exist. One such cap-and-trade program for reducing heat-trapping emissions is the European Union's Emission Trading Scheme. It includes 27 countries and all large industrial facilities generating electricity; refining petroleum; and producing iron, steel, cement, glass, and paper. The Northeast Regional Greenhouse Gas Initiative is a cap-and-trade program that covers electricity generation in 10 northeastern and mid-Atlantic States in the United States. The Western Climate Initiative includes seven western states in the United States and four Canadian provinces. In 2013, Quebec formally started operating a cap-and-trade program for CO_2 emissions that required businesses to consider the cost of emitting greenhouse gases in their decision-making processes. A year later, Quebec linked its system with California, creating the largest carbon market in North America (Government of Quebec, 2013). More recently, Ontario, Canada, proposed a comprehensive plan to reduce greenhouse gas emissions with carbon pricing as a component, with the aim of transitioning the province to a low-carbon economy.

The rest of this chapter is organized as follows. In Section 2, we briefly discuss green computing. Section 3 describes methods of enhancing green computing in cities and communities. In Section 4, we provide details of determining the metric energy usage profile (EUP) and a tool for the economic analysis of green computing initiatives. Section 5 highlights trends and challenges faced by developing countries, and Section 6 concludes.

GREEN COMPUTING

Academics have defined and discussed the terms "green IT (information technology)," "green technology," "green computing," "environmentally sustainable IT," and "green IS (Information Systems)" based on the type or scope of technology use. Green computing consists of the use of IT to accomplish sustainability goals (IT for Green—IT for efficiency such as use of automation or other technologies in manufacturing) as well as the efficient design and use of technology to reduce the negative impact on the environment (Green

IT—efficient IT). Steve Elliot (2011) provides a definition of green IT that consists of activities to minimize the negative impact and maximize the positive impact of human behavior on the environment. This is achieved ***through the design, production, application, operation, and disposal of IT and IT-enabled products and services throughout their life cycle***.

Computing has a direct impact on the environment. As we are enticed by new technologies that are ubiquitous in modern life and permeate our daily lives, in addition to globalization and interconnectivity, today's organizations cannot survive without using computers to carry out their operations. It is well accepted that technology usage is barraged with a variety of environmental issues, including the high consumption of electricity which places a heavy burden on electric grids; greenhouse gas emissions; and the production and disposal of hazardous IT hardware components (Murugesan, 2008). As users of technology, we have an obligation to reduce the negative EI of technology. More specifically, we need to create a sustainable IT environment such as green computing or green IT, an environment in which businesses "green" their IT products, applications, services, and practices through legal, ethical, and social appeals (Murugesan, 2008; Zarrella, 2008). Green IT and green computing are now a growing trend. A 2009 Boston Consulting Group report highlighted that senior managers are now considering the economic, social, and even political impacts of sustainability-related changes in the business landscape (Berns et al., 2009).

What does green computing entail? It is the efficient and eco-friendly computing practiced by organizations with the objective of achieving reduced energy cost and enhancing the organization's corporate, city, and community image. Broadly, these issues include minimizing the environmental footprint by the proper disposal of e-waste items (old computers, mobile devices, peripheral devices, and appliances, etc.); use of virtualization, cloud computing, and other computer architecture focusing on efficiency; technology designs for reduced energy consumption (hardware and software); and futuristic initiatives such as carbon-free computing, solar computing, lead-free computing, and nanotechnology.

METHODS OF ENHANCING GREEN COMPUTING IN CITIES AND COMMUNITIES

Green design, green manufacturing, green use, and green disposal criteria can be used to inform our IT choices. This can be done, for example, by reducing the energy consumption of computerized information systems through the selection of energy-efficient devices and IT architectures, by the reduction of e-waste, and by the reuse and refurbishment of computing equipment and peripherals. We describe these options in more detail in the rest of this section.

Design and Use of Power-Efficient Devices and Applications

One green computing option is the use of equipment that features low power consumption and computer power management (CPM) capabilities, for example, devices that use Energy Star—qualified products and energy-efficient monitors. Traditional LCD monitors are back illuminated with fluorescent tubes, and most of the energy consumption in these monitors is a result of the back lighting. LED monitors, on the other hand, are back illuminated by LED diodes. Consequently, LEDs use only a fraction of the electricity that fluorescent tubes use. Furthermore, LEDs can use three times less energy when they are active than LCD displays use, resulting in lower power consumption and lower EI. According to the US Environmental Protection Agency (EPA), around 30% to 40% of computers are kept on during the weekends and even after office hours and around 90% of these computers remain idle. CPM is a set of features that help reduce power consumption by placing monitors and computers into a low-power "sleep mode" after a period of activity.[1]

INFRASTRUCTURE DESIGN

Virtualization

"Virtualization" refers to dividing the resources of a computer into multiple execution environments to create a virtual, rather than the actual, version of, for example, the operating system, storage, computing capability, or network resources. Virtualization can offer a way to increase physical resources to maximize the investment in hardware. System virtualization is referred to as "server virtualization" or "desktop virtualization." Depending on the role of the virtualized system, the system uses resources of a single computer as if it were a collection of separate computers ("virtual machines"). Each virtual machine can have its own virtual CPUs, network interfaces, storage, and operating system. In this way, a single computer can be used for different computing purposes, or a single server can be used to provide services for many thin client machines. A thin client is a lightweight computer for which the server does most of the work, including data storage and processing, while the thin client provides an interface so that the user can interact with the data. Similarly, with application or workplace virtualization, where individual applications are isolated from the underlying operating system, one computer can be used by different computer users with their own application needs and profiles. Virtualization technologies also enable cloud services.

1. US Environmental Protection Agency (n.d.).

Cloud Computing

Cloud computing is a form of client server computing carried over the Internet using virtualization technologies. NIST (the National Institute of Standards and Technology) describes cloud computing as a model for enabling ubiquitous, convenient, on-demand network access to a shared pool of configurable computing resources. These resources may include networks, servers, storage, operating systems, applications, and services. Companies can choose a service depending on their need, for example, from a simple request of underlying infrastructure support and hosting their own operating systems and applications to requesting an entire system with underlying infrastructure, operating systems, and applications. Models of these services are as follows:

- Infrastructure as a Service (IaaS)[2] delivers hardware networking capabilities, including the use of servers, networking, and storage, over the cloud. The consumer is able to deploy and run arbitrary software, which can include operating systems, middleware (software that lies between an operating system and the applications providing application programs with more services than that provided by the operating systems), and applications. Thus the consumer does not manage or control the underlying cloud infrastructure but has control over operating systems, storage, and deployed applications.
- In Platform as a Service (PaaS),[2] the consumer does not manage or control the underlying cloud infrastructure including the network, servers, operating systems, or storage but has control over the deployed applications and possibly configuration settings for the application-hosting environment.
- Software as a Service (SaaS)[2] delivers applications over the cloud. In this model, consumers use the provider's applications running on a cloud infrastructure. The applications are accessible from various client devices through either a thin client interface, such as a web browser (e.g., web-based email), or a program interface. The consumer does not manage or control the underlying cloud infrastructure including the network, servers, operating systems, storage, or even individual application capabilities, with the possible exception of limited, user-specific application configuration settings.

Cloud services use a pay-per-use revenue model. Since cloud services are measured, companies have to pay for only what they use. Given the positive characteristics such as flexibility, scalability, and cost-efficiency, cloud computing has been discussed as a viable option for all organizations, large, medium, and small. For developing countries, cloud computing may be used in e-education, e-health, e-government, and e-business.

2. Mell and Grance (2011).

Cloud computing reduces the carbon footprint[3,4] through the following features:

- Dynamic provisioning—Reducing the over-allocation of infrastructure resources. IT managers typically deploy far more servers, networking, and storage infrastructure than is actually needed to accommodate fluctuations in demand, lack of prediction ability, and need for scalability. Cloud computing can overcome this excess use of resources by optimizing the number of active servers.
- Multitenancy—Sharing application instances between multiple organizations. Just as multitenant developments such as apartment complexes use less power per person than individual households, shared applications provided through cloud services are more energy efficient.
- Server utilization—Operating server infrastructure at higher utilization. Virtualizations, as discussed earlier in this chapter, offer strategies to improve server utilization both for on-premise operations and for cloud services.
- Data center efficiency—Designing data centers to focus on the way facilities are physically constructed, equipped with IT and supporting infrastructure, and managed. This has a major impact on the energy use for a given amount of computing power.

In addition to the infrastructure design features listed above, for data center facilities, innovative facility design factors may also be considered, for example, using environmentally friendly solutions for data center cooling.

Reusable Devices

The common wisdom of the three R's of protecting the environment—reduce, reuse, and recycle—is also directly applicable in the context of technology use.[5] Most technology devices are manufactured to have a short product life, which makes them obsolete within a short amount of time. This not only

3. Microsoft, Accenture, and WSP Environment and Energy (2010).
4. The results of the analysis for Microsoft clearly show significant decreases in CO_2 emissions per user across the board for cloud-based versus on-premises delivery of the three applications studied in the report: the Microsoft Exchange, MS SharePoint, and MS Dynamics CRM (Customer Relationship Management) systems. The analysis revealed as much as a 90% reduction in carbon emissions for deployment size of about 100 users (small size), 60% to 90% for medium-sized deployments of about 1000 users, and 30% to 60% for large deployments of about 10,000 users. The Microsoft report notes that cloud computing can help reduce carbon emissions significantly by enabling new solutions for smart grids, smart buildings, optimized logistics, and dematerialization. The Smart 2020 report (The Climate Group, 2008) estimates the potential impact of information and communication technologies (ICT)-enabled solutions to be as much as 15% of total global carbon emissions (or 7.8 billion tons of CO_2 equivalents per year).
5. Share-IT: Computers for Community (2014).

increases the cost of acquisition of new devices but also results in a serious e-waste problem. Companies can identify reuse opportunities for the older but still usable devices. For example, computers purchased for more computing-intensive tasks can be used in other tasks that are not computing-heavy internally within the organization, or those computers can be shared with community partners through donations. Donating older equipment to other users in need not only reduces the environmental e-waste burden but is also socially responsible to the entire community. For example, Share-IT.can makes donated devices available to low-income families, economically challenged citizens, seniors on marginal income, students in need, qualified start-up ventures, and nonprofit and community-based agencies.

The "reduce" strategy may focus on repairing and upgrading existing computer systems, replacing only components that have served out their useful life, or adding to the memory. While some products may allow such transformation, many small devices are designed to be discarded. These short-life cycle products should be designed and manufactured with modular designs, which would be an important consideration for companies seeking to reduce their environmental footprint when they acquire their devices. In recent years, refurbished products have appeared on the market. Refurbished products differ from simply used items in that the refurbished products are normally tested for functionality and defects before they are sold.[6]

ENERGY USAGE PROFILE AND ECONOMIC ANALYSIS OF GREEN COMPUTING INITIATIVES

Making the business case for any investment is an operational necessity. Environmentally sustainable infrastructure design focuses on both environmental resource constraints and more traditional IT business goals, with the objective of reducing EI and power consumption. Hence green computing should focus, for example, on data center design with comprehensive sustainability and the integration of environmental systemic quality metrics in the IT architectural process (Curtis, 2008).

Focusing on both energy efficiency and total cost of ownership (TCO) is required in the application of architecture evaluation criteria. The reduction in the amount of energy required to handle the needs of employees and customers is termed "energy efficiency." TCO, on the other hand, focuses on considering the full costs of a decision over the entire life cycle rather than focusing on the initial purchase price (such as only on the cost of hardware and software). What does TCO mean in the context of green computing? In the context of green computing, the TCO criteria should include keeping the costs low and minimizing energy consumption. Many organizations are beginning to pay

6. Electronics Take Back Coalition (2014).

attention to the positive effects of reducing their energy consumption because they are realizing that the energy consumption contributes significantly to greenhouse gas emissions. Curtis (2008) formalizes this relation as follows: reduced energy consumption equals reduced greenhouse gas emissions equals reduced operational cost for data centers and businesses. There are four main areas where a data center consumes energy: critical computational systems (servers, networks, and storage); cooling systems; power distribution units; and lighting (Curtis, 2008).

Francis and Richardson (2008) provide a green maturity model for virtualization. Virtualization is the consolidation of servers, more broadly defined as sharing. The focus is on the reduction of energy consumption over the equipment's entire life cycle as the main motivator of green application design. The primary idea is reducing energy consumption, which is highlighted as the best measure of greenness. The authors use the concept of embodied energy, which is defined as the quantity of energy required to manufacture and supply to the point of use, for a product or service along with operational energy consumption when considering the energy emissions impact of IT equipment. Francis and Richardson (2008) note that improving energy efficiency of a data center will reduce economic cost, as energy consumption cost (ECC) is a significant contributor to operating costs. Virtualization is defined at different levels in the article by Francis and Richardson (2008): Level 0 refers to zero virtualization, where all applications are resident on individual PCs; Level 3 is defined as the highest level of virtualization (cloud virtualization), where not only are resources shared but also the approach to technology planning and usage focuses on location and infrastructure ownership through cloud computing.

Although the reduction in ECC is a direct benefit of green technology, as pointed out by Chheda et al. (2008), another direct benefit which is often not considered but is significant is the EI. Consequentially, per Chheda et al. (2008), both ECC and EI are critical to evaluating application architecture design alternatives. More specifically, enterprises should assess the EUP, which consists of ECC and EI in environmentally sustainable infrastructure design applications. The development of an EUP forces the organizations not only to measure the ongoing energy cost but also to assess the EI of specific application design. It is important for organizations to quantify the EUP because it envisages understanding the energy usage of hardware, the operating system, applications, and the external (environmental) impact of the application architecture design. The energy consumption reduction and external impact of the application architecture design provide an organization the opportunity to consider investments from the perspective of corporate, community, and city environmental responsibility. That is, the organization can invest in more energy-efficient application architecture, which will reduce ECC and the EI, by considering alternatives with lower CO_2 emissions.

Several recent articles suggest that in the measurement of the effectiveness of application architecture, both energy consumption and EI should be taken into consideration (Herath et al., 2017; Chheda et al., 2008). Fundamental to both application architecture design and the evaluation of green computing alternatives is the EUP. The EUP measures the ongoing ECC and EI. It is important that an annual EUP be developed over the entire economic life of the project, not just over the initial year, when evaluating green computing alternatives. As per Herath et al. (2017) and Chheda et al. (2008), suppose for any year t that (E_t^U) is the EUP; (E_t^C) is the ECC; and (E_t^I) is the environmental energy impact. Then the relationship between ECC and environmental energy impact and energy usage is given by the following:

$$E_t^U = E_t^C + E_t^I$$

Herath et al. (2017) and Chheda et al. (2008) provide specific formulas to compute the annual ECC and the environmental energy impact, which can be used to assess the EUP. We use the same notation to be consistent. Suppose, for any period t, that e_t is the energy consumed by computing equipment in watts; c_t^e is the electricity charge in dollars per kilowatt-hour (kWh); h_t is the number of hours of equipment used per day; d_t is the number of working days in a year; $C_t^{CO_2}$ is the carbon tax (carbon price) in dollars per short ton of CO_2 emissions; and ε_t is the quantity of CO_2 emitted into the atmosphere in pounds per kWh. The annual ECC can be computed as $E_t^C = \frac{c_t^e e_t h_t d_t}{1000}$. The total energy consumed in kWh can be computed using $\gamma_t = \frac{e_t h_t d_t}{1000}$. Finally, the annual cost of CO_2 emissions, which is the environmental energy impact, can be computed as $E_t^I = \frac{C_t^{CO_2} \varepsilon_t \gamma_t}{(2000)}$, since there are 2000 pounds per short ton.

There are several traditional measures that can be used to evaluate the effectiveness of application architecture. These measures include cost, scalability, and testability, among others. Two of the contemporary measures that can be used are the ECC and environmental energy impact. The EUP, which combines ECC and environmental energy impact, is a contemporary measure of effectiveness that is ideally suited to assess the effectiveness of green computing. Hence, in the future, an EUP would be fundamental to the application architecture design, as cities and communities focus on minimizing the adverse effects of environmental pollution caused by CO_2 emissions from computing.

Making the Business Case for Virtualization

In a recent article on profit-driven corporate social responsibility (CSR), Herath et al. (2017) propose a novel approach based on Bayesian analysis to make the business case for investing in green IT. The article demonstrates how real option theory can be combined with the Bayesian framework, which allows for a more flexible approach, as investment decisions can be fine-tuned for new information. The intent of the article is to provide managers an analytical tool to evaluate investments in green computing, thereby creating economic, social, and ecological solutions to benefit stakeholders. A case study and an appendix in that article (see Herath et al., 2017) provide details of how the model was applied in a university setting. The Bayesian real options approach is general and can be applied in C^2ER (city and community environmental responsibility) investment scenarios attempting to minimize CO_2 emissions. The real option is similar to a financial option in the sense that it is modeled as a contract to purchase a physical asset rather than a traded asset on a stock exchange. From a capital investment angle, it gives managers the right to make an investment in physical assets when it is optimal to do so, hence integrating the concept of investment flexibility, which gives managers the ability to delay an investment, expand the scope and scale of an investment, abandon an investment, and switch between resource inputs, among other options (see, e.g., Herath and Jahera, 2002; Herath and Herath, 2009).

Herath et al. (2017) define environmental CSR as ecologically friendly actions and activities over and beyond those required by law and the narrow financial interest of firms. This approach aligns with the focus on C^2ER. It is based on the principles of sustainability, which concentrate on the business decision impact on current and future generations (UN World Commission on Environment and Development, 1987) vis-à-vis the ethical responsibility to act in an environmentally friendly manner. Herath et al. (2017) model the focus on energy reduction through more efficient infrastructure design using virtualization to support environmental sustainability. It is well documented that technology can play a role in reducing energy consumption and hence CO_2 emissions. Watson et al. (2010) argue that there is an economic motive to reduce energy consumption because such reduction is more profitable and lowers CO_2 emissions. The emergence of government regulations on CO_2 emissions is another factor that contributes to reductions in CO_2 emissions, as more and more countries attempt to address global warming. Although CO_2 emission regulations are evolving, there is uncertainty with regard to the level of CO_2 tax that will be imposed as well as the extent to which the tax will be imposed. There is also uncertainty as to whether such a tax will be strictly imposed or not, due to political and industry pressure.

Unlike capital investments undertaken to grow and expand business operations, generally it is quite difficult to make the business case for investing in sustainability and social responsibility initiatives aimed at reducing detrimental environmental effects such as CO_2 emissions. The real options approach is more suitable for evaluating such initiatives because the value of the project is enhanced by the option value of exploiting the economic upside. The Herath et al. (2017) model compares operating cost savings arising from migrating to a more energy-efficient computer system and the reduction in CO_2 emissions from lower energy consumption with the incremental cost of the migration to the new systems. The approach is based on the idea that if there is uncertainty with regard to the price of the CO_2 tax and the level of enforcement, then although it might be worthwhile to invest immediately if the operating cost savings and the energy impact exceed the cost of migrating to the new systems, it may be much better to delay such an investment until more information is available with regard to the level of enforcement. The strategic value of the green investment in this example arises due to the positive EI from energy savings, which can be ascertained using the formulas provided in Section 2. The value of this strategic option is shown to be equal to the present value of the energy EI over the life of the green IT project and is primarily the savings that results from the energy reduction and associated carbon tax.

The model then considers the arrival of new information from a Bayesian analysis angle and shows how the initial probabilities associated with the CO_2 tax can be revised using Bayes' Theorem. The updated probabilities change the prior probabilities and thus the economic value of the green IT investment. The advantage of the revision of probabilities is that it allows the organization to learn more about the uncertainties surrounding the CO_2 tax and integrate those uncertainties into the model. The proposed approach for making the business case for green IT investments is an active risk management approach to decision-making, since information is proactively obtained. In the Herath et al. (2017) example, it is shown that although the option value to delay the green investment is positive both when a carbon tax is strictly enforced and when it is not strictly enforced, in both these instances the project value is lower than the option value of immediately investing.

The Herath et al. (2017) model is novel in the sense that it integrates a Bayesian decision framework with environmental uncertainty to model the impact of green computing alternatives. From a green computing perspective, the key idea is that reduction in energy consumption reduces CO_2 emissions, which in turn translates into cost savings if CO_2 emissions are taxed. These cost savings justify the investment in green computing initiatives because those initiatives translate directly into cash flow savings from the project. From a C^2ER perspective, whether a carbon tax is imposed or not, in general, low energy consumption translates into low CO_2 emissions, and thus, as discussed in Section 2, assessing the energy impact and developing an energy use profile facilitates making the business case for green computing initiatives. A generic approach evaluating green computing investments is to use discounted cash flow (DCF) analysis, since net benefits and costs should be evaluated over the entire planning horizon. More specifically, let B_t be the net benefit excluding energy cost savings at time t, and let I be the capital expenditure (Capex). Then, using the DCF analysis, the net present value (NPV) of the total benefits and costs can be computed. The total benefits may include the quantifiable net benefits of the project or a product over the entire product life cycle, such as increased cash inflow from revenue generation or cost savings, annual energy cost savings due to low energy consumption (ΔE_t^C), and the resulting net cost savings from reduced energy impact (ΔE_t^I). The NPV of the total benefits and costs can be computed as follows:

$$NPV = \sum_{t=1}^{n} \frac{B_t + \Delta E_t^C + \Delta E_t^I}{(1+k)^t} - I$$

where $k$is the cost of capital or the discount rate. Life cycle costing tracks and accumulates costs attributable to each product or service from start to finish, which includes initial research and development (R&D) to final customer service and support.

The approach can be used to evaluate any type of green computing initiatives discussed in Section 3, either using a traditional capital budgeting approach or modeled as real options as in Herath et al. (2017). In the case of virtualization, as discussed in detail in Herath et al. (2017), the benefit stream includes incremental hardware support cost, incremental software support cost, incremental data center cost savings for rack and floor space, incremental cost savings for ports, incremental ECC savings, and incremental administration cost savings. The investment cost includes hardware cost of the new server system, savings from initial software license transfers, migration cost on new servers, net cost of disposal of the old server system (i.e., salvage value or environmental cleanup cost for e-waste disposal), opportunity cost of space savings, and data storage cost. A project to establish a state-of-the-art green data center would require modeling the benefits and costs including in-house power generation, renewable energy use, and multipurpose cooling systems. In the case of new product development to reduce the environmental footprint, new design ideas to reduce power loss, increased efficiency, and the reuse of waste energy would be considered some of the benefits in the NPV formula.

TRENDS AND CHALLENGES FOR DEVELOPING COUNTRIES

Developing countries tend to face special challenges when organizations in those countries try to invest in green computing technologies. Investing in more efficient computer technologies (e.g., PCs, servers, and cloud computing) is as important to developing countries as it is to the countries in the developed world, with a view to reducing CO_2 emissions from the use of computers and improving conservation of energy. Developing countries that choose environmentally friendly options must deliberate more than the economic considerations. Specifically, because developing countries tend to have a scarcity of resources, they might not be able to develop or update their IT infrastructure as frequently as developed countries can. Thus, for developing countries, it is even more important to explore reduce and reuse strategies. Even though their progress may be slower, organizations in the developing countries are steadily moving into designing and implementing energy conservation mechanisms. For example, consolidation of servers, cloud computing, and various other energy conservation mechanisms such as virtual

LAN (VLAN) are widely used by organizations in developing countries. The most interesting aspect of this progress, however, is that in many cases, the intention was not energy conservation but rather savings in terms of capital expenditures and other related costs.

Cloud computing provides a relatively new business model. With its ability to increase flexibility, accessibility, and cost-effectiveness, cloud computing is an attractive option for companies not only in developed countries but also in developing countries. Studies carried out in Ghana,[7] Tunisia,[8] and Pakistan[9] have found that in addition to scalability and flexibility, reductions in capital and operating expenditures for IT infrastructure are key drivers of cloud computing adoption in those countries. Cloud computing enables users to access a scalable and elastic pool of data storage and computing resources, as and when required. With many people in developing countries having access to mobile phones[10] and thus access to the Internet, businesses can leverage this wave of telecommunication growth to provide their services to the citizens. Armed with such portability and flexibility in the computing environment, businesses can reach their customers 24/7 anywhere in the world. On demand, scalable service without capital investment to acquire or implement IT infrastructure, which reduces capital and operating expenses, becomes an attractive feature.

Even though organizations in developing countries tend to move into renewable energy sources, such as solar, wind, and bio gas, at a slower rate than developed countries do, the whole purpose in the movement into renewable energy sources by developing countries may be to cut down the energy cost rather than to address the sustainability concerns. This is particularly true in the private sector, as the cost associated with energy adversely affects the bottom line. On the other hand, this is not the case in the public sector, as the cost is borne by the general public. It appears that organizations in both the public and the private sectors in developing countries are uneducated or unconcerned about the power consumption by computers. This lack of awareness or concern can be a barrier to developing green computing mechanisms. One possible reason for such ignorance or apathy could be the availability of cheap electricity. Even if top management wants to invest in green technologies, a majority of private sector organizations in developing countries have not grown enough to allocate enough capital from their annual budgets on information and communication technology (ICT) infrastructure design and implementation. Most of these organizations also lack the skills necessary to develop those technologies. Funding constraints and a lack of interest and support from top management in developing countries also

7. Yeboah-Boateng and Essandoh (2014), 13–20.
8. Boukhris (2017), 205.
9. Ahmad and Waheed (2015).
10. Poushter (2016).

prevent public sector organizations in those countries from investing in green computing technologies. Further, in developing countries, the infrastructure that is needed for technological development, such as network lines, has not been developed to facilitate energy conservation mechanisms. In addition to these challenges, in certain industries, such as telecommunication, the security of information limits virtualization. People's suspicions in developing countries about sophisticated technologies prevent sharing, which further impedes the development of green computing technologies there.

While individual users in developing countries are extensively using applications provided through the cloud such as email and maps, there are significant challenges at the organizational level for developing countries to adopt these new IT infrastructure design alternatives.[11,12] Recent studies in developing countries indicate that cloud adoption in these countries is comparatively low and that the majority of developing countries encounter significant obstacles to participate effectively in the cloud economy. It has been documented that external barriers to adoption include inadequate infrastructure (lack of reliable power and broadband connectivity), lack of adequate legal and regulatory frameworks for electronic commerce and cybersecurity, and lack of skills to make effective use of ICT.

The following list describes in detail the obstacles that developing countries face in adopting cloud computing and other technological advancements:

1. Knowledge, resource, and skill constraints—Access to internal expertise and knowledge to implement and use these technologies remains extremely limited in the developing countries. Even the general awareness of availability of these technological options is sometimes limited. According to a Gartner survey[13] conducted among large enterprises in 2009, half of the respondents in emerging markets either had not heard of cloud computing or did not know what it meant. Training is required for understanding cloud features and how to use them efficiently, and this training requires experts and financial resources.
2. Loss of control—One of the inherent issues related to cloud options is loss of control since cloud systems are characterized by hardware, networking, operating systems, and/or applications, and more importantly, cloud systems are managed and controlled by the service provider. Depending on the service model being used, client organizations do not have direct or complete control over these aspects of infrastructure. The control issues may relate to storage, transfer, and processing of data; functionality of the applications; and access to the digital assets. Clouds can be seen as "black

11. SciDevNet (2013).
12. United Nations Conference on Trade and Development (2013).
13. Ksherti (2010), 47–55.

boxes" in that the customer cannot directly monitor or manage the resources.[14] Most IT managers find loss of control to be one of their greatest difficulties. If managers cannot monitor closely or control their technology, it is even more challenging for them to deal with the risks associated with that technology.

3. Security and privacy issues—The cloud is bogged down with security and privacy issues. Because Internet technologies and application programming interfaces (APIs) depend on interacting with data in the cloud, those technologies and APIs are prone to malicious attacks.[15] Data stored in cloud may also be subjected to a variety of malicious attacks, thus making the cloud vulnerable to data leakage and data loss. Sensitive information and data stored in the cloud can be subjected to misuse, theft, or illegal uses. Moreover, cloud services provide scalability by sharing infrastructure, platforms, or applications by the cloud service providers. The sharing of infrastructure and services among many parties (tenants) makes cloud clients vulnerable to attack. Such sharing and dependencies will affect decisions as to whether or not to buy cyber insurance and will influence the pricing of that insurance (Herath and Herath, 2011).
   Hence, due diligence on the part of the service providers is necessary to ensure that those providers are taking appropriate precautions. In addition, the client organization must perform due diligence on its part to evaluate its technologies and service providers. The loss of control and security privacy issues are multiplied for governmental agencies that adopt cloud services because most cloud services are currently hosted in only a handful of countries, which raises concerns related to secrecy and national security.
4. Lack of adequate legal and regulatory frameworks for cybersecurity—Compounding the inherent issues of security and privacy in cloud computing for developing countries is the lack of a regulatory framework. In most developing and emerging economies, legal and regulatory frameworks are not in place to adequately address issues related to data protection and privacy.
5. Reliance on service providers—Cloud services are not completely standardized. The lack of standards makes it difficult to integrate with other applications or to switch between vendors. This situation leads to "vendor lock-in." In vendor lock-in, a client becomes dependent on a particular vendor for products and services, resulting in high switching costs and a lack of interoperability. In addition to the loss of complete control over data, another concern is the issue of migrating data, in which a decision is reached to switch to a new cloud provider. While cloud services are attractive for their flexibility and scalability, in lock-in situations, service

14. Sheppard (2014).
15. Violino (2018); Wilson (2016).

providers often demand higher fees from their clients when those clients seek upgrades for their cloud computing (Ahmad and Waheed, 2015).

6. Inadequate infrastructure—Use of cloud services requires reliable and redundant power suppliers. The feasibility of cloud computing also depends to a great extent on the availability of reliable, affordable, high-quality communications networks. Developing countries suffer from both limited and inadequate access to high-quality and affordable broadband infrastructure and power supplies. Cloud computing depends on constant availability of electricity and a running network or reliable Internet connection. Unfortunately, this is not a characteristic of most developing countries, which have unstable and slow network bandwidths and frequent power outages.[16] Power outages and slow network bandwidth may lead to immediate interruption of cloud services, increasing business downtime and reducing productivity and customer satisfaction.

How to Reduce the Barriers to Green Computing

Given that green computing is needed in our time and that technology trends such as cloud computing are recognized as the future direction for technology, developing countries have to focus on reducing the abovementioned barriers. Challenges and opportunity considerations need to be incorporated into holistic national ICT policies or a framework for development. Governmental agencies, cities, and communities can create and encourage sustainable initiatives such as Share-IT[17] that address community enrichment through green computing as well as community enrichment through broader access to technology. More focus should be placed on infrastructure development for reliable Internet, power access, an increase in broadband coverage, and access to fast Internet speeds at reasonable prices. Among developing countries, Vietnam and China (Kshetri, 2010), for example, have considered technology as a precursor for the country to build skilled workforce, which highlights the necessity for meaningful government interventions to facilitate the cloud sector. Additional emphasis should be placed by governmental and international agencies on security and privacy regulations. With security, privacy, and control questions at hand, and a general lack of regulations and standardization, significant effort will be required to create reliable regulatory frameworks to deal with these issues.

16. Veigas et al. (2012), 89–95; Yeboah-Boateng and Essandoh (2014); Ahmad and Waheed (2015); Boukhris (2017).
17. Share-IT (2014).

CONCLUSION

Global developments and international laws, regulations, and guidelines directly influence the development of CSR and environmental sustainability practices in the developing countries. Seventeen sustainable development goals and 169 associated targets were adopted by the United Nations member states in 2015 and have been incorporated into the Sri Lanka Sustainable Development Act in 2017. It is hoped that with these developments, organizations in developing countries will pay more attention to sustainability issues including precautions on green computing. Cities and communities can make a positive impact on climate change by encouraging renewable energy, using green material, recycling, and refurbishing hardware peripherals so that e-waste is diverted from landfills. Encouraging greener computing through incentive mechanisms will reduce energy consumption and the EI. It will also lessen the burden placed on power generation and electricity grids. Focusing on green computing hence will become an important aspect of minimizing the CO_2 footprint. While we have discussed approaches to minimizing the carbon footprint by individuals and organizations both public and private, cities and communities should take a proactive approach to measuring EI. Measuring, verifying, and disclosing our carbon footprint and CO_2 emissions will allow us to ensure constant improvements so that we will leave the planet a better place for future generations.

REFERENCES

Ahmad, T., Waheed, M., June 2015. Cloud computing adoption issues and applications in developing countries: a qualitative approach. International Arab Journal of e-Technology 4 (2), 84–93.

Berns, M., Towncnd, A., Khayat, Z., Balagopal, B., Reeves, M., Hopkins, M., Khruschwitz, N., 2009. The Business of Sustainability: Imperatives, Alternatives and Actions. The Boston Consulting Group, Boston.

Boukhris, Y., 2017. Cloud computing and virtualization in developing countries. In: Emerging Technologies for Developing Countries: First International EAI Conference, AFRICATEK 2017, Marrakech, Morocco, March 27–28, 2017. Proceedings, edited by Fatna Belqasmi, Hamid Harroud, Max Agueh, Rachida Dssouli, and Faouzi Kamoun, 205. Cham, Switzerland: Springer International Publishing.

Carr, N., February 20, 2008. How many computers does the world Need? Fewer than you think. Guardian. https://www.theguardian.com/technology/2008/feb/21/computing.supercomputers.

Chheda, R., Shookowsky, D., Stefanovich, S., Toscano, J., 2008. Profiling energy usage for efficient consumption. In: Microsoft (Ed.), The Architecture Journal: Green Computing Issue. https://msdn.microsoft.com/en-us/library/dd393312.aspx.

Curtis, L., 2008. Environmentally sustainable infrastructure design. The Architecture Journal 18 (1), 2–8.

Electronics Take Back Coalition, June 25, 2014. Facts and Figures on E-waste and Recycling. http://www.electronicstakeback.com/wp-content/uploads/Facts_and_Figures_on_EWaste_and_Recycling.pdf.

Elliot, S., 2011. Transdisciplinary perspectives on environmental sustainability: a resource base and framework for it-enabled business transformation. MIS Quarterly 35 (1), 197–236.

Francis, K., Richardson, P., 2008. Green maturity model for virtualization. The Architecture Journal 18 (1), 9–15.

Government of Quebec, 2013. A Brief Look at the Quebec Cap-and-trade System for Emission Allowances. http://www.mddelcc.gouv.qc.ca/changements/carbone/documents-spede/in-brief.pdf.

Herath, H.S.B., Herath, T.C., 2009. Investments in information security: a real options perspective with bayesian post-audit. Journal of Management Information Systems 25 (3), 337–375.

Herath, H.S.B., Herath, T.C., 2011. Copula-based actuarial model for pricing cyber-insurance policies. Insurance Markets and Companies: Analyses and Actuarial Computations 2 (1), 40–53.

Herath, H.S.B., Jahera Jr., J.S., 2002. Real options: valuing flexibility in strategic mergers & acquisitions as an Exchange ratio swap. Managerial Finance 28 (12), 44–62.

Herath, H.S.B., Herath, T.C., Dunn, P., 2017. Profit-driven corporate social responsibility as a Bayesian real option in green computing. Journal of Business Ethics 1–16. https://doi.org/10.1007/s10551-017-3705-1.

Ksherti, N., 2010. Cloud computing in developing economies. IEEE Computer 43 (10), 47–55.

Mell, P., Grance, T., September 2011. The NIST Definition of Cloud Computing (Technical Report). U.S. Department of Commerce, National Institute of Standards and Technology. https://doi.org/10.6028/NIST.SP.800-145. Special Publication 800–145.

Microsoft, Accenture, and WSP Environment & Energy, 2010. Cloud Computing and Sustainability: The Environmental Benefits of Moving to the Cloud. Accenture. https://download.microsoft.com/download/A/F/F/AFFEB671-FA27-45CF-9373-0655247751CF/Cloud%20Computing%20and%20Sustainability%20-%20Whitepaper%20-%20Nov%202010.pdf.

Murugesan, S., 2008. Harnessing green it: principles and practices. IT Professional 10 (1), 24–33.

Poushter, J., February 22, 2016. Smartphone Ownership and Internet Usage Continues to Climb in Emerging Economies: But Advanced Economies Still Have Higher Rates of Technology Use. Pew Research Center, Washington, DC. http://www.pewglobal.org/2016/02/22/smartphone-ownership-and-internet-usage-continues-to-climb-in-emerging-economies/.

SciDevNet, December 20, 2013. Poor Countries Struggling to Harness Cloud Computing. https://www.scidev.net/global/communication/news/poor-countries-struggling-to-harness-cloud-computing.html.

Share-IT: Computers for Community, 2014. Services. Community Environment Alliance information page. http://share-it.ca/services/.

Sheppard, D., July 2, 2014. Is Loss of Control the Biggest Hurdle to Cloud Computing? IT World Canada. https://www.itworldcanada.com/blog/is-loss-of-control-the-biggest-hurdle-to-cloud-computing/95131.

The Climate Group, 2008. Smart 2020: Enabling the Low Carbon Economy in the Information Age. N.p.: Creative Commons. https://www.theclimategroup.org/sites/default/files/archive/files/Smart2020Report.pdf.

United Nations Conference on Trade and Development, 2013. Information Economy Report 2013: The Cloud Economy and Developing Countries. New York: United Nations. http://unctad.org/en/PublicationsLibrary/ier2013_en.pdf.

U.S. Environmental Protection Agency. n.d., Put Your Computers to Sleep: Save up to $50 per Computer Annually. Energy Star Certified Products, Ways to Reduce IT Energy Costs information page. https://www.energystar.gov/products/low_carbon_it_campaign/put_your_computers_sleep.

Veigas, J.P., Naik, V., Chandrasekaran, K., 2012. Cloud computing adoption and impact in emerging markets. IRNet Transactions on Computer Science and Engineering 1 (1), 89–95.

Violino, B., January 5, 2018. The Dirty Dozen: 12 Top Cloud Security Threats for 2018. CSO Online. https://www.csoonline.com/article/3043030/security/12-top-cloud-security-threats-for-2018.html.

Watson, R.T., Boudreau, M.-C., Chen, A.J., 2010. Information systems and environmentally sustainable development: energy informatics and new directions for the is community. MIS Quarterly 34 (1), 23–38.

Wilson, M., April 1, 2016. What are the 12 biggest cloud computing security threats? IBM Cloud Computing News. https://www.ibm.com/blogs/cloud-computing/2016/04/12-biggest-cloud-computing-security-threats/.

World Bank. n.d., World Development Indicators and EconStats™. http://www.econstats.com/wdi/wdiv_597.htm.

World Commission on Environment and Development, 1987. Our Common Future. Oxford University Press, Oxford.

Yeboah-Boateng, E.O., Essandoh, K.A., 2014. Factors influencing the adoption of cloud computing by small and medium enterprises in developing economies. International Journal of Emerging Science and Engineering 2 (4), 13–20.

Zarrella, E., 2008. Sustainable IT: The Case for Strategic Leadership. KPMG IT Advisory. KPMG International. http://www.greenprof.org/wp-content/uploads/2010/05/Sustainable-IT-The-case-for-strategic-leadership.pdf.

Index

'*Note*: Page numbers followed by "f" indicate figures, "t" indicate tables.'

J

L

M

N

O

P

Q

R

S

V

W

Z

Made in the USA
Monee, IL
13 September 2026

62977014R00275